Solid Waste Management

This book provides a holistic picture of waste and its management techniques, with all the recent advancements and necessary projections for the future, which aim to maximize the value-added products for environmental sustainability on a cost-effective basis. It emphasizes the practices, problems, and management of a broad variety of industrial solid waste and facilitates a major understanding of the utilization of sustainable tools to combat all types of problems.

The book:

- Provides a holistic approach toward the topic to channelize waste management globally.
- Discusses waste minimization and regulation in conjunction with other integrated solutions and equipment.
- Reviews updated information and data for use to modify the system for advanced waste management.
- Explores innovative methods for defining, sorting, and treating solid waste.
- Includes case studies in each chapter for analyzing the concepts explored in the real world.

This book is aimed at graduate students and researchers in civil and environmental engineering, and waste management.

Solid Waste Management

Challenges and Recent Solutions

Edited by
Rajeev Pratap Singh, Ibha Suhani, Norli Ismail,
and Vaibhav Srivastava

CRC Press
Taylor & Francis Group
Boca Raton London New York

CRC Press is an imprint of the
Taylor & Francis Group, an **informa** business

Designed cover image: Shutterstock

First edition published 2025
by CRC Press
2385 NW Executive Center Drive, Suite 320, Boca Raton FL 33431

and by CRC Press
4 Park Square, Milton Park, Abingdon, Oxon, OX14 4RN

CRC Press is an imprint of Taylor & Francis Group, LLC

ISBN: 9781032534183 (hbk)
ISBN: 9781032895055 (pbk)
ISBN: 9781003543176 (ebk)

DOI: 10.1201/9781003543176

Typeset in Times
by Newgen Publishing UK

Dedication

---

*This book is dedicated to our parents and
other family members.*

Contents

Foreword xvii
Preface xix
Acknowledgments...xxiii
About the Editors xxv
List of Contributors...xxvii

Chapter 1   Understanding Waste: Type, Characterization, and Environmental Impacts.. 1

Asha Kumari Kumawat, Supriya Vaish, and Bhawana Pathak

1.1   Introduction... 1
    1.1.1   Present Status of Waste Worldwide...........................2
    1.1.2   Present Status of Waste in India 3
1.2   Types of Waste 3
    1.2.1   Solid Waste...4
    1.2.2   Liquid Waste...5
    1.2.3   Gaseous Waste 6
1.3   Sources of Waste/Waste Generation 6
    1.3.1   Municipal Waste..6
    1.3.2   Electronic Waste (e-Waste) 6
    1.3.3   Biomedical Waste...7
    1.3.4   Construction and Demolition Waste...........................7
    1.3.5   Industrial Waste 8
    1.3.6   Domestic Waste 8
    1.3.7   Radioactive Waste 9
    1.3.8   Agricultural Waste..9
    1.3.9   Hazardous Waste 10
1.4   Waste Collection 11
1.5   Waste Characterization 12
    1.5.1   Characteristics of Waste 12
1.6   Environmental Impact of Waste..15
    1.6.1   Effect on Soil...16
    1.6.2   Effect on Water..17
    1.6.3   Effect on Air 17
    1.6.4   Global Warming 18
1.7   Rules and Policies for Waste Management..............................18
1.8   Conclusion 20

Chapter 2   Waste and Its Current Status: A Middle East Perspective 26

Ghassan F. Al Samarrai, Rana I. Khaleel, and Norli Ismail

2.1   Introduction 26
2.2   Challenges Facing Solid Waste Management 27
2.3   The Main Challenges 27
2.4   Opportunities 28
      2.4.1   Reduction by Source 28
      2.4.2   Reuse 29
      2.4.3   Recycling 29
      2.4.4   Energy Production from Waste 30
2.5   Technical Methods for Heat Treatment of Waste 30
2.6   Convert Waste into Energy 30
2.7   Dumping Waste in Landfills 31

Chapter 3   Challenges and Opportunities in Waste: Indian Aspect 33

Chaitanya Narala and Bhagawan Dheravath

3.1   Introduction 33
3.2   Major Challenges of Waste Management in India 34
      3.2.1   Generation and Characterization of Waste
              in India 34
      3.2.2   Current Waste Management in India 34
      3.2.3   Waste Collection and Transport 35
      3.2.4   Environmental and Health Impacts of
              Waste Dumping 37
3.3   Rules and Regulations of Waste Management in India 38
3.4   Informal Sector's Role in Waste Materials' Reuse
      and Recycling 38
3.5   Waste-to-Energy in India 39
3.6   Improvements Needed in Waste Management in India 39
3.7   Alternative Option for Waste Management 41
3.8   Conclusion 41

Chapter 4   Waste Minimization, Management, and Resource Recovery:
               Livestock Waste 44

*Ellin Harlia, Eulis Tanti Marlina, Yuli Astuti Hidayati,
and Norli Ismail*

4.1   Introduction 44
4.2   Why Solid Waste Must Be Minimized? 45
4.3   Solid Waste Management 47
4.4   Resource Recovery 53
4.5   Conclusion 56

Chapter 5 Industrial, Construction, and Demolition Waste:
Management and Resource Recovery 60

*Nurhanim Abdul Aziz, Putri Anis Syahira Mohamad Jamil,
Muhammad Hasnolhadi Samsudin, and Norli Ismail*

5.1 Overview of Industrial, Construction, and
Demolition Waste...60
    5.1.1 Industrial Waste Ash and Geopolymers...................60
    5.1.2 Construction and Demolition Waste........................62
5.2 Overview of Waste and Health Hazards 63
    5.2.1 Industrial Waste: Ash and Its Risks.........................63
    5.2.2 Construction and Demolition Waste and Its Risks.....65
5.3 Importance of Waste Management for Public Health.............67
5.4 Waste Generation and Composition.................................68
    5.4.1 Industrial Waste: Ash..68
    5.4.2 Construction and Demolition Waste (CDW).............73
5.5 Resource Recovery Technology......................................79
    5.5.1 Chemical Treatment 79
    5.5.2 Physical Treatment..79
    5.5.3 Biological Treatment..82
5.6 Resource Recovery Technology by Physical
Treatment for Industrial Waste Ash as a Geopolymer
Precursor 83
    5.6.1 Suitability of Waste Ash for Geopolymer
Production 83
    5.6.2 Geopolymerization Techniques..............................84
    5.6.3 Performance and Properties of Geopolymer
Materials..85
    5.6.4 Successful Geopolymer Projects Utilizing
Waste Ash 86
5.7 Resource Recovery by Biological Treatment for
Construction and Demolition Waste 86
    5.7.1 *Bacillus subtilis* as Sulfate Reduction Bacteria..........86
    5.7.2 Capability of *B. subtilis* in Concrete and Mineral
Technologies...88
    5.7.3 Metabolite of *B. subtilis* 90
    5.7.4 Mechanism Adhered to *B. subtilis*..........................91
    5.7.5 Utilization of *B. subtilis* in Waste Concrete
Aggregate Treatment for Sulfate Reduction...............92
    5.7.6 Waste Concrete Aggregate Treated with
B. subtilis as Potential Recycled Aggregate 93
5.8 Conclusion and Recommendations................................100
    5.8.1 Summary of Key Findings..................................100
    5.8.2 Recommendations for Waste Minimization by
Promoting Resource Recovery Technologies...........101

Chapter 6   Health Hazards and Waste Management 111

*Siti Norabiatulaiffa Mohd Yamen, Mohd Saiful Samsudin,
and Muhammad Izzul Fahmi Mohd Rosli*

    6.1   Introduction 111
    6.2   Health Hazards of Different Types of Solid Waste 113
    6.3   Municipal Waste 113
    6.4   Industrial Waste 114
    6.5   Construction/Demolition Waste 115
    6.6   Medical Waste 115
    6.7   Household Waste 116
    6.8   Innovative Waste Management Methods for
           Reducing Health Hazards 116
    6.9   Source Reduction and Waste Minimization 117
    6.10   Recycling and Composting 118
    6.11   Waste-to-Energy Technologies 118
    6.12   Landfill Design and Management Strategies 120
    6.13   Advanced Treatment Technologies for Hazardous Waste 121
    6.14   Case Studies of Successful Implementation of
           Effective Waste Management Methods 122
    6.15   Conclusion 123

Chapter 7   Biological Hazardous Waste Management of Animal
Husbandry and Food Processing Industry 132

*Eulis Tanti Marlina, Yuli Astuti Hidayati, Ellin Harlia,
and Norli Ismail*

    7.1   Introduction 132
    7.2   Characteristics of Livestock Industry Waste 133
           7.2.1   Moisture Levels 133
           7.2.2   Organic Matter 133
           7.2.3   pH 134
           7.2.4   Odor 134
           7.2.5   Pathogens 135
           7.2.6   Texture 138
    7.3   Livestock Waste Management 140
           7.3.1   Collecting 140
           7.3.2   Transporting 140
           7.3.3   Processing 141
    7.4   Biological Hazards from Waste of Processing Livestock
           Products 141
           7.4.1   Milk Processing Waste 141
           7.4.2   Meat Processing Waste 143
           7.4.3   Leather Processing Waste 144

7.5   Reducing the Impact of Biological Hazards from
Livestock Waste 145
     7.5.1   Composting 145
     7.5.2   Anaerobic Digestion...147
7.6   Conclusion 150

Chapter 8   Ecological and Human Health Risk Assessments in
Water and Soil Samples from Beris Lalang Waste Dumpsite,
Kelantan, Malaysia..157

*Florence Rutselin Kelanit, Norli Ismail, Mohd Hafiidz Jaafar,
Nurul Ilyana Sansudin, Hasmah Abdullah, and Widad Fadhullah*

8.1   Introduction..157
8.2   Methodology 159
     8.2.1   Study Area..159
     8.2.2   Data Collection...160
     8.2.3   Estimation of Pollution Indices in Water Samples...160
     8.2.4   Heavy Metal Pollution Index..............................161
     8.2.5   Heavy Metal Assessment Index...........................161
     8.2.6   Soil Samples' Metal Risk Assessments...............161
     8.2.7   Index of Geo-accumulation................................161
     8.2.8   Potential Ecological Risk Index 162
     8.2.9   Health Risk Assessment 162
     8.2.10 Statistical Analyses..164
8.3   Results and Discussion 164
     8.3.1   Relationship between Metals in Soil and
Water of Beris Lalang Waste Dumpsite 164
     8.3.2   Groundwater and Surface Water Assessment
Using Pollution Evaluation Indices,
Heavy Metal Pollution Index, and
Heavy Metal Evaluation Index..................................165
     8.3.3   Soil Metal Assessments Using Index of
Geo-accumulation and Potential Ecological
Risk Index..168
     8.3.4   Health Risk Assessment of Heavy Metals in Soil....171
     8.3.5   Health Risk Assessments of Groundwater
and Surface Water in Beris Lalang
Waste Dumpsite..172
8.4   Recommendations and Limitations of the Study 175
8.5   Conclusion 175

Chapter 9   Wealth from Municipal Solid Waste 179

Ramnarayan Singh and Neha Verma Gour

9.1   Introduction..179
9.2   Generation of Municipal Solid Waste..................................180

9.3 Classification of Municipal Solid Waste 182
9.4 Characterization of Municipal Solid Waste 184
9.5 Current Status of Waste Management Worldwide 185
9.6 Challenges in Waste Management in India..........................188
9.7 Opportunities in Waste Management in India......................190
9.8 Plants with Waste-to-Energy in India.................................193
9.9 Conclusion 195

Chapter 10 Plastic Waste and Its Management Strategies: Recent Advances 197

Aryadeep Roychoudhury, Shreyosi Pahari, and Alen D. Rozario

10.1 Introduction...197
10.2 Chemicals Involved in Plastics 198
    10.2.1 Bisphenol A (BPA)...198
    10.2.2 Phthalates 198
10.3 Sources of the Plastic Waste 198
10.4 Types of Environmental Hazards Related to Plastics 199
10.5 Nature of Plastics 200
10.6 Physical States and Molecular Morphologies......................201
10.7 Properties of Plastics...202
    10.7.1 Colorants 202
    10.7.2 Stabilizers 202
    10.7.3 Plasticizers..203
10.8 Chemical Resistance of Plastics...203
    10.8.1 Dimensional Stability 203
    10.8.2 Durability of Plastics...203
    10.8.3 Electric Insulation..203
    10.8.4 Resistance to Fire 203
    10.8.5 Melting Point..203
    10.8.6 Optical Property 204
    10.8.7 Sound Absorption..204
    10.8.8 Strength 204
10.9 Categorization of Plastics 204
10.10 Microplastics...207
10.11 What Are Nanoplastics?..208
    10.11.1 Nanoplastics from Tire Wear..................................208
    10.11.2 Nanoplastics from Laundry Wastewater...................209
10.12 The Vulnerability of Aquatic Life to Ingesting Microplastics...209
10.13 Impact of Microplastics on Terrestrial Ecosystem 210
10.14 Impact of Plastics on Human Health 212
10.15 Strategies for Managing Plastic Waste................................213
    10.15.1 Recycling..213
    10.15.2 Incineration...214

10.15.3 Landfills 215
10.15.4 Pyrolysis 215
10.15.5 Concrete Production 216
10.16 Using Biotechnology to Get Rid of Microplastics 217
10.16.1 Degradation of Microplastics by Algae 218
10.16.2 Degradation of Microplastics by Fungi 219
10.16.3 Modern Biotechnology Techniques to
Accelerate the Breakdown of Microplastics 219
10.16.4 Metagenomic Investigation of Microbes
That Degrade Plastic for Biotechnological Use 220
10.17 Biotechnological Imputation on Metagenomic
Information 221
10.17.1 Changing the Composition of the Microbial
Community 221
10.18 Engineering Microbiological Genetic Organization 222
10.19 Employing Bioinformatics Strategy 222
10.20 Conclusion 223

Chapter 11 Nanowaste and Its Management: New Waste Management
Paradigm 237

Ekta Tiwari, Nitin Khandelwal, and Nisha Singh

11.1 Introduction to Nanowaste 237
11.2 Nanowaste Classification and Sources 238
11.2.1 Primary Sources 238
11.2.2 Secondary Sources 239
11.3 Environmental and Human Health Risks of Nanowaste 244
11.4 Challenges of Nanowaste Management 246
11.5 Conventional Management Methods and Nanowaste 247
11.5.1 Physical, Chemical, and Biological Recycling
Approaches 248
11.5.2 Bioremediation Approaches 248
11.6 Nanowaste Management Paradigm: Emerging
Approaches and Challenges 249
11.6.1 Magnetic Separation 249
11.6.2 Adsorption-Based Approach for Removal
and Recovery of Nanoparticles 250
11.6.3 Thermal Processes 251
11.6.4 Antisolvent Technique by Using CO_2 251
11.6.5 Microemulsion-Based Phase Separation
Techniques 251
11.6.6 Nanoparticles Recovery Using
Electrodeposition 252
11.6.7 Sludge Treatment before Incineration 252
11.6.8 Use of Nanomembranes 253

11.6.9 Use of Ionic Liquids 253
11.6.10 Cloud Point Extraction 253
11.7 Conclusion 254

Chapter 12 Algal Nanoparticles as a Green Approach for
Bioremediation of Heavy Metals from Wastewater 261

*Sujata, Anuj Sharma, Bansal Deepak, Rachna Bhateria, and
Sharma Mona*

12.1 Introduction 261
12.2 Synthesis of Nanoparticles 262
12.3 Conventional Practices for Remediation of
Heavy Metals 263
12.3.1 Electrochemical Process 264
12.3.2 Electro-Flotation Process 264
12.3.3 Coagulation/Flocculation 265
12.3.4 Activated Carbon Adsorption 266
12.3.5 Biological Treatment 266
12.3.6 Membrane Filtration 266
12.3.7 Oxidation Processes 266
12.4 Synthesis of Microalgal Nanoparticles 266
12.5 Application of Algae-Mediated Nanoparticles in the
Remediation of Contaminants from Wastewater 267
12.6 Conclusion and Future Outlook 270

Chapter 13 From Crisis to Opportunity: Advancing Plastic Waste
Management in the Age of Nanowaste and
COVID-Related Waste 276

*Muhammad Izzul Fahmi Mohd Rosli, Mohd Saiful Samsudin,
and Siti Norabiatulaiffa Mohd Yamen*

13.1 Introduction 276
13.2 Plastic Waste Management Techniques 278
13.3 Recycling 279
13.4 Pyrolysis 280
13.5 Mechanical Treatment 281
13.6 Waste-to-Energy Conversion 283
13.7 Challenges and Drawbacks of Existing Plastic Waste
Management Techniques 284
13.8 Introduction to Chemical Recycling as an Innovative
Approach 284
13.9 Biodegradable Plastics and Their Potential Impact
on Waste Management 285
13.10 Case Studies Highlighting Successful Implementation of
Innovative Approaches 285

13.11 Examples of Plastic Waste Management Practices
from Different Regions 286
13.12 Case Studies from Various Industries (e.g., Manufacturing,
Packaging, and Retail) 287
13.13 Successful Strategies and Approaches for Managing
Plastic Waste 288
13.14 Overview of Nanowaste and COVID-Related Waste.............288
13.15 Challenges and Risks Associated with Managing
These Types of Waste..289
13.16 Techniques for Safe Handling and Disposal of
Nanowaste and COVID Waste 289
13.17 Case Studies Demonstrating Effective Management of
Nanowaste and COVID Waste 290
13.18 Conclusion and Recommendations..290

Chapter 14 *Pinus roxburghii* Sarg: Needle Management to Combat Forest
Fire: Status, Prospects, and Constraints 295

*Prashant Sharma, Kamlesh Verma, Daulat Ram Bhardwaj,
Manoj Kumar Singh, Dhirender Kumar, and Pankaj Thakur*

14.1 Introduction...295
14.2 Indian Himalayan Region and Forest Fire 296
14.3 Role of Chir Pine in Forest Fire...297
14.4 Chir Pine Needle Management 299
14.4.1 Bio-Energy Source 300
14.4.2 Pulp, Paper, and Fiber Boards 301
14.4.3 Small-Scale Supplementary Utilization of
Chir Pine Needles...303
14.5 Constraints in the Utilization of Chir Pine Needles...............303
14.6 Future Perspectives 305
14.7 Conclusion 306

Chapter 15 Waste Management Practices of Riverside Communities at the
Limbang River Sarawak, Malaysia 312

Siti Noor Aliza Apandi and Norli Ismail

15.1 Introduction...312
15.2 Methods...313
15.2.1 Study Area 313
15.2.2 Respondents 313
15.2.3 Data Collections 314
15.2.4 Data Analysis...314
15.2.5 Ethical Considerations...314
15.3 Results and Discussion 314
15.3.1 Demographic Profile of Respondents........................314

15.3.2 Anthropogenic Activity in the Limbang River.........316

15.3.3 Household Drinking Water Supply 318

15.3.4 Household Waste Management...............................319

15.3.5 Health Status and Impact of the Unsustainability Waste Disposal in the River.....................................321

15.3.6 Variables Correlation..322

15.3.7 Comparison of the KSK Communities' Health Impact with Other Riverside Dwellers in Southeast Asian Countries....................................323

15.3.8 Challenges in Waste Management Practices of Riverside Communities 323

15.4 Conclusion 325

Index..329

Foreword

I am extremely pleased to have been asked to write the Foreword to the book *Solid Waste Management: Challenges and Recent Solutions* edited by Dr. Rajeev Pratap Singh. One of the major environmental concerns haunting society today is improper waste management that inevitably contributes to greenhouse gas emissions which in turn influence climate change. A demand has arisen where integration of several technologies is needed which are ecologically sustainable and economically viable to achieve overall sustainability.

This book provides a wholesome picture of waste and its management techniques with several recent advancements, also providing necessary projections for the future, to maximize economic stability with environmental sustainability. To enable these, the book has been designed into 15 chapters, which delve into several aspects of waste management, waste utilization, and the impact of improper waste management on human health as well as ecosystem health. To facilitate proper functioning of waste management, the most important step is understanding waste, its types, characterization, and environmental impacts. As waste types can range from hazardous to non-hazardous, proper waste characterization is crucial to ensure efficient waste management, and this characterization must take into account physical, chemical, and biological properties of the waste that is generated.

The need to upgrade waste management systems is the main challenge due to urbanization and industrialization. To combat this ensuing war on wastes, waste minimization, management, and resource recovery from solid and liquid waste are a must. Waste from human settlements, agriculture, and livestock; industrial, construction, and demolition waste; and electronic waste continue to threaten the environment as these could unintentionally release hazardous substances into the environment, which can serve as a haven for pathogenic microbes and vermin, thereby posing serious health risks to all biota. This book discusses aspects of resource recovery technologies through physical and biological treatment to improve carbon footprint.

The chapters on health hazard and waste management give insights into how improper management of different types of waste can lead to exposure to pathogens, chemicals, and other contaminants that can have adverse effects on human health. In fact, effective biohazardous waste management practices can minimize environmental impacts and protect the health and safety of workers and the public.

The authors have taken care to highlight plastic waste and its management strategies including diverse applications, chemical compositions, and properties. With technological development, the editor has wisely incorporated a chapter on nanowaste and its management as is, in fact, essential to the current status and challenges associated with the safe disposal and recycling of nanomaterials.

The book though has emphasized several technologies to handle waste and its efficient management and resource recovery, it also discusses green technology using algal nanoparticles for bioremediation of heavy metals from wastewater using microalgae.

This book, *Solid Waste Management: Challenges and Recent Solutions*, focuses on every issue and concern and importantly reviews issues related to compliance and regulation. It also provides well-consolidated information and knowledge for the benefit of not only the faculty and students, but also everyone who wishes to seek knowledge about waste management. I wish to not only congratulate but also offer my best wishes to the editors, the contributing authors, and the publishers for bringing out a consolidated book.

Sultan Ahmed Ismail, PhD, DSc,
Soil Biologist and Ecologist
Member, State Planning Commission, Government of Tamil Nadu, India

Preface

Solid waste generation is a phenomenon which is directly related to population growth and the amount of waste produced. With the passing decades, urbanization, industrialization, and uncontrolled exploitation have added complexity to this phenomenon. Waste management is a global requirement to achieve a sustainable environment. Consumer patterns, socio-economic status, and living standards all impact waste generation per capita. Solid waste management is also influenced by several driving factors, such as environmental, socio-economic, and political factors, which vary according to regions and countries. Due to the lack of proper waste management techniques and practices, the condition is more serious in developing countries as when compared with developed countries. Improper waste management practices contribute significantly to greenhouse gas emissions which in turn impact climate change. However, not much attention has been given in this regard as it is always linked with direct pollution sources. In this respect, waste-to-energy resources seem to be the most effective and efficient solutions. The demand has arisen where integration of various technologies along with economic benefits is needed to achieve sustainability with maximum utilization of waste.

This book provides the wholesome picture of the waste and its management techniques with all the recent advancements, also providing the necessary projections for the future outlook, to maximize the value-added products for environmental sustainability on a cost-effective basis. This book will be a crucial guide for managers, environmental consultants, engineers, and remediation specialists. It emphasizes the practices, problems, and management of a broad variety of industrial solid waste. This book has suggested using a global lens to facilitate a better understanding for the utilization of sustainable tools to combat all types of problems. Innovative ideas and valuable information regarding all the necessary efficient technologies are provided, which can further help government and other organizations to manage solid waste-related issues. In addition, much more information for waste minimization and regulation is provided that can be utilized in conjunction with other integrated solutions and equipment. Chapters in this book provide detailed information for the readers to understand the topic from the grassroots level. Each chapter provides sufficient details to help readers relate to the problems and solutions in a practical manner. Case studies included in the chapters give appropriate practical information to analyze the concept in the real world.

Chapter 1 entitled "Understanding Waste: Type, Characterization, and Environmental Impacts" describes that waste is a universal problem that impacts every part of our life. Waste types can range from hazardous to non-hazardous, and therefore proper waste characterization is crucial to ensuring good waste management. This characterization must take into account physical, chemical, and biological properties of waste. Waste management is critical not only for public health and safety, but also for long-term environmental sustainability.

Chapter 2 entitled "Waste and Its Current Status: A Middle East Perspective" focuses on the Middle East, which is one of the most fertile places for waste generation in the world. The wasteful lifestyle, ineffective legislation, obstacles to providing infrastructure, indifferent public behavior, and lack of environmental awareness are among the main factors that lead to the growth of waste problems in the Middle East.

Chapter 3 entitled "Challenges and Opportunities in Waste: Indian Aspect" describes that every country needs to upgrade by urbanization, industrialization, etc. We cannot stop the growth of population and development of megacities. So, the process of rapid urbanization definitely increases waste generation, but we must improve and make advanced steps in the waste management systems. This is the main challenge for India. The government must focus and make efforts on the improvement of existing waste management.

Chapter 4 entitled "Waste Minimization, Management, and Resource Recovery: Livestock Waste" states that production activities in animal husbandry and animal food consumption always generate by-products, namely, solid and liquid waste, which pollute the environment. So far, conventional waste management strategies still tend to be reactive, namely, responding after the formation of waste, not carrying out prevention or prevention from the start, but making improvements after pollution occurs. As a result, it requires high costs to repair environmental damage, and even environmental damage continues to increase.

Chapter 5 entitled "Industrial, Construction, and Demolition Waste: Management and Resource Recovery" deals with an overview of uncontrolled industrial, construction, and demolition waste, which could unintentionally release hazardous substances into the environment, , where they serve as a haven for pathogenic microbes and pose serious health risks. Systematic and efficient management of solid waste from the construction and processing industries is needed. Resource recovery technologies through physical and biological treatment are becoming a promising technology when it comes to improving the carbon footprint of both industries.

Chapter 6 entitled "Health Hazards and Waste Management" gives insights into how the improper management of different types of waste can lead to exposure to pathogens, chemicals, and other contaminants that can have adverse effects on human health. Innovative waste management methods are increasingly being adopted to mitigate the health risks associated with solid waste management. This chapter focuses on the intersection of health hazards and waste management.

Chapter 7 entitled "Biological Hazardous Waste Management of Animal Husbandry and Food Processing Industry" focuses on several solutions for handling waste from the livestock and food processing industries. Adopting effective biohazardous waste management practices, the livestock and food processing industries can minimize environmental impacts and protect the health and safety of workers and the public.

Chapter 8 entitled "Ecological and Human Health Risk Assessments in Water and Soil Samples from Beris Lalang Waste Dumpsite, Kelantan, Malaysia" states that open dumpsites present ecological and human risk in relation to exposure to heavy metals. The condition of water in Beris Lalang was found to be unsuitable due to the HP1> 100, while the soil is classified as not contaminated because PERI is less than 40. The population around the vicinity of Beris Lalang Dumpsite are at low risk of developing cancer and non-cancer health problems as the overall health and cancer-related indices are at a lower level.

Chapter 9 entitled "Wealth from Municipal Solid Waste" describes that generation of municipal organic waste (MOW) is continuously increasing throughout the world. Effective management of this waste has become a critical environmental and public health issue globally. Proper waste management is necessary to prevent the adverse effects on the environment and human health caused by improper disposal of MOW. This chapter provides information about the different types and characterization of MOW. It also discusses the current status of waste management worldwide, challenges and opportunities in waste management in India, and various waste management technologies used in different industries.

Chapter 10 entitled "Plastic Waste and Its Management Strategies: Recent Advances" offers the idea of diverse applications, chemical compositions, and properties of plastics being extensively used around the globe. The more plastics are used, the higher the generation of plastic waste. Both the marine and terrestrial life are seriously affected by plastic waste. Conventional plastics, which are non-biodegradable, require a longer time for degradation, show high resistance to aging, and get converted to small debris that accumulates in the environment.

Chapter 11 entitled "Nanowaste and Its Management: New Waste Management Paradigm" provides an overview of the current status and challenges associated with the safe disposal and recycling of nanomaterials. We begin with understanding the different nanomaterial types from day-to-day and industrial applications that end in landfills and the environment as nanowaste. We discuss advanced methods to characterize and classify various types of nanowaste, and the related environmental and health risks.

Chapter 12 entitled "Algal Nanoparticles as a Green Approach for Bioremediation of Heavy Metals from Wastewater" focuses on the benefits of green nanomaterials for bioremediation of heavy metals as compared to conventional practices. Using microalgae, production of nanoparticles offers a greater boost to biotechnology for sustainable production and cost reduction. Interaction between pollutants, microalgae, and nanomaterials is important. There may be positive and negative effects produced because some nanomaterials stimulate the microalgae, while others are toxic.

Chapter 13 entitled "From Crisis to Opportunity: Advancing Plastic Waste Management in the Age of Nanowaste and COVID-Related Waste" provides useful insights for policymakers, waste management practitioners, and researchers to address the global issue of plastic waste and promote a cleaner and healthier planet. Furthermore, this chapter focuses on the emerging issues of nanowaste and COVID waste management approaches. This chapter discusses the challenges and risks associated with managing these types of waste and presents techniques for their safe handling and disposal.

Chapter 14 entitled "*Pinus roxburhii* Sarg. Needle Management to Combat Forest Fire: Status, Prospects, and Constraints" addresses different measures that have been already implemented to manage these needles, such as control burning, preparation of fire lines, pine needle check dams, bailing and briquetting, bio-gasifiers, fiber, and composite wood. Simultaneously, constraints associated with the utilization of the pine needles and potential future strategies for the management of the pine needles are discussed.

Chapter 15 entitled "Waste Management Practices of Riverside Communities at the Limbang River Sarawak, Malaysia" aims to understand better the relationship between the riverside community and waste management practice by determining the socio-demographic profile, river water utilization, and waste disposal practices at the Limbang River. The riverside communities at the Limbang River use the river for many purposes, mainly for transportation. The increasing population has caused the river water quality to deteriorate, where more household waste and wastewater are disposed directly into the river and potentially cause river water contamination.

Dr. Rajeev Pratap Singh
Banaras Hindu University, Varanasi, India
Ibha Suhani
Banaras Hindu University, Varanasi, India
Prof. Norli Ismail
Universiti Sains Malaysia, Malaysia
Dr. Vaibhav Srivastava
University of Allahabad, Prayagraj, India

Acknowledgments

First and foremost, we thank the Almighty God for sustaining the enthusiasm with which we plunged into this endeavor. We avail this opportunity to express our profound sense of sincere and deep gratitude to all the people who are responsible for the knowledge and experience we have gained during this book project.

The editors extend their sincere gratitude to all contributing authors from around the globe for raising their views in the form of chapters on every current important issue of waste and its management in the present book. We would like to acknowledge the valuable contributions of all reviewers who played a pivotal role in improving the quality and presentation of manuscripts. Rajeev Pratap Singh and Ibha Suhani are extremely thankful to the Institute of Environment and Sustainable Development, Banaras Hindu University, Uttar Pradesh, Varanasi, India. Vaibhav Srivastava is thankful to the University of Allahabad, Prayagraj, India. Norli Ismail is thankful to the School of Industrial Technology, *Universiti Sains Malaysia, Malaysia.*

About the Editors

Rajeev Pratap Singh is currently working as Associate Professor in the Institute of Environment and Sustainable Development, Banaras Hindu University, Varanasi, India. He also works in the area of waste management and has worked on various kinds of waste, such as fly ash, sewage sludge, tannery sludge, palm oil mill waste, and contained water irrigation. He has published significantly on waste management and similar topics in reputed journals. Rajeev Pratap Singh has received several international awards, such as the 'Green Talent' award from Federal Ministry of Education and Research (BMBF), Germany; the Prosper.Net Scopus Young Scientist award, the DST Young Scientist Award, Prof Archana Sharma Memorial Award by ISCA, Kolkata, and Prof. C. N. R. Rao Award for Excellence in Scientific Research by Banaras Hindu University. Also, Singh has co-edited 10 books and more than 50 highly cited research and review articles on different aspects of environmental science. He has received a Water Advanced Research and Innovation (WARI) Fellowship, a fellowship supported by the Department of Science and Technology (DST), Government of India, the Indo-US Science and Technology Forum (IUSSTF), University of Nebraska-Lincoln (UNL), and the Robert Daugherty Water for Food Institute (DWFI). He is also a fellow of the National Academy of Agricultural Science (NAAS), New Delhi, India.

Ibha Suhani has completed her MSc in Environmental Science from Banasthali Vidyapith, Rajasthan, India. She has completed her Master of Philosophy in Environmental Science from the Institute of Environment and Sustainable Development (IESD), Banaras Hindu University, Varanasi, India. She is a PhD research scholar currently working in the field of environmental science, ecotoxicology, solid waste management, and sustainability at IESD. During her doctoral research she has worked on the current issue of salt-affected lands by using amelioration techniques through utilizing compost and vermicompost. Currently she is working to check the threshold of earthworms for naturally sodic soil and organic amendments.

Norli Ismail holds a BSc (honors) in Environmental Science from Universiti Putra Malaysia and completed her MSc in Chemical Processes and PhD in Environmental Technology at Universiti Sains Malaysia (USM). Norli joined the School of Industrial Technology as a lecturer attached to Environmental Technology Division, USM, in October 2003. She has research experience in various areas of environmental science and technology with emphasis on water quality, management issues, and treatability studies in relation to water, wastewater, and analytical testing. She is actively involved in the technical aspects of biological and physicochemical treatability studies, bioremediation and bioaugmentation, environmental analytical techniques, sampling and data validation, and kinetic studies of the water and wastewater treatment. Norli leads a contract research project on wastewater treatment and acclimatization studies for a semiconductor industry, and food and beverages for local and international companies. She is a consultant to a company for sewage treatment plant development and modification and introduced a bioremediation and bioaugmentation concept for latex effluent treatment. Norli collaborated with Osaka Gas Ltd., Japan, on hydrothermal gasification of palm oil mill effluent (POME) attached to Eco-energy Department. She also collaborated on research work with Hohenheim University, Germany. Her research focuses on low-cost treatment technologies of different types of wastewater that may be used in developing countries. Her ongoing studies are related to hybrid sewage treatment plants associated with biological nitrogen removal of sewage, enhancement and purification of biogas, bio-flocculant and biocarrier development for wastewater, and solid waste treatment. Her research interests also extend to micro-biology of wastewater, pharmaceutical waste, heavy metals bio-sorption, recovery, and XOC biodegradation.

Vaibhav Srivastava is currently working as Assistant Professor in the Department of Botany, Faculty of Science, University of Allahabad, Prayagraj, Uttar Pradesh, India. Srivastava obtained his masters (MSc) from the University of Allahabad, Prayagraj, India, and his PhD from Banaras Hindu University, Varanasi, India, both in Botany. During his doctoral research , he has presented innovative ideas and tried to develop an economically cheaper and sustainable method of solid waste management and performed vermicomposting of municipal solid waste (MSW) as a circular economy-based management approach. Moreover, he characterized and quantified MSW of Varanasi city for the first time. He also assessed the impact of MSW vermicompost on soil health and plant growth responses. During his postdoctoral research at IIT Gandhinagar, he worked on wastewater based epidemiology surveillance of COVID-19 in Ahmedabad. His research interests include solid waste management, organic waste recycling, nutrient and energy recovery from different biowaste, compost science, plant stress physiology, emerging contaminants, and sustainability science. He has received National Post-Doctoral Fellowship Award-2021 from the Science and Engineering Research Board (SERB); Outstanding Researcher Award 2019 from the International Institute of Organized Research, Chandigarh, India; and International Travel Support Award from the Science and Engineering Research Board (SERB) and Council of Scientific and Industrial Research, New Delhi.

Contributors

Hasmah Abdullah
Biomedicine Program, School of Health
  Sciences, Health Campus
Universiti Sains Malaysia, Malaysia

Ghassan F. Al Samarrai
Department of Biology, College of
  Education, University of Samarra,
  Samarra, Iraq

Siti Noor Aliza Apandi
SMART College (Faculty of Health and
  Science), Kuala Lumpur,
Federal Territory of Kuala Lumpur,
  Malaysia

Nurhanim Abdul Aziz
Department of Environmental
  Engineering, Faculty of Engineering
  and Green Technology (FEGT),
  Universiti Tunku Abdul Rahman,
  Malaysia

Daulat Ram Bhardwaj
Department of Silviculture and
  Agroforestry, Dr. YS Parmar
  University of Horticulture and
  Forestry, Solan, Himachal
  Pradesh, India

Rachna Bhateria
Department of Environmental Science,
  Maharshi Dayanand University,
  Rohtak, Haryana, India

Bansal Deepak
JBM Group, Gurugram, Haryana,
  India

Bhagawan Dheravath
Department of Environmental Science
School of Earth Sciences, Central
  University of Rajasthan,
  Bandarisindri, Ajmer,
  Rajasthan, India

Widad Fadhullah
Environmental Technology Division,
  School of Industrial Technology,
  Universiti Sains Malaysia, Malaysia
Renewable Biomass Transformation
  Cluster, School of Industrial
  Technology, Universiti Sains
  Malaysia, Malaysia

Neha Verma Gour
School of Energy and Environmental
  Studies, Devi Ahilya
  Vishwavidyalaya, Indore, India

Ellin Harlia
Department of Animal Products
  Technology, Faculty of Animal
  Husbandry, Universitas
  Padjadjaran-Indonesia

Yuli Astuti Hidayati
Department of Animal Products
  Technology, Faculty of Animal
  Husbandry, Universitas
  Padjadjaran-Indonesia

Norli Ismail
Environmental Technology Division,
  School of Industrial Technology,
  Universiti Sains Malaysia, Penang,
  USM, Malaysia
Renewable Biomass Transformation
  Cluster, School of Industrial
  Technology, Universiti Sains
  Malaysia, Penang, Malaysia

Mohd Hafiidz Jaafar
Environmental Technology Division,
School of Industrial Technology,
Universiti Sains Malaysia,
Malaysia
Renewable Biomass Transformation
Cluster, School of Industrial
Technology, Universiti Sains
Malaysia, Malaysia

**Putri Anis Syahira Mohamad
Jamil**
Department of Environmental
Engineering, Faculty of Engineering
and Green Technology (FEGT),
Universiti Tunku Abdul Rahman,
Malaysia

Florence Rutselin Kelanit
Environmental Technology Division,
School of Industrial Technology,
Universiti Sains Malaysia,
Malaysia

Rana I. Khaleel
Department of Architecture, College of
Engineering, University of Samarra,
Samarra, Iraq

Nitin Khandelwal
Department of Hydrology, Indian
Institute of Technology Roorkee,
Roorkee, Uttarakhand, India

Dhirender Kumar
Department of Silviculture and
Agroforestry, Dr. YS Parmar
University of Horticulture and
Forestry, Solan, Himachal
Pradesh, India

Asha Kumari Kumawat
Central University of Gujarat,
Gandhinagar, India

Eulis Tanti Marlina
Department of Animal Products
Technology, Faculty of Animal
Husbandry, Universitas
Padjadjaran-Indonesia

Sharma Mona
Department of Environmental Studies,
School of Interdisciplinary and
Applied Sciences, Central University
of Haryana, Mahendragarh,
Haryana, India

Chaitanya Narala
University College of Engineering,
Science & Technology Hyderabad,
Jawaharlal Nehru Technological
University Hyderabad,
Hyderabad, India

Shreyosi Pahari
Department of Biotechnology, St.
Xavier's College (Autonomous),
Kolkata, West Bengal, India

Bhawana Pathak
Central University of Gujarat,
Gandhinagar, India

Muhammad Izzul Fahmi Mohd Rosli
Environmental Technology Division,
School of Industrial Technology,
Universiti Sains Malaysia,
Malaysia

Aryadeep Roychoudhury
Discipline of Life Sciences, School of
Sciences, Indira Gandhi National
Open University, New Delhi, India

Alen D. Rozario
Department of Biotechnology,
St. Xavier's College (Autonomous),
Kolkata, West Bengal, India

Mohd Saiful Samsudin
Environmental Technology Division
School of Industrial Technology
Universiti Sains Malaysia, Penang,
Malaysia
Renewable Biomass Transformation
Cluster School of Industrial
Technology, Universiti Sains
Malaysia, Penang, Malaysia

Muhammad Hasnolhadi Samsudin
Department of Environmental
Engineering, Faculty of Engineering
and Green Technology (FEGT),
Universiti Tunku Abdul Rahman,
Malaysia

Nurul Ilyana Sansudin
Environmental and Occupational Health
Program, School of Health Sciences,
Health Campus, Universiti Sains
Malaysia, Malaysia

Anuj Sharma
Department of Environmental Science
& Engineering, Guru Jambheshwar
University of Science & Technology,
Hisar, Haryana, India

Prashant Sharma
Department of Silviculture and
Agroforestry, Dr. YS Parmar
University of Horticulture and
Forestry, Solan, Himachal
Pradesh, India

Manoj Kumar Singh
Department of Agronomy, Institute
of Agricultural Sciences, Banaras
Hindu University, Varanasi, Uttar
Pradesh, India

Nisha Singh
Japan Agency for Marine-Earth Science
and Technology (JAMSTEC),
Yokosuka, Kanagawa, Japan

Ramnarayan Singh
School of Energy and Environmental
Studies, Devi Ahilya
Vishwavidyalaya, Indore, India

Sujata
Department of Environmental Science
& Engineering, Guru Jambheshwar
University of Science & Technology,
Hisar, Haryana, India

Pankaj Thakur
Department of Agribusiness
Management, A.N.D. University
of Agriculture and Technology,
Kumarganj-Ayodhya, Uttar
Pradesh, India

Ekta Tiwari
Natural Resources Management &
Environmental Sciences, College of
Agriculture, Food & Environmental
Sciences, California Polytechnic State
University, CA, USA

Supriya Vaish
Central University of Gujarat,
Gandhinagar, Gujarat, India

Kamlesh Verma
ICAR-Central Soil Salinity
Research Institute, Karnal,
Haryana, India

Siti Norabiatulaiffa Mohd Yamen
Environmental Technology Division,
School of Industrial Technology,
Universiti Sains Malaysia,
Malaysia

1 Understanding Waste
Type, Characterization, and Environmental Impacts

Asha Kumari Kumawat, Supriya Vaish, and Bhawana Pathak

1.1 INTRODUCTION

Waste is a complex and multifaceted problem that has potential consequences for our environment, economy, and society. Waste, in its broadest sense, refers to any unwanted or unusable material or substance produced by human activities. Everything from domestic waste to industrial by-products to hazardous chemicals and electronic waste falls into this category. The production of waste is directly influenced by urbanization, and improper waste management degrades the urban environment and poses health risks. The volume of garbage produced from various sources, including home waste, institutional and commercial waste, and industrial waste, has significantly increased as a result of increased production and consumption, solid material rejection, and regular waste generation in urban life. The majority of waste that emerge from a distinctive urban society consist of waste packaging materials, hazardous wastes, construction and destruction waste, medical waste, plant litter, etc. (Rajput et al., 2009). The amount of waste produced globally is tremendous. In 2018, the global production of municipal solid waste (MSW) was 2.01 billion tonnes, an average of 1.2 kg per person in a day. Globally, only 14% of plastic packaging is recycled; the remainder ends up in landfills, incinerators, or openly in the environment (MacArthur, 2013). The ratio of hazardous to nonhazardous medical waste is 4:1. However, approximately 15% of medical waste is both infectious and radioactive, rendering it the most perilous for our environment (Chu et al., 2023). According to World Bank report, East Asia and the Pacific generate 23% of global waste. Households generate the majority of waste; however, commercial and industrial waste is also included. Population growth and urbanization are anticipated to result in a 70% increase in waste generation by 2050. Rapid urbanization, industrialization, and population growth have led to an increase in waste production, but the expansion of medical facilities has added hazardous refuse and biomedical waste to the waste stream, which has detrimental effects on the environment. UNEP estimated that over 90% of waste is burned or thrown in the open in low-income countries, having a substantial negative environmental and human impacts (UNEP, 2015). Since 2000, the World Bank has given more than $4.7 billion to more than 340 programs of solid waste management globally due to the urgency of the waste challenge.

DOI: 10.1201/9781003543176-1

India, the world's second most populous country, has increased its urbanization rate from 27.81% in 2001 to 31.16% in 2011 as per the census of India report, 2011. Increase in human populations, industrialization and development of new technologies, and waste management has become increasingly complicated (Akinbile et al., 2019). In developing countries, because of over population and urbanization occurrence in the past two decades, the development of exemplary waste management systems is required. Uncontrolled open dumping is a widespread practice in Indian cities that has a severe impact on the ecosystem and contaminates the groundwater, air, and soil (Gupta et al., 1998). Greater than 90% of the MSW produced in India is not properly disposed of directly in open spaces (Das et al., 1998). The greater the growth of population and economic development, the greater the amount of MSW generated (Krishnamurti & Naidu, 2003). The proper management of waste is obligatory for the protection of the environment and the public health. By instituting efficient waste management processes, we can reduce the detrimental effects of waste on our landscapes and preserve natural resources for future generations, and it would be a source of income, if proper research and development will be conducted on useful by products of waste such as gas production, composting, reusable products, and integrated waste management system.

1.1.1 PRESENT STATUS OF WASTE WORLDWIDE

In 2012, the world produced about 1.3 billion tons of waste, and by 2025 and 2050, respectively, it was predicted that waste output would increase to 2.2 and 3.4 billion tons annually (Kumari & Raghubanshi, 2023). 2.01 billion total waste is generated annually (Kaza et al., 2018). At least 33% of the world's 2.01 billion tonnes of MSW is not processed safely. Over the next 30 years, increased urbanization, population expansion, and economic development will increase worldwide garbage by 70% to 3.40 billion tons. (Kaza et al., 2018). Cities, which serve more than half of the world's population and produce more than 80% of the world's GDP, are key to solving the problem of global waste. As per the data provided by the World Bank, 2019, the global waste production in various regions has been identified and the projected increase in waste generation in the future is as follows (Table 1.1).

TABLE 1.1
Region-wise waste generation and prediction

	Millions of tons per year		
Region-wise waste generation	In 2016	In 2030	In 2050
North Africa and Middle East	129	177	255
Sub-Saharan Africa	174	269	516
North America	289	342	396
Latin America and Caribbean	231	290	369
Europe and Central Asia	392	440	490
South Asia	334	466	661
East Asia and Pacific	468	602	714

1.1.2 PRESENT STATUS OF WASTE IN INDIA

The amount of solid waste produced in India each year is 62 million tonnes; however, only 43 million tonnes are collected and 11.9 million tonnes are treated. This indicates a significant gap in the nation's solid waste management (CPCB, 2020). According to estimates, the urban areas of India produce a daily amount of MSW ranging from 130,000 to 150,000 metric tonnes. This translates to a daily per capita waste generation of 330–550 grams for urban residents. This results in an approximate annual production of 50 million metric tonnes. Based on current trends, it is projected that this amount will increase to approximately 125 million metric tonnes per year by 2031 (NITI Aayog Report, 2021). Indian cities generate around 15,000 tonnes of plastic waste daily, yet only a fraction is recycled. It pollutes landfills, rivers, and oceans (Kumar, 2019). China and India will contribute one-third of global waste by 2050 because population and lifestyle increase urban waste. By 2036, urban India would generate 131.2 million tonnes of waste, five times more than in 2011 (Das, 2020).

Ahmedabad, Greater Mumbai, Delhi, Kanpur, Jaipur, Lucknow, Surat, and Pune account for 36% (8 out of 22) of total waste generation. In total, 13.6% (3 cities out of 22) create garbage in the 500–1000 TPD range (Ludhiana, Indore, and Vadodara). Asansol, Agartala, Faridabad, Chandigarh, Guwahati, Kochi, Jamshedpur, Mangalore, Kozhikode, Mysore, and Shimla generate less than 500 TPD (11 out of 22 cities). Ahmedabad, Chandigarh, Asansol, Delhi, Greater Mumbai, Faridabad, Jaipur, Kanpur, Jamshedpur, Ludhiana, Lucknow, Pune, Mangalore, and Vadodara supply more than 75% of their garbage to dumpsites (63.6%, 14 out of 22 cities). Out of the 17 class I cities, 47.05% (8) have a single dumpsite, 29.4% (5) have two, 5.88% (1) have three, and 11.76% (2) have four. Greater Mumbai and Ludhiana furnish the dumpsite with 100% of the waste collected.

1.2 TYPES OF WASTE

The categories of waste and site of generation are crucial factors in waste management. Commonly waste is classified into two types: biodegradable and non-biodegradable waste. These two types of waste are explained below: another classification of waste is divided into three types: solid waste, liquid waste, and gaseous waste shown in Figure 1.1. These types of waste can originate from diverse sources. Each form of waste has distinctive properties that determine its effect on environmental health, the economy, and public health.

(A) Biodegradable Waste
Examples of waste include garden and park garbage, food and kitchen waste from homes, restaurants, mass catering services, retail establishments, and food processing facilities. Biodegradable waste refers to any organic matter that can be broken down by biological processes into simpler, non-toxic substances. These substances can then be assimilated by the environment without causing harm. Biodegradable waste contains high moisture content, around 60%–90% and this type of waste can come from various sources, including households, industries, and agricultural activities (Garcia

FIGURE 1.1   Types and sources of waste.

et al., 2005). Examples of biodegradable waste include food waste, yard waste, paper products, and some types of plastics, such as biodegradable plastics made from plant-based materials. Proper management of biodegradable waste involves separating it from non-biodegradable waste, composting it, or using it as a source of renewable energy through processes such as anaerobic digestion.

(B) Non-biodegradable waste

Non-biodegradable waste refers to any material that cannot be broken down by biological processes into simpler, non-toxic substances within a reasonable timeframe. Such waste can linger in the environment for hundreds or even thousands of years, polluting the environment and harming ecosystems as well as human health. Non-biodegradable waste includes a wide variety of materials, such as plastics, metals, glass, and some chemicals. These materials are often used in everyday products such as packaging, electronics, and construction materials, and their improper management can lead to pollution and harm to the environment and human health (Bharadwaj et al., 2015). Landfills, for example, can release toxic gases and leachate into the soil and groundwater, contaminating nearby ecosystems. To address the issue of non-biodegradable waste, it is essential to reduce its production and promote the use of more sustainable materials. Recycling and upcycling can also help to reduce the amount of waste that ends up in landfills and incinerators, prolonging the lifespan of non-biodegradable materials.

1.2.1 Solid Waste

Solid waste is a complex mix of inorganic and organic substances with different biodegradability, toxicity, and reuse potential. Solid waste includes garbage from domestic, businesses, building sites (construction and demolition [C&D]), industrial, municipal, agricultural, biomedical, and electronic waste. It can be hazardous or non-hazardous based on its qualities and environmental and health effects. A massive amount of complicated solid waste has been produced as a result of the continued overuse of abundant resources (Singh et al., 2011). The amount and variety of solid waste produced by industrial, mining, residential, and agricultural operations have increased because of rising urbanization, rising population, and rising standards of living carried on by technological advancements. Sugarcane bagasse, grain and wheat stalks and husk, vegetable waste, food goods, tea, oil production, jute fiber, groundnut shell, wood mill waste, coconut husk, cotton stalk, and other agricultural waste are major sources of garbage. Coal, combustion leftovers, bauxite red clay, and tailings from aluminum, iron, copper, and zinc are major industrial non-hazardous inorganic solid waste. Solid waste management is a matter of concern for both national and municipal governments due to its impacts on the environment and human health, as well as its capacity to significantly enhance resource preservation (Ghinea et al., 2016). Insufficient skilled labor, inconsistent waste collection services, inadequate waste collection equipment, insufficient legal regulations, and limited resources are contributing factors to ineffective solid waste management in developing nations (Al-Khatib et al., 2015).

1.2.2 Liquid Waste

Liquid waste is generated by a variety of sources, including commercial establishments such as retail locations, fuel depots, mines, and quarries, along with marine vessels, manufacturing plants, workplaces, and private residences. The mixture consists of wastewater from, laundry effluent, water containing fats, oils and greases (FOGs), chemicals, sewage sludge lavatory waste, and kitchen sink waste. The chemical makeup of the substance is typically comprised of approximately 99.9% water and 0.1% organic and inorganic contaminants. The categorization of liquid waste can be primarily based on three sources, namely, domestic, industrial, and agricultural. Liquid waste generated from household sources can be distinguished into two types: black water, consisting of human excreta, and greywater, which does not contain excreta. The nature of liquid chemical waste can vary depending on the raw materials, products, and processes used in the respective industries. Due to accelerated urbanization, municipal wastewater has increased, resulting in 0.1–30.8 kg of sewerage per population equivalent per year (Syed-Hassan et al., 2017). The petroleum and refining industries generate a substantial volume of liquid waste in the form of oil. The splitting of oils in the aquatic environment poses a significant risk to the survival of aquatic species, beneficial microorganisms, and aquatic flora, as noted by Tangchirapat et al. (2007). The liquid waste produced by hospitals and laboratories consists of infectious materials, including blood and bodily fluids, cultures of infectious agents, and biological waste, as well as discarded liquid vaccines and cultures from dishes (Biswal, 2013). Additionally, the waste may contain chemically hazardous substances such as formaldehyde, which is obtained from pathology labs, autopsy, dialysis, and preserving, as well as mercury from wrecked thermometers, sphygmomanometers, and dental amalgams, and solvents from pathology and preserving procedures. Furthermore, the waste may contain radioactive isotopes

1.2.3 GASEOUS WASTE

Any waste that is in a gaseous state, such as poisonous fumes, greenhouse gases, and air pollution, is referred to as "gaseous waste." Agricultural operations, transportation, and industrial processes are just a few of the causes of gaseous waste. The majority of gaseous waste generated from chemical industries is a result of the industrial processes as byproducts, which include CO_2, CH_4, NOx, SO_2, volatile organic compounds (VOCs), and particulates, these byproducts cause respiratory issues, acid rain, global warming, and other environmental issues (Hussain et al., 2021). Cleaner energy, pollution control, industrial improvements, and sustainable agriculture reduce gaseous waste. Proper implementation of government policies and regulations also plays a key role in the reduction of gaseous waste emissions.

1.3 SOURCES OF WASTE/WASTE GENERATION

Waste generation increases with the increase in population because basic needs (food, shelter, and cloth) and other necessities are required for survival. It can be generated from various sources which are difficult to divide in a systematic manner but crucial in waste management. Therefore, here are some waste generation sources in broad categories, as shown in Figure 1.1.

1.3.1 MUNICIPAL WASTE

Waste produced by human settlements, small industries, commercial, and municipal operations is referred to as MSW. Plastics, organic (compostable) materials, textiles, wood, and others are MSWs. Besides these, according to Singh et al. (2011), wastewater treatment plants (sewage sludge), domestic waste (glass, metals, paper, etc.), and waste from public areas such as parks and streets are also sources of MSW. The composition of MSW varies depending on the source and location. Millati et al. (2019) stated that MSW includes recyclables (biodegradable plastic, paper waste, metals, glass, etc.), compostable organic substances (green waste, vegetable scraps, food waste, food spoiled paper, manure, yard trimming, non-hazardous wood waste, etc.), and medical waste (blood-stained cotton, disposable needles, sanitary napkins, broken glass, body part, tissues, etc.). Population, income, consumption, and geography affect MSW production. Population and income are the biggest contributors to MSW. The main source of MSW generation in the world is domestic or household activities of increased population. MSW volume is much higher in urban areas as compared to rural areas. However, India's 70% population lives in villages.

1.3.2 ELECTRONIC WASTE (E-WASTE)

E-waste refers to unwanted electronic items, including devices that are no longer working and have reached or surpassed the end of their useful life. Electrical items that rely on electricity to function include computers, televisions, cellphones, office and medical equipment, VCRs, stereos, copiers, microwaves, fans, fax machines, etc. E-waste stream is considered hazardous because there are serious risks to human and environmental health from e-waste because of the presence of high quantities of heavy metals (mercury, lead, and cadmium), toxic chemicals, polychlorinated biphenyls, halogens, and bromine (Ilankoon et al., 2018). E-waste grows between 3% and 5% each year, making it one of the fastest-growing waste streams worldwide (European Parliament, 2020). Therefore, it gained significant interest from the scientific community, business leaders, members of the media, and governmental bodies worldwide (Chen et al., 2015). E-waste recycling, including electrical waste and electronic assembly components like mercury switches, accumulators and other batteries, active glass bridges from cathode-ray tubes and other active glass, PCB capacitors, etc., need to be controlled due to hazardous components. Eco-friendly e-waste recycling guidelines have already been issued. Extended producer responsibility may oblige electronics manufacturers to centralize their e-waste.

1.3.3 Biomedical Waste

The waste generated from healthcare practices such as veterinary hospitals, dental practices, laboratories, hospitals, clinics, blood banks, and vaccinations of humans or animals is called bio-medical waste (BMW). BMW contains sharps: needles, scalpel blades, anatomical body parts, microbiology cultures, blood samples (pathological in nature), dressings, catheters, and I.V. lines (infectious in nature) and these require special handling and disposal (Babu et al., 2009). Radioactive trash, mercury-containing devices, and polyvinyl chloride (PVC) plastics are other healthcare waste. Approximately 75%–95% of BMW is classified as nonhazardous, while the remaining 10%–25% is considered hazardous (Datta et al., 2018). The perilous component of biomedical waste poses potential physical, chemical, and microbiological threats to both the general public and healthcare personnel involved in the management, processing, and elimination of said waste (Li & Jenq, 1993). Handling, storing, transporting, and dispersing of biomedical waste in healthcare facilities is governed by stringent rules and regulations. This involves separation, sterilization, and disinfection, as well as disposal by incineration, internment, or other approved methods. Biomedical waste management is crucial for preventing the transmission of infectious diseases, safeguarding public health, and preserving the environment. Biomedical waste improperly handled can spread dangerous diseases, pollute the environment, and risk public health. Therefore, suitable methods and standards for biomedical waste disposal are needed.

1.3.4 Construction and Demolition Waste

The term C&D waste pertains to the substances produced during the process of constructing, repairing, restoring, remodeling, rehabilitation, renovating, and demolishing buildings, including residential properties, commercial establishments, and infrastructure. C&D waste encompasses a range of materials, including but not limited to metals, bricks, concrete, wood, glass, and plastics. Some components of demolition waste are toxic or hazardous like plasterboard breaking down in landfills, releasing poisonous hydrogen sulfide (Ponnada & Kameswari, 2015). According to a 2021 UNEP report, C&D trash would reach 2.2 billion tonnes by 2025, rising 25%

from 2018. C&D waste accounts for 20%–30% of garbage in most nations, making it a significant environmental issue. Governments and industry stakeholders should work together to encourage sustainable practices, create and uphold laws and regulations, and invest in C&D waste management infrastructure. This entails enhancing garbage collecting and sorting infrastructure, boosting the use of recycled supplies in construction, and fostering public understanding and education on sustainable practices.

1.3.5 INDUSTRIAL WASTE

Industrial waste is the byproduct of various industrial processes, including but not limited to manufacturing, construction, mining, cleaning, and agriculture. It may produce in the form of solid, liquid, or gaseous and is comprised of a diverse range of substances, encompassing chemicals, metals, plastics, hazardous and non-hazardous waste, radioactive substances, wastewater sludge, and other materials (Godswill, 2017). Most of these waste is minimally toxic and generated by a single source in enormous volumes. Coal-burning solids, including bottom debris, flying ash, and desulfurization of flue gas sediment, are types of industrial waste streams, paper and pulp processing sector, the steel and iron production sector, and the chemical industry are common contributors to industrial waste (Pichtel, 2014). Each year, billions of tonnes of industrial solid waste are produced and managed locally at industrial sites; this quantity is approximately four times that of MSW (Tammemagi, 1999). Worldwide, industrial waste poses a serious environmental challenge, and different nations and areas are currently dealing with it differently. Global industrial waste is predicted to be approximately 7.6 billion tonnes per year, and this number is expected to climb to 13.5 billion tonnes by 2050 (International Solid Waste Association, 2020). High-income countries generate a disproportionate quantity of industrial waste compared to low- and middle-income nations. However, low- and middle-income countries frequently lack the infrastructure, technology, and financial resources required to effectively manage industrial waste. Environmental and health effects of industrial waste include soil and water contamination, greenhouse gas emissions, and exposure to hazardous substances and pollutants (Godswill et al., 2023).

1.3.6 DOMESTIC WASTE

Domestic waste generated as a result of domestic activities is used to describe the garbage produced by households, including the objects thrown away during regular domestic tasks like cooking, sweeping, garden waste, cleaning, cloth washing, and other chores (Mohammed & Elias, 2017). Organic solid waste is generated from domestic sources, including food scraps, paper, cardboard, plastics, glass, metal, and other items. It constitutes a significant contributor to the presence of microplastics in the environment (Shi et al., 2023). Yoada et al. (2014) found in their research that food debris was discarded as waste by 93.1% of households, while 77.8% of households discarded plastic-based items as waste. Inappropriate domestic waste management causes malaria and diarrhea. To avoid health risks, pollution, and environmental harm, proper waste management is crucial. Some strategies for managing household waste include recycling, composting, and proper disposal. Separating recyclables from non-recyclables and correctly disposing of hazardous waste, such as batteries and chemicals, are also crucial. In India, the majority of MSW consists of domestic waste, 50% of which is biodegradable waste, and sustainable technologies such as incineration, pyrolysis/gasification, microbial fuel cells and bio methanation are being used for its treatment (Mondal & Yadav, 2018). Numerous methodologies and remedies have been proposed for the sustainable management of household waste, including the generation of biogas, heat, power, fuel for vehicles, and other related alternatives (Ajay et al., 2021). Governments and municipalities frequently have special regulations for handling household waste, which may include regular curbside collection, recycling initiatives, and facilities for disposing of hazardous waste. Using reusable bags, containers, and utensils, composting organic trash, and avoiding single-use plastics are some further actions people may take to reduce their home waste.

1.3.7 RADIOACTIVE WASTE

Materials that release ionizing radiation are considered to be radioactive materials. Waste containing radioactive material and showing radioactivity is called radioactive waste (Deng et al., 2020). As per the General Plan for Radioactive Waste, any substance or by-product that shows measurable levels of radioactivity and lacks any functional utility is classified as radioactive waste. This encompasses waste materials that are contaminated with liquids and gases. The primary sources of power generation include nuclear power plants that produce electricity, facilities that reprocess nuclear waste, and establishments that manufacture nuclear weapons. Research and medical procedures, such as pharmacology, also generate radioactive waste (Pichtel, 2014). Producing an annual output of 1 gigawatt of electrical power involves the presence of approximately 400 metric tonnes of hazardous heavy metals, which include radioactive uranium and thorium (Chapman & Hooper, 2012). These substances pose a risk to living organisms due to the potential for extended exposure to radiation, which can result in harm. Radioactive materials are a cause for concern due to their long-lasting nature. The duration of which radiation is emitted is commonly quantified and denoted as the half-life, which denotes the time taken for a specific quantity of radioactivity of a substance to decrease by half of its initial value. For example, the radioactive decay of uranium compounds show varying half-lives, with U232 having a half-life of 72 years and U236 having a half-life of 23,420,000 years.

1.3.8 AGRICULTURAL WASTE

Agricultural waste includes crop and animal waste. Agricultural waste is produced through various means such as the remains of crops, agro-industrial processes, livestock farming, and aquaculture (Koul et al., 2022). Crop residue, including the stalks, leaves, and other plant parts left in the field after harvesting, is a major agricultural waste component. Approximately 998 million metric tonnes of agricultural residue waste is generated annually (Agamuthu et al., 2009). Rural areas generate an estimated 2 metric tonnes of agricultural waste per day. In addition, cow dwellings and the sugar manufacturing sector generate an estimated 20 million metric tonnes of waste, which represents a valuable reservoir of nutrients and fertilizers (Iqbal et al., 2020). According to Searchinger et al. (2008), inadequate management of agricultural by-products (like rice residue combustion) results in the production of greenhouse gases (GHGs), including carbon dioxide (CO_2), carbon monoxide, nitrous oxide (N_2O), particulate matter, and methane (CH_4). This phenomenon poses a risk to human health (by causing respiratory issues), air quality, water quality, and the natural environment (UNEP, 2022).

In the context of waste-to-wealth and sustainable agriculture, agricultural waste is defined as biodegradable organic waste. Therefore, it is capable of undergoing composting. In rural areas, cow dung is readily available for sowing. Adding it to garbage boosts microbial activity and accelerates organic matter oxidation and stabilization. Compost gets nutrients from cow dung and boosts compost quality (Lokeshwari & Swamy, 2010).

1.3.9 HAZARDOUS WASTE

Any product, other than domestic and radioactive waste, that is incorrectly managed, stored, transported, and disposed of because of its bulk and corrosive, reactive, ignitable, poisonous, cancerous, and infectious properties pose severe threats to human health or the environment (Misra & Pandey, 2005). Hazardous waste comes from several sources, including businesses, manufacturing, agriculture, homes, MSW, and the bio-medical sector. Globally, an estimated 400 million tonnes of hazardous waste is generated annually, equating to approximately 60 kg per capita (Rucevska et al., 2015). Between 5% and 7% of the MSW is classified as hazardous (Couto et al., 2013). Hazardous waste may harm ecosystems and humans due to their non-degradability, biological amplification, severe toxicity, and lethality even at low concentrations (Kanagamani et al., 2020). Toxicity, phytotoxicity, genetic activity, and bio-concentration determine the hazard of it.

Classification of Hazardous Waste

For correct management of hazardous waste, USEPA categorized hazardous waste on the basis of its sources into lists—F, K, P, and U lists.

(i)   **F List:** The F list comprises hazardous waste that originates from unspecified sources, namely, diverse industrial procedures that may produce such waste. It comprises solvents that are frequently employed in metal immersion baths and sludge, effluent generated from metal plating procedures, and chemical compounds or their antecedents that contain dioxins. Examples of hazardous waste materials include benzene (F005), carbon tetrachloride (F001), and chrysalis acid (F004).

(ii) **K List:** The K list is a compilation of dangerous waste materials produced through particular manufacturing procedures. Industries that produce K-listed waste comprise of wood preservation, pigment production, petroleum

refining, chemical production, iron and steel production, explosives manufacturing, and pesticide production.

(ii) **P and U Lists:** Commercial chemicals, off-specification compounds, container residues, and spill residues are on the P and U lists. These lists include
formulations using the chemical as the only active ingredient, technical
grades, and commercial pure grades. A pesticide that is past its expiration
date and needs to be disposed of in bulk is an example of a P or U classified
hazardous waste.

1.4 WASTE COLLECTION

Local governments typically collect waste from residences through routine rubbish
collection or through special recycling pickups. If growing economies want to achieve
sustainable solid waste management, waste collection and transportation costs must
be reduced. Therefore, it is important to collect and transport solid waste in ways that
are both economical and environmentally friendly (Sulemana et al., 2018). The scents
produced by decomposition and the accumulated volumes should also be considered
when choosing the frequency of collection.

In Table 1.2, descriptions of the major categories of collection systems are
provided.

TABLE 1.2
The major categories of collection systems

	System	Description
I	Dumping at nominated location	Residents and other generators must dispose of their garbage in a masonry enclosure or at a designated area.
II	Shared container	Residents and other waste producers fill a container with their waste, which is then taken or emptied.
III	Block collection	Collector waits at designated spots while sounding horn or ringing a bell while residents deliver rubbish to the collection van.
IV	Curbside collection	Waste is either swept up and collected by a sweeper or left outside the property in a container to be picked up by a passing vehicle.
V	Door-to-door collection	Waste collector rings doorbell or knocks on each door and waits for residents to bring out their waste.
VI	Yard collection	A collection worker enters a building to remove waste.
VII	Transfer stations	Transfer stations are facilities where waste is consolidated and transferred to larger transportation vehicles for long-distance transport to disposal facilities.
VIII	Automated Collection	This system uses specially designed trucks equipped with automated arms to collect and empty standardized waste and recycling carts. This system is designed to increase efficiency and reduce the need for manual labor.

1.5 WASTE CHARACTERIZATION

Waste characterization is the process of identifying and analyzing the types and quantities of waste streams generated in a particular area or location. This information is used in the planning of waste management and reduction strategies, recycling programs, and the conservation of money and resources. Accurate characterization of waste is imperative for the formulation of sustainable long-term plans and development strategies.

Waste characterization involves several steps, including waste sampling, sorting, and analysis.

    (i)   **Waste sampling:** Selecting a representative sample of the waste produced in a specific location is known as waste sampling. The sample is often taken over a predetermined time period, like a week or a month, and it depends on the volume, kind, and location of the waste generated. Usually, the gathered samples are weighed, recorded, and taken to the lab for additional examination.

    (ii)   **Waste sorting:** Following collection, the waste sample is divided into various groups according to its makeup, composition, size, and shape. Typically, the waste is divided into different categories, including hazardous, recyclable, organic, and non-recyclable waste. Depending on the type and volume of garbage produced, the sorting process may utilize human or automated techniques, such as magnetic or optical sorting devices.

    (iii)   **Waste analysis:** After sorting, the waste is analyzed to determine its composition and properties. Measurements of physical, biological, and chemical qualities of waste, such as moisture content, pH, bulk density, particle size distribution, and calorific value, are often included in the analysis that can influence waste management decisions. The waste may also be analyzed for the presence of pollutants or hazardous compounds.

1.5.1 CHARACTERISTICS OF WASTE

1.5.1.1 Physical Characteristics

Physical properties of waste are those that may be measured or observed without changing its chemical makeup. Some of the common physical properties of waste include:

    (i)   **Odor:** The odor of waste is referred to its smell. VOCs and other gases, such as methane, ammonia, and hydrogen sulfide, can contribute to the smell of trash. These gases, which can be released into the atmosphere, are created as the organic material in the waste breaks down. It may have an impact on the way of life of those who live close to waste management facilities and may also reveal the existence of dangerous or hazardous materials in the waste.

(ii) **Appearance:** The aesthetic qualities of garbage, such as color and texture, are referred to as its appearance. It can have an impact on the aesthetic value of the environment while also indicating the sort of waste and its source.

(iii) **Moisture Content:** The amount of water that is contained in garbage is referred to as its moisture content. Waste that has a lot of moisture can be heavy and challenging to manage, transport, and store. The amount of moisture in waste is a key factor in determining whether the combustion of that waste is economically feasible or not. Additionally, it may result in the production of leachate, which endangers the environment and the human health.

(iv) **Density:** The mass of waste per unit volume is referred to as the density of waste. Low-density waste, such as plastics and paper, requires more space and larger storage and transportation facilities, but high-density waste, such as metals and glass, is heavy and challenging to carry.

The weight of a material per unit volume is referred to as its density. It can be determined by applying the formula

$$D = W_1 - W_2/V$$

where
D = Density.
W_1 = Weight of fresh waste and container used to measure waste
W_2 = Weight of waste container as measured
V = Volume of container used to measure waste

Density is important for waste storage, collection, and transportation as well as for designing hygienic landfills. Compaction of waste to the ideal density is necessary for the landfill to operate efficiently.

(v) **Particle Size:** The variety of particle sizes contained in the garbage is referred to as its particle size distribution. Measuring the size distribution of waste stream particles is crucial because it affects how mechanical separators and shredders are designed. It may have an impact on how garbage is handled and treated because large particles may need additional processing to be turned into a usable state.

(vi) **Calorific Value:** The quantity of heat that may be produced by the combustion of waste is referred to as its calorific value. Waste has a different calorific value based on its content and characteristics. In general, garbage that contains a lot of organic material, like food scraps, paper, and wood, has a high calorific value, whereas waste that contains a lot of inorganic material, like metals and glass, has a low calorific value. Waste with high calorific values can be used as a fuel source; it is an important quality in the waste-to-energy conversion process.

1.5.1.2   Chemical Characteristics

The chemical qualities of waste pertain to its chemical composition and reactivity. Some of the common chemical properties of waste include:

(i)   **pH:** The acidity or alkalinity of waste is indicated by its pH. The stability and reactivity of waste can be impacted by pH levels, which can provide risks to the environment and public health. Additionally, high alkalinity and acidity of waste can affect soil and water quality near dumping sites.

(ii)   **Chemical Composition:** The characteristics and behavior of waste are governed by its chemical makeup. The kinds and concentrations of chemical components in trash can have an impact on its toxicity, reactivity, and applicability for various waste treatment techniques. Chemical composition of waste contains organic and inorganic matter. The amount of carbon-containing chemicals found in waste is referred to as its organic matter content. It has the potential to influence the rate of biodegradation and methane production in landfills. On the other hand, inorganic waste is made up of substances devoid of carbon-hydrogen bonds. It consists of substances like metals, glass, plastic, and ceramics. Small amounts of organic material, such as plasticizers or additives, may also be present in inorganic trash.

(iii) **Reactivity:** Reactivity is the capacity of waste to engage in chemical processes like combustion or deterioration. It easily erupts or experiences strong reactions; therefore, when heated, exposed to water, compressed, or in typical handling circumstances, it bursts or burns violently, emitting hazardous fumes or vapors. For example, lithium-sulfur batteries and explosives.

(iv) **Toxicity:** Toxicity refers to a substance's capacity to harm living things. It causes long-term illness like cancer. Toxic waste (such as those containing lead, mercury, etc.) consumed or absorbed by human beings or animals can be dangerous or even lethal. When toxic waste is dumped on land, polluted liquid may seep out and contaminate the groundwater and constitute a serious threat to both the environment and public health.

(v)   **Volatility:** The potential of waste to evaporate or emit gases is known as volatility. This can occur when the waste is exposed to heat, such as during incineration or hot weather conditions. Organic waste materials that include VOCs are frequently linked to volatility. These consist of substances like fuels, chemicals, and solvents. These substances can emit dangerous gases and compounds into the air when they are heated or subjected to other circumstances that encourage vaporization.

(vi) **Ash Content:** The weight of residue left over from burning in an open crucible became visible in the ash content. Depending on the composition of the waste, ash content varies. For instance, waste that has a lot of organic material, like food scraps and paper, has a lower ash level than waste that contains a lot of inorganic material, like metals and glass.

(vii) **Biodegradability:** The ability of microorganisms to breakdown and break down waste is referred to as biodegradability. It influences the rate and degree of waste decomposition and can indicate whether waste is suitable for composting or anaerobic digestion.

(viii) Heavy Metal Content: Heavy metals are highly persistent in the environment and challenging to break down or remove because they are dense and have high atomic weights. Toxic metals including lead, mercury, chromium, arsenic, and cadmium are often present in electric waste, batteries, and other waste materials which leach into the environment and cause hazards to it. Heavy metal exposure can result in a variety of health problems, including brain damage, cancer, and developmental disorders. Pregnant women and children are more vulnerable to the consequences of heavy metal exposure.

According to Han et al. (2018), some natural factors like climate and geography of a particular waste generation location affect characteristics of waste. The varied quantities of rainfall, temperature, humidity, and agricultural activities that take place during the year all have an impact on how residential waste behaves. The factors that directly affect the moisture content of household waste are rainfall and humidity. In general, more humidity and more rainfall will cause the home waste to have more moisture. The methods and resources used for residence heating are mostly influenced by the outside temperature, which alters the properties of household waste. Agricultural and aquacultural activities will add their equivalent solid waste to the home waste stream during harvest seasons, which might increase the volume and make-up of the household waste generated. Waste characterization information can be utilized to build waste management plans that are suited to the specific demands of the location. Composting or anaerobic digestion, for example, may be recommended as a means of distracting organic waste from landfills if waste characterization reveals that a large amount of organic waste is being generated. If hazardous waste is discovered, steps can be taken to ensure that it is handled and disposed of properly to reduce the risk of harm to human health and the environment.

1.6 ENVIRONMENTAL IMPACT OF WASTE

Waste can have serious environmental consequences, both in short and long term. The management of all produced waste must be done in a proper way. Open dumping is a persistent issue both domestically and internationally. Disposing of garbage produced by communities and industry on land, in the sea, or in low-lying areas is a highly widespread practice. Sites may be entirely or primarily made up of home or industrial garbage. The open disposal of solid waste creates a number of environmental and health risks. Methane is created through the degradation of organic materials and contributes to global warming by igniting and exploding. Significant environmental effects are caused by waste. When waste is not managed properly, it can pollute our air, water, and soil, causing damage to plants, animals, and humans. For instance, landfills release methane gas, a powerful greenhouse gas that accelerates climate change. Many gases are created while the waste is being dumped. It contains greenhouse gases like methane (CH_4) (30%–60%) and carbon dioxide (CO_2) (34%–60%) that contribute to global warming and progressively worsen the environment (Swachh Bharat Mission Urban, 2020). Unsafe disposal of hazardous trash can contaminate the land and groundwater, endangering both human and animal health.

1.6.1 EFFECT ON SOIL

Soil is defined as a "dynamic biological system on the outermost layer of the earth where vegetation grows, comprising of minerals, organic components, and living organisms, provides construction materials" (Brady et al., 2008). Additionally, it also serves as a layer of protection and filtration that is placed over groundwater to lessen the effects of various hazardous pollutants (Venkatesan & Swaminathan, 2009). The weight of MSW on land has increased due to urbanization and industrialization, which is negatively affecting the soil's yield and biotic and abiotic qualities. Compared to the control sites, the soils at the disposal sites had higher pH, total dissolved solids (TDS), and electrical conductivity (EC) regimes. Lead (Pb), copper (Cu), nickel (Ni), chromium (Cr), and zinc (Zn) were among the heavy metal concentrations that were found to be greater at the dumping sites, except for cadmium (Cd), which had a higher value at the control site. Salinity is another major danger to soil; it results from incorrect human land use, including forest destruction, construction work, and industrialization. Salinity has an impact on the physical and chemical composition of the soil, contributes to soil degradation, and raises the salinity level of underground water (Oo et al., 2015).

Different metals are carried by waste and then transported to plants in various ways (Voutsa et al., 1996). Depending on the pollutants' propensity, water retained in the soil or underground water is where they end up. According to Shayler et al. (2009), metals such as Cd, Cu, Ni, Pb, and Zn can change the chemistry of the soil and have an effect on the organisms and plants that rely on the soil for sustenance. The reason is that the vegetation diversity is directly influenced by soil properties.

Additionally, Biro et al. (2013) showed in their research that the degradation of the land caused by changes in land use and land cover may have contributed to the change in soil characteristics caused by waste, which in turn has caused a drop in the productivity of soils. In ecological systems, the restoration of degraded soil is a challenging and protracted process that involves the creation of a vegetative community and the resumption of microbial activity (De Souza et al., 2013). Total nitrogen, pH, and the degree of erosion are critical variables that affect soil growth. According to Pallavicini et al. (2014), soil richness dropped as these parameters rose. Quantifying the eco-toxicological risk of composted residuals is a topic of growing attention. The review by Kapanen and Itavaara (2001) discussed the wide range of eco-toxicity test techniques for compost that use microbes, enzymes, soil organisms, and plants. Increases in the fungal and bacterial populations in amended soil as well as an increase in CO_2 production were effects of MSW-compost incorporation on soil biological characteristics.

1.6.2 EFFECT ON WATER

For human survival, water is a precious natural resource. On the one hand, there is a serious shortage of fresh water in the world, and on the other hand, any kind of groundwater supplies that are still left, are under severe quality stress from inappropriate waste generated and poor management due to the urbanization and industrialization processes.

Nagarajan et al. (2012) stated in their study that TDS, total hardness (TH), total alkalinity (TA), sodium (Na^+), magnesium (Mg^{2+}), chloride (Cl^-), fluoride (F^-), and nitrate (NO_3^-) concentrations were found to be higher than the upper permissible limit for drinking water in Erode city, Tamil Nadu, India when compared to Bureau of Indian Standards (BIS) and World Health Organization (WHO) standards. According to Nagarajan et al. (2012), this leachate affects the quality of groundwater by changing its EC, TDS, chloride (Cl^-), and sulfate (SO_4^{2-}), among other factors. Therefore, before designing a landfill site, it is important to consider both the availability of resources and the risk of subsurface water contamination. Similar findings were made by Vasanthi et al. (2008) when they examined the quality of the groundwater in various wells near the Perungudi dumping site in Chennai. They discovered elevated concentrations of pollutants such as TDS, EC, TH, Cl^-, chemical oxygen demand (COD), NO_3^-, and SO_4^-. Mor et al. (2006) revealed that groundwater samples taken from the Gazipur landfill site and its surrounding areas in Delhi had somewhat high concentrations of Cl^-, NO_3^-, SO_4^-, NH_4^+, phenol, Fe, Zn, and COD. This suggested that leachate percolation had contaminated the groundwater.

1.6.3 EFFECT ON AIR

In spite of the fact that methane-rich gas released from open dumping or landfills presents a chance for energy recovery, it is frequently viewed as a risk due to its flammability, capacity to combine with air to form explosive mixtures, and propensity to diffuse and advent outside the landfill's boundaries. As the migratory gas penetrates buildings and underground facilities situated on or near dump sites, it creates gas pockets and presents an explosion risk. Waste disposal or landfilling or its treatment may release air-polluting gases such as CO_2, SO_2, CH_4, NO_2, particulate matter, and many more. Particulate matter serves as the primary air pollution indicator, also called as particle pollution (Gangwar et al., 2019). Incineration process of MSW releases fly ash, mixture of other gaseous residues which are toxic for air quality (Rani et al., 2008).

There is evidence that the gas released from landfills and open waste combustion contains a variety of additional constituents (PM_{10}, $PM_{2.5}$, CO_2, and toxic substances) in trace proportions that are big enough to raise environmental and health issues, even though methane and carbon dioxide are its two main constituents (Wiedinmyer et al., 2014). Both CH_4 and CO_2 are greenhouse gases, but CH_4 has a global warming potential that is more than 25 times greater than CO_2's, given that it has an atmospheric residence time of 12–13 years (IPCC, 2023). The release of VOCs has the potential to raise cancer risks in the surrounding area as well as contribute to the production of ambient ozone. Trace gases may also influence methane production by limiting the growth of methanogens, as well as corroding gas recovery equipment/ instruments.

1.6.4 GLOBAL WARMING

Several researchers have measured the rates at which trash emits gases into the atmosphere. Methane and carbon dioxide emissions from landfill surfaces have a big impact on the greenhouse effect and global warming. Extensive characterization studies of the world's methane sources and sinks have been stimulated by recent increases in atmospheric methane concentrations. The IPCC study states that methane concentrations in the atmosphere have been rising on average between 1% and 2% annually. It is concluded that methane contributes about 18% toward total global warming (Saadatlu et al., 2022). Methane is created through the degradation of organic materials and contributes to global warming by igniting and exploding.

In addition to its environmental effects, waste also has economic repercussions. Waste management expenses, such as those for collection, transportation, and disposal, can be very expensive. As a result, many governments and businesses are seeking to reduce waste and improve their waste management practices to save money and increase productivity.

1.7 RULES AND POLICIES FOR WASTE MANAGEMENT

The obligations for hygienic waste management for Indian cities and residents were outlined in the country's solid waste policy. Based on the Supreme Court's submission of the Committee for Solid Waste Management's March 1999 Report on Class 1 Cities in India, which asked legislative bodies to follow the report's proposals and recommendations, this policy was created in September 2000. These also act as instructions on how to follow the MSW regulations.

    (i)   **Municipality Solid Waste Rules:** To stop the current unplanned open dumping of waste outside of city limits, the MSW rules have established a strict timetable for compliance: by the end of 2001, existing landfill sites must be improved; by the end of 2002, landfill sites must be identified and ready for operation; by the end of 2003, waste-processing and disposal facilities must be established; and by the end of 2004, a buffer zone must be established around such sites. Composting, vermicomposting, and other methods should be used to handle biodegradable waste, and landfilling should only be used for non-biodegradable inert waste and compost rejects. The regulations also call on towns to encourage recycling or reuse of segregated materials and to assure community involvement in waste segregation (by preventing the mixing of "wet" food waste with "dry" recyclables including paper, plastics, glass, and metal). The combustion of waste and dried leaves is prohibited. Industrial and biomedical waste cannot be combined with municipal garbage. Residents must deliver waste in accordance with the city's collection and segregation system, preferably by house-to-house collection at predetermined times in multi-container handcarts or tricycles (to avoid manual handling of waste), or directly into trucks stopping at street corners at regularly scheduled, predetermined timings. Additionally, residents must store their dry (recyclable) and wet (food) waste inside their properties until collection. Whether or not such a practice amounts to unfair competition is a matter of debate, but many governments take action against dumping to safeguard domestic industry. There are no conclusions drawn in the WTO agreement. It is often called the "anti-dumping agreement" and focuses on what actions governments can and cannot take in response to dumping. The following are various sections of government law in India concerning waste and its disposal:

- The Water Act (prevention and control of pollution), 1974
- The Air (prevention and control of pollution) Act, 1981
- The Environment (Protection) Act, 1986
- The Hazardous Waste Rules (management and handling), 1998
- The Biomedical Waste Rules (management and handling), 1998
- The Biomedical Waste Rules Amendment (management and handling), 2000 and 2003
- Municipal Solid Waste Rules (management and handling), 2000
- The Biomedical Waste Rules (Management and Handling), 2011

(ii) **Challenges and Opportunities:** India, being the fastest growing economy in the world, leads to urbanization, industrialization, increased population, R&D activities, and increased institutional and organizational activities which ultimately increase the production of waste. In 2025, India's metropolitan areas will produce 0.7 kilograms of waste per person each day, which is four to six times more waste than they did in 1999 (Kumar et al., 2017). Solid waste is a key issue for the sustainable model in metropolitan cities, as it encompasses the collection, storage, transportation, segregation, processing, and disposing of rubbish to reduce its negative impact on human health and the environment (Pal & Bhatia, 2022). Increased waste needs to be disposed and managed properly; therefore, many technological and sustainable methods are being developed across the world.

Waste has significant uses in energy security of the nation through waste to energy conversion processes like biological and thermal processes. It can help in not only energy security of the world but also in waste to wealth conversion. Because of their advantages for both the environment and the economy, these advancements can also be beneficial to society (Vaish et al., 2016). Urban, local, and commercial waste could be proven to be a helpful source of energy with ensured safe disposal of waste (Rafey et al., 2020). Agriculture waste such as crop residues, slaughterhouse waste, livestock, etc., industry waste (paper/pulp, bagasse, press sludge, polluted petroleum or synthetic oil, linger residue waste, coal tar, etc.) (Shin et al., 2004), and domestic waste are alterable to wealth through sustainable technologies resulting valuable biogas, hydrogen gas bio-alcohol. Waste-to-energy technologies that promote sustainable waste management in India include pyrolysis, incineration, gasification, bio-methanation, anaerobic digestion, composting, and landfill gas recovery (Annepu, 2012).

1.8 CONCLUSION

Waste is a universal problem that affects every part of our lives. Waste types can range from dangerous to non-hazardous, and proper waste characterization is crucial to ensuring good waste management, and this characterization must consider physical, chemical, and biological properties of waste. Waste management is critical not just for public health and safety, but also for long-term environmental sustainability.

The effects of waste on the air, water, land, and climate can be severe and long-lasting. Climate change is exacerbated by the production of greenhouse gases from waste, and the wrong disposal of hazardous waste can contaminate soil and groundwater, endangering both human and ecological health. Long-term environmental deterioration can result from the buildup of non-biodegradable garbage in landfills.

These effects can be lessened with the use of efficient waste management techniques including recycling, composting, and waste-to-energy technology. These remedies, while helpful, are not a cure-all, and the best method for minimizing the harm to the environment still involves cutting waste production at the source. Individuals, governments, and industries must all collaborate to reduce trash production and enhance sustainable waste management practices such as recycling and proper disposal.

Waste management is a serious issue that requires the attention of scientists, governments, and the general public. We can work toward a more sustainable and resilient future by knowing the types and characteristics of waste and applying good waste management practices.

REFERENCES

Agamuthu, P., Khidzir, K. M., & Hamid, F. S. (2009). Drivers of sustainable waste management in Asia. *Waste Management & Research, 27*(7), 625–633.

Ajay, C. M., Mohan, S., & Dinesha, P. (2021). Decentralized energy from portable biogas digesters using domestic kitchen waste: a review. *Waste Management, 125,* 10–26.

Akinbile, C. O., Eze, R. C., Yusuf, H., Ewulo, B. S., & Olayanju, A. (2019). Effect of some selected soil properties, moisture content, yield and consumptive water use on two Cassava (TMS 0581 and TME 419) varieties. *Journal of Agricultural Engineering, 50*(4), 166–172.

Al-Khatib, I. A., Kontogianni, S., Nabaa, H. A., & Al-Sari, M. I. (2015). Public perception of hazardousness caused by current trends of municipal solid waste management. *Waste Management, 36,* 323–330.

Annepu, R. K. (2012). Sustainable solid waste management in India. *Columbia University, New York*, 2(1), 1–189.

Babu, B. R., Parande, A. K., Rajalakshmi, R., Suriyakala, P., & Volga, M. (2009). Management of biomedical waste in India and other countries: A review. *Journal of International Environmental Application & Science*, 4(1), 65–78.

Bharadwaj, A., Yadav, D., & Varshney, S. (2015). Non-biodegradable waste–its impact & safe disposal. *International Journal of Engineering and Advanced Technology*, 3(1), 184–191.

Biro, K., Pradhan, B., Buchroithner, M., & Makeschin, F. (2013). Land use/land cover change analysis and its impact on soil properties in the northern part of Gadarif region, Sudan. *Land Degradation & Development*, 24(1), 90–102.

Biswal, S. (2013). Liquid biomedical waste management: an emerging concern for physicians. *Muller Journal of Medical Sciences and Research*, 4(2), 99.

Brady, N. C., Weil, R. R., & Weil, R. R. (2008). *The nature and properties of soils* (Vol. 13, pp. 662–710). Upper Saddle River, NJ: Prentice Hall.

Central Pollution Control Board (CPCB). (2020). Annual Report on Solid Waste Management (2020-21), CPCB, Delhi. https://cpcb.nic.in/uploads/MSW/MSW_AnnualReport_2020-21.pdf

Chapman, N., & Hooper, A. (2012). The disposal of radioactive wastes underground. *Proceedings of the Geologists' Association*, 123(1), 46–63.

Chen, M., Huang, J., Ogunseitan, O. A., Zhu, N., & Wang, Y. M. (2015). Comparative study on copper leaching from waste printed circuit boards by typical ionic liquid acids. *Waste Management*, 41, 142–147.

Chu, Y. T., Zhou, J., Wang, Y., Liu, Y., & Ren, J. (2023). Current state, development and future directions of medical waste valorization. *Energies*, 16(3), 1074.

Couto, N., Silva, V., Monteiro, E., & Rouboa, A. (2013). Hazardous waste management in Portugal: An overview. *Energy Procedia*, 36, 607–611.

Das, D., Srinivasu, M. A., & Bandyopadhyay, M. (1998). Solid state acidification of vegetable waste. *Indian Journal of Environmental Health*, 40(4), 333–342.

Das, P. K. (2020). Present scenario of waste management in India. *American International Journal of Social Science Research*, 5(1), 22–32.

Datta, P., Mohi, G., & Chander, J. (2018). Biomedical waste management in India: critical appraisal. *Journal of Laboratory Physicians*, 10(01), 006–014.

Deng, D., Zhang, L., Dong, M., Samuel, R. E., Ofori-Boadu, A., & Lamssali, M. (2020). Radioactive waste: A review. *Water Environment Research*, 92(10), 1818–1825.

De Souza, R. G., Da Silva, D. K. A., De Mello, C. M. A., Goto, B. T., Da Silva, F. S. B., Sampaio, E. V. S. B., & Maia, L. C. (2013). Arbuscular mycorrhizal fungi in revegetated mined dunes. *Land Degradation & Development*, 24(2), 147–155.

European Parliament (2020). E-waste in the EU: Facts and figures (infographic). Available online: www.europarl.europa.eu/news/en/headlines/society/20201208STO93325/e-waste-in-the-eu-facts-and-figures-infographic

Gangwar, C., Choudhari, R., Chauhan, A., Kumar, A., Singh, A., & Tripathi, A. (2019). Assessment of air pollution caused by illegal e-waste burning to evaluate the human health risk. *Environment International*, 125, 191–199.

Garcia, A. J., Esteban, M. B., Marquez, M. C., & Ramos, P. (2005). Biodegradable municipal solid waste: Characterization and potential use as animal feedstuffs. *Waste Management*, 25(8), 780–787.

Ghinea, C., Drăgoi, E. N., Comăniţă, E. D., Gavrilescu, M., Câmpean, T., Curteanu, S., & Gavrilescu, M. (2016). Forecasting municipal solid waste generation using prognostic tools and regression analysis. *Journal of Environmental Management*, 182, 80–93.

Godswill, A. C. (2017). Industrial waste management: brief survey and advice to cottage, small and medium scale industries in Uganda. *International Journal of Advanced Academic Research, 3*, 26–43.

Godswill, A. C., Gospel, A. C., Otuosorochi, A. I., & Somtochukwu, I. V. (2023). Industrial and community waste management: Global perspective. *American Journal of Physical Sciences, 1*(1), 1–16.

Gupta, S., Mohan, K., Prasad, R., Gupta, S., & Kansal, A. (1998). Solid waste management in India: Options and opportunities. *Resources, Conservation and Recycling, 24*(2), 137–154.

Han, Z., Liu, Y., Zhong, M., Shi, G., Li, Q., Zeng, D., ... & Xie, Y. (2018). Influencing factors of domestic waste characteristics in rural areas of developing countries. *Waste Management, 72*, 45–54.

Hussain, C. M., Paulraj, M. S., & Nuzhat, S. (2021). *Source reduction and waste minimization.* Amsterdam: Elsevier.

Ilankoon, I. M. S. K., Ghorbani, Y., Chong, M. N., Herath, G., Moyo, T., & Petersen, J. (2018). E-waste in the international context—a review of trade flows, regulations, hazards, waste management strategies and technologies for value recovery. *Waste Management, 82*, 258–275.

International Solid Waste Association. (2020).The future of the waste management sector. www.iswa.org/wp-content/uploads/2022/06/ISWA-Digital-Future-of-Waste-Management-Sector.pdf

IPCC (2023). *Climate Change 2023: Synthesis Report.* Contribution of Working Groups I, II and III to the Sixth Assessment Report of the Intergovernmental Panel on Climate Change [Core Writing Team, H. Lee and J. Romero (eds.)]. IPCC, Geneva, Switzerland, pp. 35–115.

Iqbal, N., Agrawal, A., Dubey, S., & Kumar, J. (2020). Role of decomposers in agricultural waste management. In *Biotechnological applications of biomass.* London: IntechOpen.

Kanagamani, K., Geethamani, P., & Narmatha, M. (2020). Hazardous waste management. In *Environmental issues and sustainable development*, Sarvajayakesavalu, S., Charoensudjai, P., Eds. London: IntechOpen.

Kapanen, A., & Itävaara, M. (2001). Ecotoxicity tests for compost applications. *Ecotoxicology and Environmental Safety, 49*(1), 1–16.

Kaza, S., Yao, L., Bhada-Tata, P., & Van Woerden, F. (2018). *What a waste 2.0: a global snapshot of solid waste management to 2050.* World Bank Publications.

Koul, B., Yakoob, M., & Shah, M. P. (2022). Agricultural waste management strategies for environmental sustainability. *Environmental Research, 206*, 112285.

Krishnamurti, G. S., & Naidu, R. (2003). Solid–solution equilibria of cadmium in soils. *Geoderma, 113*(1–2), 17–30.

Kumar, S. (2019). India's environmental challenges and their solutions. In *Green initiatives and sustainable development* (pp. 1–36). Meerut: Learning Media Publications.

Kumar, S., Smith, S. R., Fowler, G., Velis, C., Kumar, S. J., Arya, S., ... & Cheeseman, C. (2017). Challenges and opportunities associated with waste management in India. *Royal Society Open Science, 4*(3), 160764.

Kumari, T., & Raghubanshi, A. S. (2023). Waste management practices in the developing nations: Challenges and opportunities. In *Waste management and resource recycling in the developing world* (pp. 773–797). Amsterdam: Elsevier.

Li, C. S., & Jenq, F. T. (1993). Physical and chemical composition of hospital waste. *Infection Control & Hospital Epidemiology, 14*(3), 145–150.

Lokeshwari, M., & Swamy, C. N. (2010). Waste to wealth—agriculture solid waste management study. *Pollution Research, 29*(3), 513–517.

MacArthur, E. (2013). Towards the circular economy. *Journal of Industrial Ecology, 2*(1), 23–44.

Millati, R., Cahyono, R. B., Ariyanto, T., Azzahrani, I. N., Putri, R. U., & Taherzadeh, M. J. (2019). Agricultural, industrial, municipal, and forest wastes: an overview. In *Sustainable resource recovery and zero waste approaches* (pp. 1–22). Amsterdam: Elsevier.

Misra, V., & Pandey, S. D. (2005). Hazardous waste, impact on health and environment for development of better waste management strategies in future in India. *Environment International, 31*(3), 417–431.

Mohammed, A., & Elias, E. (2017). Domestic waste management and its environmental impacts in Addis Ababa City. *African Journal of Environmental and Waste Management, 4*(3), 206–216.

Mondal, P., & Yadav, A. (2018, January). An overview on different methods of domestic waste management and energy generation in India. In *2018 International Conference on Smart City and Emerging Technology (ICSCET)* (pp. 1–5). Mumbai: IEEE.

Mor, S., Ravindra, K., Dahiya, R. P., & Chandra, A. (2006). Leachate characterization and assessment of groundwater pollution near municipal solid waste landfill site. *Environmental Monitoring and Assessment, 118*, 435–456.

Nagarajan, R., Thirumalaisamy, S., & Lakshumanan, E. (2012). Impact of leachate on groundwater pollution due to non-engineered municipal solid waste landfill sites of Erode city, Tamil Nadu, India. *Iranian Journal of Environmental Health Science & Engineering, 9*, 1–12.

Nations, U. (2015). Transforming our world: The 2030 agenda for sustainable development. *New York: United Nations, Department of Economic and Social Affairs, 1*, 41.

Niti Aayog. (2021). Reforms in urban planning capacity in India. Ministry of Housing and Urban Affairs & Ministry of Education. www.niti.gov.in/sites/default/files/2021-09/UrbanPlanningCapacity-in-India-16092021.pdf

Oo, A. N., Iwai, C. B., & Saenjan, P. (2015). Soil properties and maize growth in saline and nonsaline soils using cassava-industrial waste compost and vermicompost with or without earthworms. *Land Degradation & Development, 26*(3), 300–310.

Pal, M. S., & Bhatia, M. (2022). Current status, topographical constraints, and implementation strategy of municipal solid waste in India: A review. *Arabian Journal of Geosciences, 15*(12), 1176.

Pallavicini, N., Engström, E., Baxter, D. C., Öhlander, B., Ingri, J., & Rodushkin, I. (2014). Cadmium isotope ratio measurements in environmental matrices by MC-ICP-MS. *Journal of Analytical Atomic Spectrometry, 29*(9), 1570–1584.

Pichtel, J. (2014). *Waste management practices: municipal, hazardous, and industrial* (2nd edn.). Boca Raton: CRC Press.

Ponnada, M. R., & Kameswari, P. (2015). Construction and demolition waste management—a review. *Safety, 84*, 19–46.

Rafey, A., Prabhat, K., & Samar, M. (2020). Comparison of technologies to serve waste to energy conversion. *International Journal of Waste Resources, 10*, 372.

Rajput, R., Prasad, G. A. K. C., & Chopra, A. K. (2009). Scenario of solid waste management in present Indian context. *Caspian Journal of Environmental Sciences (CJES), 7*(1), 45–53.

Rani, D. A., Boccaccini, A. R., Deegan, D., & Cheeseman, C. R. (2008). Air pollution control residues from waste incineration: Current UK situation and assessment of alternative technologies. *Waste Management, 28*(11), 2279–2292.

Rucevska, I., Nellemann, C., Isarin, N., Yang, W., Liu, N., Yu, K., ... & Nilsen, R. (2015). Waste Crime–Waste Risks: Gaps in Meeting the Global Waste Challenge. A UNEP Rapid Response Assessment. United Nations Environment Programme and GRID-Arendal, Nairobi and Arendal.

Saadatlu, E. A., Barzinpour, F., & Yaghoubi, S. (2022). A sustainable model for municipal solid waste system considering global warming potential impact: A case study. *Computers & Industrial Engineering, 169*, 108127.

Searchinger, T., Heimlich, R., Houghton, R. A., Dong, F., Elobeid, A., Fabiosa, J., ... & Yu, T. H. (2008). Use of US croplands for biofuels increases greenhouse gases through emissions from land-use change. *Science, 319*(5867), 1238–1240.

Shayler, H., McBride, M., & Harrison, E. (2009). *Sources and impacts of contaminants in soil.* Cornell Waste Management Institute. http://cwmi.css.cornell.edu

Shi, Y., Chai, J., Xu, T., Ding, L., Huang, M., Gan, F., ... & Yang, J. (2023). Microplastics contamination associated with low-value domestic source organic solid waste: A review. *Science of the Total Environment, 857*, 159679.

Shin, H. S., Youn, J. H., & Kim, S. H. (2004). Hydrogen production from food waste in anaerobic mesophilic and thermophilic acidogenesis. *International Journal of Hydrogen Energy, 29*(13), 1355–1363.

Singh, R. P., Singh, P., Araujo, A. S., Ibrahim, M. H., & Sulaiman, O. (2011). Management of urban solid waste: Vermicomposting a sustainable option. *Resources, Conservation and Recycling, 55*(7), 719–729.

Sulemana, A., Donkor, E. A., Forkuo, E. K., & Oduro-Kwarteng, S. (2018). Optimal routing of solid waste collection trucks: A review of methods. *Journal of Engineering, 2018*(1), 4586376.

Syed-Hassan, S. S. A., Wang, Y., Hu, S., Su, S., & Xiang, J. (2017). Thermochemical processing of sewage sludge to energy and fuel: Fundamentals, challenges and considerations. *Renewable and Sustainable Energy Reviews, 80*, 888–913.

Swachh Bharat Mission Urban (SBMU). (2020). Management manual on municipal solid waste. Central Public Health and Environmental Engineering Organisation (CPHEEO). Available on https://mohua.gov.in/upload/uploadfiles/files/Part1(1).pdf

Tammemagi, H. Y. (1999). *The waste crisis: Landfills, incinerators, and the search for a sustainable future.* New York: Oxford University Press.

Tangchirapat, W., Saeting, T., Jaturapitakkul, C., Kiattikomol, K., & Siripanichgorn, A. (2007). Use of waste ash from palm oil industry in concrete. *Waste Management, 27*(1), 81–88.

United Nations Environment Programme (2022). Synthesis Report on the Environmental and Health Impacts of Pesticides and Fertilizers and Ways to Minimize Them. Geneva. www.fao.org/fileadmin/user_upload/soils/publications/pesticides.pdf

Vaish, B., Srivastava, V., Singh, P., Singh, A., Singh, P. K., & Singh, R. P. (2016). Exploring untapped energy potential of urban solid waste. *Energy, Ecology and Environment, 1*, 323–342.

Vasanthi, P., Kaliappan, S., & Srinivasaraghavan, R. (2008). Impact of poor solid waste management on ground water. *Environmental Monitoring and Assessment, 143*, 227–238.

Venkatesan, G., & Swaminathan, G. (2009). Review of chloride and sulphate attenuation in ground water nearby solid-waste landfill sites. *Journal of Environmental Engineering and Landscape Management, 17*(1), 1–7.

Voutsa, D., Grimanis, A., & Samara, C. (1996). Trace elements in vegetables grown in an industrial area in relation to soil and air particulate matter. *Environmental Pollution, 94*, 325–335.

Wiedinmyer, C., Yokelson, R. J., & Gullett, B. K. (2014). Global emissions of trace gases, particulate matter, and hazardous air pollutants from open burning of domestic waste. *Environmental Science & Technology, 48*(16), 9523–9530.

Yoada, R. M., Chirawurah, D., & Adongo, P. B. (2014). Domestic waste disposal practice and perceptions of private sector waste management in urban Accra. *BMC Public Health, 14*(1), 1–10.

2 Waste and Its Current Status
A Middle East Perspective

Ghassan F. Al Samarrai, Rana I. Khaleel, and Norli Ismail

2.1 INTRODUCTION

The Middle East is one of the most fertile places for waste generation in the world, and the wasteful lifestyle, ineffective legislation, obstacles to providing infrastructure, indifferent public behavior, and lack of environmental awareness are among the main factors leading to the growth of waste problems in the Middle East. The high standard of living, coupled with the lack of waste collection and disposal facilities, also contributes to transforming the issue of "garbage" into a responsibility (Ahsan et al., 2014).

The general perception of waste is centered around indifference, and waste is treated as "waste" and not as a "resource". There is an urgent need to raise public awareness about environmental issues, solid waste management practices, and sustainable living. Public participation in community initiatives for waste management is characterized by inactivity due to the low level of environmental awareness and public education. Unfortunately, none of the countries in the region have an effective waste separation mechanism at the source (Vishnu and Singh, 2020).

Solid waste management in the Middle East suffers from deficiencies in legislation and poor planning. There is no legislation available in many countries to deal with the issue of waste. In addition to inadequate funding, the absence of waste management plans, weak coordination between the relevant authorities, the lack of qualified human resources, as well as weak technical and administrative decisions, in general, lead to those countries' inability to implement an integrated waste management strategy in the region. The powers of waste management are assigned in many countries to companies affiliated with the government and municipalities, and this in turn leads to discouraging the participation of the private sector and entrepreneurs in this sector (Hussein et al., 2021).

The per capita waste generation rates in many developing countries in Arab countries are among the highest in the world. Due to the lack of proper facilities for waste collection and disposal, it is common in the region to see waste dumped randomly in open spaces, deserts, and water bodies. Moreover, the lack of awareness and failure

DOI: 10.1201/9781003543176-2

of people to reduce their waste, separate it at the source, and manage it in a proper manner constitute another critical issue.

A sustainable waste management system requires broad public participation, effective legislation, adequate funding, as well as modern waste management practices and technologies. The region hopes to improve waste management scenarios by separating waste at the source, encouraging the participation of the private sector, applying recycling systems and generating energy from waste, and developing a strong and influential legislative and institutional system (Elagroudy et al., 2011; Zafar, 2021).

2.2   CHALLENGES FACING SOLID WASTE MANAGEMENT

Governments and local communities face great and varied challenges arising from the production of solid waste. The situation is the same, if not greater, in developing countries, where per capita production rates of solid waste are high compared to other countries. This is because of several reasons, the most important of which are the rapid pace of industrial growth, construction boom, population growth, rapid urbanization, and an improved lifestyle coupled with unsustainable consumption patterns that have all contributed to creating a waste crisis facing the governments there (Taghipour et al., 2021). The major part of it is biodegradable and is possible to be recycled. Nevertheless, the widely practiced method is to dump the waste in landfills, due to the abundance of vast lands in many countries. The waste management sector in Middle Eastern countries faces several major challenges, including Arab countries (Engelhard and Schulz, 2017).

2.3   THE MAIN CHALLENGES

1. The absence of a clear work structure in the management of the solid waste sector, which includes collection and transportation as well as proper waste treatment.
2. The lack of effective and comprehensive legislative frameworks that regulate the solid waste sector and the inefficiency of application mechanisms, which are no less important than these legislations themselves.
3. Most of the waste management activities are directly related to state institutions, i.e. the public sector. This arrangement in waste management is considered weak and lacks market mechanisms. In this case, economic incentives cannot be used to improve and develop waste treatment services.
4. Inadequate human and organizational capabilities and specialized competencies in this field.
5. Scarcity of accurate and reliable basic data as well as lack of information on solid waste and waste in general, such as generation rates for various types of solid waste, assessment needs for natural resources, land use, transportation, treatment scenarios, and solid waste growth scenarios. It is obvious that data and information are crucial elements for developing a waste management system including adequate monitoring of the sector.

6. **Inadequate infrastructure and waste management strategies:** Currently, in most Middle East countries, waste treatment capacities are insufficient and recycling rates are low. Moreover, in the absence of local recycling facilities, there is no alternative to recycling them other than landfilling them.
7. **Waste recycling is costly:** Although recent years have witnessed an increase in the number of waste recycling facilities, the economics of recycling are not very favorable and, in many cases, it is costly compared to purchasing the product which can be attributed to the lack of recycling facilities.
8. Inadequate demand for recycled products in the local market is another reason that impedes the growth of the waste recycling industry.
9. **Prevailing public opinion:** The economy in the Gulf Cooperation Council countries depends primarily on oil and gas, with high levels of land reserves. Accordingly, several decades ago, renewable energy alternatives, such as solar energy, wind energy, etc., did not receive much attention, and oil was the easiest option. However, in the last few years, due to the drop in oil prices and the subsequent shortage of imports, more consideration has been given to alternative and renewable energy sources.

2.4   OPPORTUNITIES

Middle East countries urgently need an ambitious waste management plan that includes reducing, reusing, and recycling it. It must be recalled that waste has a set of harmful and negative environmental effects on air, water, and land. It is also a major draining factor on the economy, especially on local city budgets, where it is estimated that 50% of the city budget is spent on waste management. Waste not only poses a challenge but also provides a golden opportunity that is largely untapped. Correct and programmed waste management provides an opportunity not only to avoid the harmful effects associated with waste but also to recover basic resources and benefit from the environmental, economic, and social benefits that result from its proper management and take a step on the path to a sustainable future (Schubeler et al., 1996). Integrated management means reducing the amount of waste and its danger at the source through the process or technology that is taken or considered during the life stages of the product, starting from the design stage of the product, through its manufacture, sale, and purchase, and ending with its use. The integrated treatment of waste includes five circles with the aim of reducing, as much as possible, damage to the environment and human health, and converting waste into an economic and social resource through reuse and recycling (Teshome, 2021).

2.4.1   REDUCTION BY SOURCE

Reducing, reusing, and recycle reduction are done at the source by encouraging producers and consumers to produce, buy, and use various products in a way that reduces, as much as possible, the amount of waste before it reaches the waste stream. The aim of source reduction is to reduce pollution, conserve natural resources, and protect human health. The reduction begins with the design of products and the reduction in the purchase of unnecessary things, continuing with the separation of waste in the home, the factory, and the commercial store, where the local authority responsible for waste treatment can convert the separated waste materials into recycling and reuse, and reduce the materials that must be landfilled in the end (Safeguarding the Plastic Recycling Value Chain, 2022).

Reducing waste at source is the simplest and least expensive way to reduce waste. Each person can contribute to the reduction of the source in many ways that, over time, become a new approach to daily life:

- Preferring products without excess wrap.
- Buyimg bulky reams and bags for refilling.
- Reducing the use of single-use bags.
- Preferring to buy products bearing the green badge.

2.4.2 REUSE

People should be encouraged to reuse products that have been put together in some way and use them again for the same purpose or use them for new purposes, which brings about economic and environmental benefits, thus encouraging reuse.

Examples of classic reuse are as follows:

- Using empty cans or boxes as a piggy bank to collect money or as gift boxes with wrapping and decorating them.
- Refilling bottles and glass containers.
- Buying various life necessities from shops selling used goods.

2.4.3 RECYCLING

The process of recycling waste is known as the process of treating consumable materials, so that they are returned to the raw form of this material, to be manufactured again and reused and used again. It can also be defined as the process of using the natural resource again after it has been recycled. In recycling, we convert waste into products that are useful to humans. This process also helps to reduce (1) the consumption of raw materials, (2) energy consumption, (3) water and air pollution, and (4) greenhouse gas emissions.

- Based on the Ministry's policy for environmental protection, and because of increasing awareness of the importance of reducing waste as much as possible, many local authorities have begun to treat waste through separation from the source and waste sorting at transitional stations.
- Sorting the waste and reducing the remaining waste that must be landfilled may contribute to reducing the expenses of the local authority in transporting the waste to the landfill sites and paying the landfill fees, which gradually increase.
- The Ministry of Environmental Protection supports the local authorities regarding waste management and conducts an educational campaign for

citizens about the importance of waste separation and transferring what is suitable from it to recycling:

- Support for transferring waste to recycling plants.
- Support for setting up facilities for sorting waste and separating materials suitable for recycling.
- Public consultation of local authorities on the development of plans for waste treatment infrastructure.
- Support and guidance on citizen education campaigns.

- Local authorities must submit reports
- There are some stations for sorting residues and waste.

2.4.4 ENERGY PRODUCTION FROM WASTE

Waste to energy or energy from waste refers to any processing of waste to generate energy in the form of electricity or heat. Waste is converted to energy in many ways. Producing energy from waste is also a way to reduce landfilling as much as possible. The policy for materials management in the waste market in Israel calls for the reduction of waste and final waste through four means: reduction at source, reuse, recycling, and energy generation. The decision on how to use the materials found among the waste is taken based on the economic benefit of each material. The materials that cannot be recycled are recommended by the Ministry of Environmental Protection to be converted into energy with the appropriate technology, with an emphasis on the principles of Best Available Technology. Facilities to generate energy from waste and household waste may generate electricity, power generation, or combined heat and electricity (Breaking the Plastic Wave, PEW, 2020).

2.5 TECHNICAL METHODS FOR HEAT TREATMENT OF WASTE

There are some technical means for the heat treatment of waste:

- **Mass Burn:** When the waste is burned in its raw state without any prior treatment except for the removal of unwanted objects.
- **Extracting Fuel from Waste, Refuse-Derived Fuel:** Waste is treated before burning, and waste that has a very high heating value, such as plastic, paper, and cardboard, is selected. The fuel extracted from the waste is used in the cement industries as an alternative to fossil fuels, thus reducing the negative impact on the environment.

2.6 CONVERT WASTE INTO ENERGY

Its objective is to convert waste into energy, which is complex and less commercially viable

- Gasification, i.e. converting waste into gases, is the process of converting materials that contain carbon in their composition, such as coal and biomass, into carbon monoxide and hydrogen, by the interaction of raw materials at high temperatures with quantities of oxygen. Gasification is an effective process for extracting energy from organic matter.

- **Thermal analysis (pyrolysis):** It is an anaerobic chemical analysis process for solid waste that takes place under high pressure and a temperature higher than 450 degrees Celsius and without the presence of oxygen. Thermal analysis takes place in a closed device, and therefore no pollutants are released into the atmosphere. This method is considered better than other methods.
- **Plasma, that is, ionized gas:** Commonly known as industrial gas, becomes a substance that needs high amounts of oxygen to decompose into raw materials. Plasma gas does not burn waste like normal incinerators, but it converts waste into fuels that still retain chemicals and thermal energy from waste, and it converts inorganic waste into a glassy substance that can be used for industrial purposes.

Obstacles to constructing waste incineration facilities:

- The high costs involved in building a waste incineration facility.
- High moisture content in the waste.
- Opposition to citizens living near incineration facilities

2.7   DUMPING WASTE IN LANDFILLS

The Ministry's policy for environmental protection calls for reducing, as much as possible, the dumping of waste and waste in landfills, to prevent damage to nature and the environment. Solid waste landfill is the final link in the integrated waste treatment chain. The waste that remains after exhausting all means of reduction at the source, creation, and recycling must be buried in regulated and licensed landfills, without causing environmental pests. Regardless of how much we contribute to reducing the amount of waste, it will always remain landfill waste.

REFERENCES

Ahsan, A.; Alamgir, M.; El-Sergany, M.; Shams, S.; Rowshon, M.; Daud, N. 2014. Assessment of municipal solid waste management system in a developing country. *Chin. J. Eng.* 1–11.

Elagroudy, S.; Elkady, T.; Ghobrial, F. 2011. Comparative cost benefit analysis of different solid waste management scenarios in Basrah. *Iraq. J. Environ. Prot.* , 2:555–563.

Engelhard, H.; Schulz, F. 2017. *Demographic Developments in the Middle East and North Africa.* University of Amberg: Amberg, Germany.

Hussein, L.; Uren, C.; Rekik, F.; Hammami, Z. 2021. A review on waste management and compost production in the Middle East–North Africa region. *Waste Manag. Res. J. A Sustain. Circ. Econ.*, 40:1110–1128.

PEW Charitable Trusts and SystemIQ. "Breaking the Plastic Wave: A Comprehensive Assessment of Pathways Towards Stopping Ocean Plastic Pollution," 2020.

Safeguarding the Plastic Recycling Value Chain: Insights from COVID-19 impact in South and Southeast Asia, Circulate Capital, 2020.

Schubeler, P.; Christen, J.; Wehrle, K. 1996, August. Conceptual Framework for Municipal Solid Waste Management in Low-income Countries. SKAT (Swiss Center for

Development Cooperation): St. Gallen. https://documents1.worldbank.org/curated/en/829601468315304079/pdf/400960Municpal1te0framework.

Taghipour, H.; Amjad, Z.; Aslani, H.; Armanfar, F. 2021. Characterizing and quantifying solid waste of rural communities. *J. Mater. Cycles Waste Manag.*, 18:790–797.

Teshome, F.B. 2021. Municipal solid waste management in Ethiopia; the gaps and ways for improvement. *J. Mater. Cycles Waste Manag.*, 23:18–31.

Vishnu, B.; Singh, L.A. 2020. Study on the suitability of solid waste materials in pavement construction: A review. *Int. J. Pavement Res. Technol.*, 14:625–637.

Zafar, S. 2021. Municipal waste management in Saudi Arabia. Bioenergy Consult Available online: www.bioenergyconsult.com/municipal-wastes-in-saudi-arabia/

3 Challenges and Opportunities in Waste
Indian Aspect

Chaitanya Narala and Bhagawan Dheravath

3.1 INTRODUCTION

Indian urbanization will present several challenges over the next decade. The progression of economic growth, including urbanization, industrialization, migration, and changes in lifestyle, leads to an increase in waste generation, which in turn creates threats to the environment (1–5). Treatment, energy recovery, reuse, recycling, and/ or disposal—as well as other releases to the environment—are all necessary means of managing waste once it is produced. An effort is being made to differentiate between the upcoming application in the field of energy recovery and solid waste management (SWM) (6, 7).

Waste is produced in India by every individual, organization, and resident. Municipal solid waste, hazardous waste, industrial nonhazardous waste, agricultural and animal waste, medical waste, radioactive waste, construction and demolition debris, extraction and mining waste, oil and gas production waste, fossil fuel combustion waste, and sewage sludge are just a few of the many different types of waste that are produced (8). Biomedical waste production is also rising because of the recent threat posed by the coronavirus disease 2019 (COVID-19) effect. Consequently, the country now faces a sparse waste management issue.

Modern technologies are not only used to produce energy but also to reduce the amount of waste that needs to be disposed of safely. To meet this challenge, India's waste management processes and systems will need to be upgraded. The Ministry is encouraging in a number of ways by providing numerous options for the establishment of various technology projects for the recovery of energy (biogas, electricity, and bio CNG) from various types of waste of an agricultural, industrial, and urban renewable nature, including municipal solid waste, vegetable, and other market waste, slaughterhouse waste, agricultural residues, and industrial/sewage treatment plant (STP) waste and effluents (9). Despite this, an immediate move toward advanced facilities and sustainable waste management systems is required, because the current systems are insufficient and waste has a negative impact on public health, the environment, and the economy. The Ministry of Environment and Forests (MoEF) introduced the Waste Management and Handling Rules in India, although compliance is inconsistent and limited. In the end, we must acknowledge that, for sustainable management, we must reduce and reuse waste as urbanization continues (10).

DOI: 10.1201/9781003543176-3

3.2 MAJOR CHALLENGES OF WASTE MANAGEMENT IN INDIA

3.2.1 GENERATION AND CHARACTERIZATION OF WASTE IN INDIA

Population growth at a rapid rate will result in waste. Each year, India generates 62 million tonnes of waste. With changing consumption patterns and rapid economic growth, it is anticipated that urban municipal solid waste generation will increase to 165 million tonnes in 2030 (11–13). Of the approximately 43 million tonnes that are collected, approximately 12 million tonnes are treated and 31 million tonnes are dumped in landfill sites (14, 15).

Globally, there are 2.01 billion tonnes of municipal solid waste generated annually, of which at least 33% are not managed in an environmentally responsible manner, according to a World Bank report. The average amount of waste produced per person worldwide is 740 g/day, but national averages range from 110 g/day to 4.54 kg/day. Despite having only 16% of the world's population, high-income nations generate 34% of all waste (16).

For further estimation of the quantity of waste produced, which is required for planning future waste management, it is necessary to calculate and characterize waste (17). The amount of waste produced is determined by the season, type of activities, and living pattern.

The rate of waste generation is also impacted by factors like the city or region, population density, level of commercial activity, culture, and economic status (18).

With more than 1.27 billion people, India is the second largest country in the world after China, accounting for 17.6% of the world's total population. Over the past few decades, urban populations have been steadily rising. India has a gross domestic product (GDP) growth rate of 7.30%, making it one of the fastest-growing economies in the world. India's GDP is expected to grow at a 10% rate by 2030. Living standards will rise because of an increase in GDP (19). With rapid industrialization and over-population, as a result of improved living standards and unchecked urbanization, the rate of waste generation per capita rises. Medical waste, radioactive waste, construction and demolition debris, extraction and mining waste, oil and gas production waste, fossil fuel combustion waste, sewage sludge, household, municipal, mining, and agricultural waste are among the many types of waste (20). Other types of waste include industrial nonhazardous waste, agricultural and animal waste, municipal hazardous waste, and industrial nonhazardous waste. The production of electronic waste will more than double as the digital economy grows. It mostly consists of large amounts of organic matter, fine earth and ash, paper and plastic, glass, and metals, all of which pose a serious threat to both the environment and living things (21).

Percentage gap in SWM has been calculated for the period 2015–21 and is listed in Table 3.1 (22).

3.2.2 CURRENT WASTE MANAGEMENT IN INDIA

Collection, segregation, transportation, recycling, and waste disposal are the main steps in waste management. Pollution of the environment is caused by waste management that is inappropriate. Waste in India is regulated and managed by the Union

TABLE 3.1
Gap in solid waste management

Year	Gap in solid waste management (%)
2015–16	40.09
2016–17	79.78
2017–18	9.38
2018–19	30.35
2019–20	25.82
2020–21	31.70

Ministry of Environment, Forests, and Climate Change (MoEFCC). The Ministry's primary goal is pollution prevention and control, and numerous regulations have been issued over time to ensure a clean environment by safely handling and disposing of waste. The MoEFCC established guidelines for proper management for the State Pollution Control Board and the Central Pollution Control Board. Based on where it comes from and what it contains, waste is typically divided into municipal, industrial, and hazardous categories. Explicit standards and compliances are to be adhered to for safe dealing with and removal of the different sorts of waste (23).

India's regulations for waste management are based on "sustainable development," "precaution," and "polluter pays." These principles require municipalities and businesses to restore equilibrium if their actions disturb it in an environmentally responsible and accountable manner. The Environment Protection Act (EPA) of 1986 serves as the umbrella for a number of subordinate laws that regulate the manner in which waste is disposed of and how it is dealt with because of the rise in waste generation as a result of economic growth. Different types of waste are covered by different regulations and must adhere to different things, mostly in the form of authorizations, keeping accurate records, and proper disposal mechanisms.

3.2.3 WASTE COLLECTION AND TRANSPORT

In cities, efficient waste collection, storage, and transportation can be significant obstacles. These steps are crucial to the management system. In India, the municipal corporations are in charge of collecting waste, and bins are typically provided for inert and biodegradable waste (24–26). Open burning is a common method of disposing of waste that is both inert and mixed with biodegradable material. India's waste collection and transportation infrastructure will be improved, which will increase tourism, job creation, and better public health (27). Around $500 to $1,000 per ton for SWM is spent by local bodies, with 70% going toward collection and 20% going toward transportation.

The Assortment and Transportation fragment of municipal solid waste management (MSWM) can be advantageously separated into seven stages (Figure 3.1), starting with isolation of waste at source and finishing with mass transportation/move

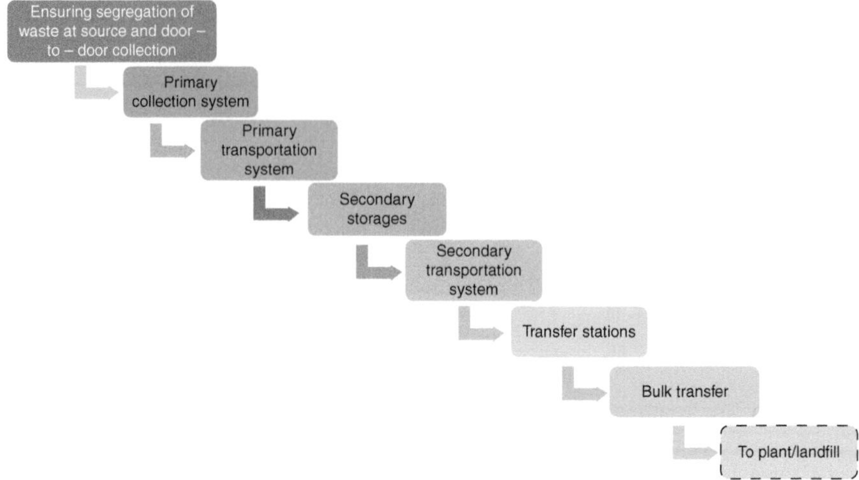

FIGURE 3.1   Seven steps to efficient collection and transportation of waste.

of waste to the handling plants/landfills. It is not necessary for each and every urban local body (ULB) to use or deploy all seven steps. Even the requirement that each of the seven steps includes waste disposal is not ideal. A few ULBs have found it helpful and proficient to utilize just a portion of the means for finishing the MSWM venture from squander generator premises to the waste handling/landfilling offices. Using case studies of successful ULBs' MSW collection and transportation models, these seven steps are explained and demonstrated (28).

The transfer of waste from the point of generation and pickup to the point of treatment or disposal in a landfill is referred to as waste collection and transportation. It includes waste collection from door to door and curbside by municipal, private, informal, and other waste collectors. In MSWM, collecting separated municipal waste is an essential step (29). The public health and aesthetics of towns and cities are negatively impacted by inefficient waste collection services. Recycling is maximized when wet, dry, and household hazardous waste are collected separately. Additionally, it makes it more cost-effective to treat such waste to meet the minimum quality requirements for various products, such as the production of compost from organic waste or refuse-derived fuel (RDF) (30).

There are two types of waste collection services: primary and secondary collection. Depending on the MSWM system in place in the city or town, primary collection is the process of collecting, lifting, and transporting separated solid waste from the source of its generation—such as households, shops, offices, markets, hotels, and other residential or nonresidential premises—to a storage depot or transfer station or directly to the disposal site. As previously mentioned, primary collection must guarantee the separate collection of waste streams or fractions. Secondary collection involves transporting waste to waste processing facilities or the final disposal site from community bins, waste storage depots, or transfer stations. Separate covered

FIGURE 3.2 Ways of waste management.

bins, containers, vats, or vaults must be used for the transportation of separated waste at the secondary collection points. For organic and biodegradable waste, the secondary storage points should be serviced and emptied daily or before they begin to overflow (28).

The accumulated garbage from cities is immediately dumped at the dump site. The waste that is dumped is mixed in nature. A method of waste disposal that is more effective than the one currently in use is needed (Figure 3.2). Landfills are used to dispose of the collected waste without supervision; additionally, it is difficult to see that the entirety of a city or town's waste collection is processed, with only leftovers going to a landfill. In fact, to realize the vision and mission of achieving "zero" landfilling, remnants that are labeled as "inert/nonrecyclable" must be transformed into other products that can be used. In addition to posing a risk of groundwater contamination and likely contributing to a decline in the quality of the surrounding air, landfills are not maintained in a scientific manner (31).

3.2.4 Environmental and Health Impacts of Waste Dumping

The improper handling of solid waste poses potential threats to human health and the environment. Workers in this industry face the greatest direct health risks, and they must be shielded from waste as much as possible. Handling hospital and clinic waste also carries particular dangers. The breeding of disease vectors, primarily flies and rats, poses the greatest indirect health risks to the general public. Methane is released into the air when biodegradable waste breaks down in open dumps under anaerobic conditions. Methane is a major contributor to global warming and is responsible for explosions and fires (32). Odor and leachate migration into receiving waters are two additional issues (33). Odor is a serious issue, especially in the summer, when India's average temperature can exceed 45°C (34). At dumps, discarded tires collect water, allowing mosquitoes to breed and raising the risk of West Nile, malaria, and dengue fever. Uncontrolled consuming of waste at dump destinations delivers fine particles which are a significant reason for respiratory infection and cause brown haze (32). Every year, around Mumbai, open burning of MSW and tires releases 22,000 tons of pollutants (35). The effects of unfortunate waste administration on general well-being are indisputable, with expanded rates of nose and throat contaminations, breathing hardships, aggravation, bacterial diseases, weakness, decreased resistance, sensitivities, asthma, and different contaminations (36).

3.3 RULES AND REGULATIONS OF WASTE MANAGEMENT IN INDIA

The EPA, which was approved in 1986, aims to establish a sufficient environmental protection system. It grants the central government the authority to regulate all forms of waste and address specific issues that may arise in various parts of India. To ensure proper waste management in India, the MoEF issued the Waste Management Rules 2000, which have since been updated and published. It is the primary piece of legislation that needs to be taken into consideration and has significant environmental provisions. These regulations and the development of infrastructure for waste collection, storage, segregation, transportation, processing, and disposal are the responsibilities of municipal authorities (37).

Some of the important legislations for environment protection are as follows:

- The National Green Tribunal Act, 2010
- The Air (Prevention and Control of Pollution) Act, 1981
- The Water (Prevention and Control of Pollution) Act, 1974
- The Environment Protection Act, 1986
- The Hazardous Waste Management Regulations, etc.

3.4 INFORMAL SECTOR'S ROLE IN WASTE MATERIALS' REUSE AND RECYCLING

India's waste management relies heavily on the informal recycling industry. The small-scale, labor-intensive, largely unregistered, low-technology manufacturing or service provision, which is characteristic of the informal sector, is to extract recyclables prior to treating mixed waste or dumping it haphazardly into landfills to cut costs associated with waste treatment and disposal (38–40). People who extract recyclable and reusable materials from mixed waste are referred to by these terms. These exercises encapsulate the casual area as this work is serious, of low innovation, low-paid, unrecorded, and unregulated, frequently finished by people or family gatherings. The documentation, comprehension, and expansion of existing informal collection and recycling systems are necessary for strategic MSWM planning (41).

3.5 WASTE-TO-ENERGY IN INDIA

An improper waste disposal might be major problem and requires an emergency solution to recovery the waste material to sustainable management. The significant view of waste-to-energy cycle is to effectively dispose of waste, reduce landfill and dump yards, and promote alternate energy source.

The central and state governments of India are becoming increasingly interested in alternative and renewable energy sources due to the country's growing energy deficit. Waste-to-energy cycle is one of these, and it is gathering expanding consideration from both the focal and state legislatures. Status of waste to energy: In 1987, the first waste-to-energy plant was built in Delhi's Timarpur. In India, 14 waste-to-energy plants have been built, 7 of which have been shut down. In the years 2019–20, the nation generated 1,50,761 tons of solid waste per day (42).

Through the use of thermal, thermochemical, biochemical, and electrochemical techniques, energy can be extracted from the organic portion of waste, which includes both biodegradable and nonbiodegradable components.

Combustion generates heat and power in the waste-to-energy technology that is utilized the most frequently for residual waste. India's dumping would be significantly reduced if an integrated waste management system implemented maximum recycling and waste-to-energy. India's industry is eager to take advantage of waste-to-energy technologies that can process unsegregated low-calorific waste. Presently, a number of waste-to-energy projects employing the combustion of unsorted low-calorific waste are being developed. Gasification, pyrolysis, the production of RDF, and gas-plasma technology are alternative thermal treatment methods to combustion (43).

The waste-to-energy technology has been recognized as a renewable technology by the Indian government, and it is supported by a variety of subsidies and incentives. The Ministry of New and Renewable Energy (MNRE) actively promotes each and every technological option for energy recovery from industrial and urban waste. In accordance with the MNRE's research and development (R&D) Policy, the MNRE is also supporting research on waste to energy by providing cost-shared financial support for R&D projects. In addition, projects involving applied research and development as well as studies on resource assessment, technology advancement, and performance evaluation receive financial support from MNRE (44).

3.6 IMPROVEMENTS NEEDED IN WASTE MANAGEMENT IN INDIA

The utilization of waste as resources with increased value extraction, recycling, recovery, and reuse are essential improvements required for India's waste management. If SWM is to advance in India, a powerful and independent authority is required to regulate waste management. Improvements will not occur unless clear regulations and enforcement are in place. Innovation can be triggered by stringent

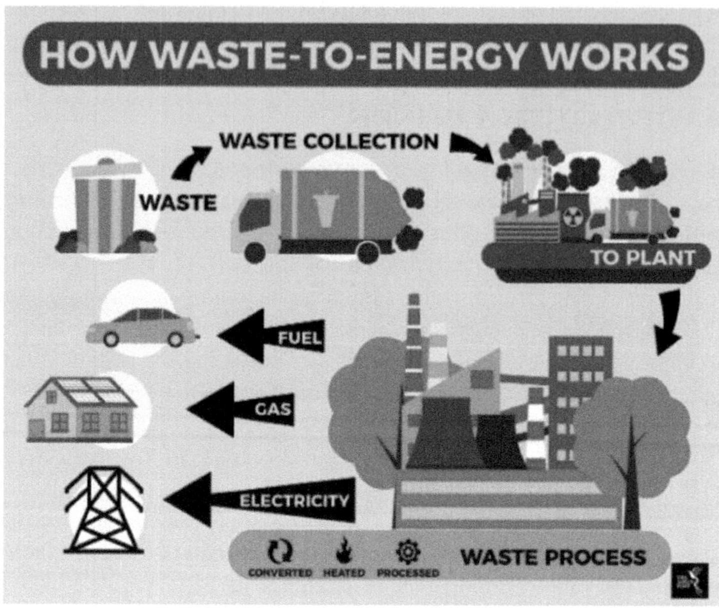

FIGURE 3.3   How waste-to-energy works.

waste regulations. Businesses that are appealing and profitable must be part of the waste management industry. The ULB must have clear performance requirements and impose financial penalties for ineffective waste management services. Through a waste tax, producers of waste must raise funds for infrastructure and waste management companies. To provide efficient waste management throughout India, an average fee of 1 rupee per person per day would likely be sufficient to generate close to 50,000 crores of dollars annually. The appropriateness of various waste management and treatment options is determined by information on future waste quantities and characteristics. For primary and secondary collection, effective systems for monitoring collection, transport, and disposal require state-level procurement of vehicles and equipment.

ULBs, the private sector, and nongovernmental organizations (NGOs) must develop visionary projects for long-term waste management planning. To deliver sustainable systems, roles and responsibilities must be defined, and progress must be monitored and evaluated. Experiences ought to be shared among various social groups and regions of India. A comprehensive approach to SWM is being developed by a number of private sector businesses, research institutes, and organizations. In the future, India's waste management system will need to include a significant amount of informal sector participation. At every level, training and capacity building must be developed. The roles and responsibilities of each individual in the waste management system, as well as the effects of poor waste management on the environment and public health, should all be understood by Indian schoolchildren. Citizens who see waste as a resource opportunity will benefit because of this.

3.7 ALTERNATIVE OPTION FOR WASTE MANAGEMENT

Waste can be depolymerized under catalysis process. This depolymerization of waste generates value-added monomers, which are chemicals, materials, edible products and fuels, etc. The natural resources might be reserved for future used by adopting depolymerization technology.

3.8 CONCLUSION

Every country needs to upgrade by urbanization, industrialization, etc. We cannot stop growth of population and development of megacities. Therefore, in the process of rapid urbanization, waste generation definitely increases, but we must improve and make proactive, advance step in the waste management systems. However, this is the main challenge for India. The government must focus and keep the effort on the existing waste management improvements. At the same time, every citizen is spirited, and without their involvement government efforts might remain ineffective. Therefore, it is desired to place more emphasis on information, planning, funding, and unified waste management along with community education. Further, the 4 Rs philosophy of Reducing, Reusing, Recycling, and Recovering Resources should be actively encouraged.

REFERENCES

1. Government notifies new solid waste management rules. www.downtoearth.org.in/. Retrieved 26 March 2019.
2. Municipal Solid Waste (MSW) Management—IAS Prep—Sept 2017 update. iascurrent.com. Retrieved 26 March 2019.
3. Solid Waste Management Rules, 2016—India Environment Portal | News, reports, documents, blogs, data, analysis on environment & development | India, South Asia. www.indiaenvironmentportal.org.in/content/427824/solid-waste-management-rules-2016/. Retrieved 26 March 2019.
4. Solid waste management rules, 2016. Civilsdaily. 15 September 2017. Retrieved 26 March 2019.
5. Annepu RK. (2012). Report on sustainable solid waste management in India. Waste-to-Energy Research and Technology Council (WTERT), 1–189. http://swmindia. blogspot.in/ (accessed 26 June 2015).
6. Annual Report on Solid Waste Management. (2020-21). CPCB, Delhi https://cpcb.nic. in/uploads/MSW/MSW_AnnualReport_2020–21.pdf
7. Bernard Galea November. (2010). Waste regulation in India: An overview. www. cppr.in/wp-content/uploads/2012/10/Waste-Regulation-in-India-An-Overview-Bern ard.pdf
8. Bhoyar RV, Titus SK, Bhide AD, Khanna P. (1999). Municipal and industrial solid waste management in India. *JIAEM* **23**, 53–64.
9. CPCB (Central pollution Control Board). (2000). Management of municipal solid waste Delhi. https://cpcb.nic.in/uploads/MSW/MSW_AnnualReport_2000-01.pdf (accessed 27 June 2015).
10. Das S, Bhattarcharya BK. (2014). Estimation of municipal solid waste generation and future trends in greater metropolitan regions of Kolkata, India. *Ind. Eng. Manage. Innov.* **1**, 31–38. http://doi.org/10.2991/jiemi.2014.1.1.3

11. Dasgupta B, Yadav VL, Mondal MK. (2013). Seasonal characterization and present status of municipal solid management in Varanasi, India. *Adv. Environ. Res.* **2**, 51–60. http://doi.org/10.12989/aer.2013.2.1.051

12. Wilson DC, Velis C, Cheeseman C. (2006). Role of informal sector recycling in waste management in developing countries. *Habitat Int.* **30**, 797–808.

13. Singh GK, Gupta K., Chaudhary S. (2014). Solid waste management: Its sources, collection, transportation and recycling. *Int. J. Environ. Sci. Dev.* **5**(4), 347–351. DOI: 10.7763/IJESD.2014.V5.507

14. Gupta S, Mohan K, Prasad R, Gupta S, Kansal A. (1998). Solid waste management in India: Options and opportunities. *Resour. Conserv. Recycl.* **24**(2), 137–154. http://doi.org/10.1016/S0921-3449(98)00033-0

15. Guria N, Tiwari VK. (2010). Municipal solid waste management in Bilaspur city (C.G.) India. *National Geographer, Allahabad* **1**, 1–16.

16. India waste to energy. www.eai.in/ref/ae/wte/wte.html

17. India's challenges in waste management. www.downtoearth.org.in/blog/waste/india-s-challenges-in-waste-management-56753

18. Jha MK, Sondhi OAK, Pansare M. (2003). Solid waste management—a case study. *Indian J. Environ. Prot.* **23**, 1153–1160.

19. Kansal A. (2002). Solid waste management strategies for India. *Indian J. Environ. Prot.* **22**(4), 444–448.

20. Kaushal RK, Varghese GK, Chabukdhara M. (2012). Municipal solid waste management in India—current state and future challenges: A review. *Int. J. Eng. Sci. Technol.* **4**, 1473–1489.

21. Kaza S, Yao LC, Bhada-Tata P, Van Woerden F. (2018). *What a waste 2.0: A global snapshot of solid waste management to 2050.* Urban Development; © Washington, DC: World Bank. http://hdl.handle.net/10986/30317

22. Mathur V. (2012). Scope of recycling municipal solid waste in Delhi and NCR—integral review. *J. Manage.* **5**, 27–36. 21.

23. Minghua Z, Xiumin F, Rovetta A, Qichang H, Vicentini F, Bingkai L, Giusti A, Yi L. (2009). Municipal solid waste management in Pudong new area, China. *Waste Manag.* **29**(3), 1227–1233.

24. Ministry of Environment and Forests (MoEF). (2015). *The Gazette of India. Municipal Solid Waste (Management and Handling) Eules.* New Delhi, India.

25. Misra V, Pandey SD. (2005). Hazardous waste, impact on health and environment for development of better waste management strategies in future in India. *Environ. Int.* **31**, 417–431.

26. Narayana T. (2009). Municipal solid waste management in India: From waste disposal to recovery of resources? *Waste Manage.* **29**(3), 1163–1166. http://doi.org/10.1016/j.wasman.2008.06.038

27. Rana PR, Yadav D, Ayub S, Siddiqui AA. (2014). Status and challenges in solid waste management: A case study of Aligarh city. *J. Civil Eng. Environ. Technol.* **1**, 19–24.

28. Rathi S. (2006). Alternative approaches for better municipal solid waste management in Mumbai, India. *Waste Management,* **26**(10), 1192–1200. https://doi.org/10.1016/j.wasman.2005.09.006

29. Ray MR, Roychoudhury S, Mukherjee G, Roy S, Lahiri T (2005) Respiratory and general health impairments of workers employed in municipal solid waste disposal at an open landfill site in Delhi. *Int. J. Hyg. Environ. Health* **208**(4), 255–262.

30. Sharholy M, Ahmad K, Mahmood G, Trivedi RC. (2005). Analysis of municipal solid waste management systems in Delhi–a review. In *Book of Proceedings for the Second International Congress of Chemistry and Environment, Indore,* 773–777.

31. Sharholy M, Ahmad K, Vaishya R, Gupta R. (2007). Municipal solid waste characteristics and management in Allahabad, India. *Waste Manage.* **27**, 490–496.
32. Solid Waste Management Rules Revised After 16 Years; Rules Now Extend to Urban and Industrial Areas: Javadekar. pib.nic.in. Retrieved 26 March 2019.
33. Sridevi P, Modi M, Lakshmi MVVC, Kesavarao L. (2012). A review on integrated solid waste management. *Int. J. Eng. Sci. Adv. Technol.* **2**, 1491–1499.
34. Sridevi P, Modi M, Lakshmi MVVC, Kesavarao L. (2012). A review on integrated solid waste management. *Int. J. Eng. Sci. Adv. Technol.* **2**, 1491–1499.
35. Srivastava R, Krishna V, Sonkar I. (2014). Characterization and management of municipal solid waste: A case study of Varanasi city, India. *Int. J. Curr. Res. Acad. Rev.* **2**, 10–16.
36. Statistics India. www.statista.com/topics/5586/waste-management-india/#topicOverview
37. Kumar S, Smith SR, Fowler G, Velis C, Jyoti Kumar S, Arya S, Rena, Kumar R, Cheeseman C. (2017). Challenges and opportunities associated with waste management in India. *R. Soc. Open Sci.* 4, 160764. http://dx.doi.org/10.1098/rsos.160764.
38. Swachh Bharat Mission—Urban guidance on efficient collection and transportation of municipal solid waste. https://sbmurban.org/storage/app/media/pdf/SBM%20CT%20Booklet%20_10%20July%20Final.pdf .
39. United States Environmental Protection Agency (USEPA). (2002). Solid waste management: a local challenge with global impacts. www.epa.gov/osw/nonhaz/municipal/pubs/ghg/f02026.pdf (accessed 1 July 2015).
40. United States Environmental Protection Agency (USEPA). www.epa.gov/report-environment/wastes
41. Waste to Energy Project. www.iasparliament.com/current-affairs/waste-to-energy-projects-in-india
42. Waste to Energy Works. www.iasgyan.in/blogs/waste-to-energy
43. Wilson DC, Velis C, Cheeseman C. (2006). Role of informal sector recycling in waste management in developing countries. *Habitat Int.* **30**, 797–808. https://doi.org/10.1016/j.habitatint.2005.09.005
44. World Energy Council Report. (2013). World energy resources: waste to energy. https://www.worldenergy.org/assets/images/imported/2013/10/WER_2013_7b_Waste_to_Energy.pdf (accessed 1 July 2015).

4 Waste Minimization, Management, and Resource Recovery

Livestock Waste

Ellin Harlia, Eulis Tanti Marlina,
Yuli Astuti Hidayati, and Norli Ismail

4.1 INTRODUCTION

Farms in the industry are highlighted, because environmental protection and issues related to land, water, and air pollution are currently receiving global attention. As a result of environmental damage, more nations worldwide have increased their commitment to enforcing pollution restrictions. Many regulations focus on preventing pollution caused by industry, animal husbandry, and agriculture (Herero et al., 2018). The importance of handling and managing dirt animals in some countries has caused the implementation of relevant laws, regulations, standards, and policies to promote handling dirt-sustainable animals. Some are implemented at the community, state, national, regional, and international levels (Gabriel et al., 2018). Some techniques start from simple, cost-effective to complex strategies available for handling animal waste properly.

Efforts to increase food security are not always in line with production capacity. There is a problem with the use of land and water, which impacts the environment from livestock, and applicable regulations limit the ability of producers to add several cattle to fulfill the request for food animals in the future. Therefore, agriculture must be carried out in a way that does not damage natural resources because they are needed for human and animal life in the future.

Farm animals are maintained especially for the objective as a source of animal protein and nonfood such as fancy, skin, and even fertilizer pens in several system productions. The waste produced during animal production, processing, shipping, and selling is the product side, which is possible if not properly managed. Several waste results potentially during operating production cattle, including waste or remainder feed, wastewater, sewage house, cut animals, and livestock feces. Dirt from production cattle often causes external issues such as contamination from bedding, urine, washing water, precipitation, feed spills, and water spills (Gabriel et al., 2018). Before introducing organic fertilizers, animal waste was central to increasing soil fertility. Despite the role of organic fertilizer in agricultural production, manure remains an

DOI: 10.1201/9781003543176-4

essential source of fertilizer, especially in blood, where organic fertilizer is unavailable or accessible to farmers.

Intensification of animal operations has led to the production of large amounts of manure concentrated in certain locations in excess of need. Intensive animal production can pose significant problems regarding waste storage and disposal. Air and water pollution associated with animal waste has become the focus of regulatory discussions around the world. Animal waste contains various microorganisms that can be a source of danger to humans and animals. These microorganisms can cause food contamination and outbreaks, making them hazardous for public health.

Worldwide, it has been linked directly or indirectly to solid waste contamination from livestock. Therefore, to limit some of the challenges associated with handling livestock solid waste, the use of sustainable management practices and strategies is recommended. A solid waste management plan from livestock is an integral part of the livestock production strategy, and based on regulations and other legal instruments as well as innovative practices to prevent and reduce the risk of exposure. Several strategies and technologies for managing livestock solid waste can be applied to various scales and production environments.

Livestock waste means livestock manure, bedding material, rainwater or other water, soil, hair, fur, or other remains, usually included in handling animal waste in cages. Inappropriate management of livestock feces causes fecal pollution of rivers that receive agricultural runoff (Parihar et al., 2019). Humans are at risk if fecal microorganisms enter water sources. Groundwater and surface waters can harbor pathogens originating from animal waste deposits. Animal manure is a solid waste from livestock that contains many valuable constituents that, if recycled effectively, can be used as plant fertilizer, animal feed, and energy production. Minimizing solid waste can be done through solid waste management as an effort to recover resources.

4.2 WHY SOLID WASTE MUST BE MINIMIZED?

Unprocessed solid waste must be reduced as much as possible, because it can negatively impact the environment (Kementerian Lingkungan Hidup dan Kehutanan, 2021). Solid waste can cause pollution to water, land, and air, and can cause health problems for humans and animals (Parihar et al., 2019) Additionally, improper management of solid waste can increase the potential for spreading disease caused by microorganisms (Adi Permana, 2021). Management of solid waste has to be done both proactively and reactively. A proactive approach is that using materials will produce minimal waste, with the lowest possible level of danger, or, in other words, reduce the waste produced as much as possible with the lowest possible level of danger. Solid waste must also be processed to minimize its impact on the environment. A proactive approach is an approach that handles upstream waste generation. This process has been introduced to the world since the 1970s in the industrial world as a clean process or clean technology, with the concept of modifying the processes and technology used so that the emissions or waste produced can be minimized, as well as the risks. The reactive approach involves handling waste after it has begun to accumulate. This is an approach that handles downstream waste generation. It has an initial concept of technology to manage waste that has arisen so that the resulting emissions and residues are safe to be released back into the environment. This concept is then improved by activities to reuse residues or waste directly (*reuse*) and through a process before reusing (*recycling*) the waste (Evita, 2019). This is then known as 3R (*reduce, reuse, recycle*). The 3 Rs alone is an effort to minimize waste, business processing, or extermination of waste. Processing livestock waste can reduce the impact on the environment when waste products are used directly on the environment. Once processed, it can be handled more readily or released into the environment without risk, but most processing changes the material's nature into waste, which is considered detrimental. This item should then be able to dispose of garbage that has not been processed and must be safely released into the environment, for example, by placing it in a landfill (Adi Permana, 2021).

Utilizing resources effectively and creating products with little trash that can be recycled will help limit the quantity of waste produced. In the long term, reducing waste can produce positive outcomes like reduced cost management and investment, reduced potency water and land pollution, and saving usage source. Power nature: However, in Indonesia, efforts to reduce waste still do not get good attention, because they are considered complicated things and do not show results in time (*immediate result*).

In Indonesia, the concept of reuse and reduction has long been known, especially in the regions of agriculture, where its people know the cycle of repeat waste, especially the nature of biological waste, like the leftover food and leaves. In managing waste, there should be early handling, which is generally the case. More recycling and simple handling will be done to get desired outcomes. They are beginning of handling for classifying garbage according to types, lowering volume, and minimizing size. Cycle efforts will succeed when sorting and separating component waste from the source until the final process.

The benefits of the 3 Rs over a period are reduced dependency on place processing, increased efficiency and effectiveness in the use of suggestions and infrastructure-related management waste, creation of opportunity business for society, creation of braid cooperation between government and community/private sector toward quality management and controlling the undesirable impacts of waste on the environment. Should all public farmers and ranchers be aware of the 3Rs and able to put them into practice daily? If 3Rs are used to reduce and avoid the buildup of waste from agriculture and animal husbandry, there will be numerous advantages. Sustainable agriculture and animal husbandry are created from cooperation between various parties, businesses, governments, and society to keep the environment awake for generations. Waste-dense ones that are not under control can harm the environment and human health. Some impacts from waste solids that can happen (Gabriel et al., 2018; Parihar et al., 2019; Adi Permana, 2021) are as follows:

1. Formation of poisonous gases, like methane (CH_4), acid sulfide (H_2S), ammonia (NH_3)
2. Cause a decrease in air quality
3. Cause pollution to soil, water, and air
4. Waste that is congested contains organic materials when entering waters which can increase biochemical oxygen demand (BOD) and chemical oxygen

demand (COD) as well as lower oxygen dissolved (DO), interfering with biotic life in water.

5. Cause the formation of greenhouse gas (GHG)
6. Pollution leads to decreasing diversity
7. Improve potency spread of diseases caused by microorganisms
8. Causes disease transmission through the food chain
9. There is a decline in environmental quality and aesthetics.
10. Causes of environmental conditions becoming less comfortable. Considering the impacts that occur, solid waste must be appropriately managed to ensure that the negatives can be minimized.

4.3 SOLID WASTE MANAGEMENT

Several methods for managing solid waste from livestock are biogas production, composting, vermicompost, biodynamic fertilizer, bio-briquettes, etc. Biogas is a clean, environmentally friendly fuel that can be obtained by anaerobic digestion of livestock waste and household waste, which is abundantly available in rural areas (Parihar et al., 2019).

Solid waste from livestock is a potential resource to be used as raw material for biogas and compost. Various methods can be used to manage waste, including:

1. Recycling is converting waste into new products to prevent energy use and consumption of fresh raw materials.
2. Biogas production technology is a process in which biofuel is produced naturally from the decomposition of organic materials, either livestock and plant solid waste or simply gas production from anaerobic biomass.
3. **Vermicomposting:** This is an organic waste processing method that utilizes the role of earthworms as decomposers.
4. **Composting:** This is the process of recycling organic materials, such as leaves and waste food, which becomes a valuable fertilizer that can fertilize the soil and plants.
5. **Collection:** This involves collecting or assembling raw materials/waste.
6. **Storage:** Waste is then stored in waste ponds/storage buildings and sewage/waste processing lagoons. Treatment involves implementing chemical reactions and biology that occur in waste-stored animals.
7. **Transfer:** As a result of processing waste and materials burn, the extracted gas from the reaction is then transferred into a cylinder or storage tank for utilization.

The general technology used to obtain biogas is fermenting livestock manure using an anaerobic digester. Biogas digesters include fixed-dome, floating drum, and balloon reactors. Solid waste consists of two parts, namely organic and inorganic parts. The average percentage of organic waste material reaches ±80%, so composting is a suitable alternative treatment. Compost is an organic fertilizer derived from plant residues and animal waste that undergoes decomposition or weathering. During composting, microorganisms grow and release carbon dioxide gas, water, other organic products, and energy. Some energy is used for metabolism, and the rest is released as heat. When the food supply runs out, microbial growth and heat generation decrease, and humus material called compost is produced in general methods or technology composting such as *static piles, windrow composting, passively aerated windrows, forced aeration static piles, enclosed composting, and vermicomposting*. System agriculture integrated or Simantri business-oriented agriculture and animal husbandry (without zero waste) produces 4 F (food, feed, fertilizer, and fuel). The activity mainly is integration of cultivation crops and livestock, waste plant processed for feed livestock, and to reserve feed in season dry. Apart from that, waste-congested livestock (feces) are processed into biogas and organic fertilizer (Wisnuardhana, 2009; Iwan et al., 2014; Gde Prima Tangkas and dan Yulinah Trihadiningrum, 2016).

Fertilizer pens contain lots of valuable components and can be recycled repeat. The physical and chemical properties of dirt animals will influence the potency of their use, especially as fertilizer and convenience in handling it. Dirt animals can be categorized based on consistency or rate. The water transforms into different types of fertilizer: liquid (up to 5% solids), porridge and half solids (between 5% and 25% solids), and a dense dense (more than 25% solids). Profit is from the use of manure because it contains nutrients, organic materials, solids, energy, and fiber (Salundik, 2019).

According to Martin (2005), compost results from decomposing biological material by microorganisms in condition-controlled aerobics and becomes material similar to relative humus stable. Contrarily, in the process of composting, organic material undergoes biological breakdown, particularly by bacteria that use organic material as a source of energy. Composting is regulated and subject to regulations so that it can form more quickly. Making a mixture of balanced components, giving enough water, controlling aeration, and adding an activator to compost are all included in this process.

Waste agriculture can be made of a mixture of compost straw and husks, rice, weeds, stems and cobs corn, all parts of vegetative plants, banana stems, and coir coconut. Compost is like a multivitamin for land agriculture. Compost will boost strong roots and raise the soil's fertility. Compost strengthens the structure of the soil and increases the ability of the soil to manage groundwater levels. The composting process will involve microorganisms for the decomposition process in the soil. This aids in the production of chemicals that can encourage plant growth and the absorption of nutrients from the soil by plants. Compost fertilization of plants results in higher quality plants than fertilizer chemistry fertilization. For instance, the harvest is heavier, fresher, and more flavorful as a result of the findings. Benefits of own composting lots are examined from several aspects:

Economy aspect:

- Saves costs for transportation and waste storage
- Reduces the volume/size of waste
- Has a higher selling value than the original material

Environmental aspects:

- Reduces air pollution due to waste burning
- Reduces the need for land for landfill

Aspects for land/plants:

- Increases soil fertility
- Improves soil structure and aeration
- Increases soil water absorption capacity
- Increases soil microbial activity
- Improves the quality of the harvest (taste, nutritional value, and number of harvests)
- Provides hormones and vitamins for plants
- Increases nutrient retention/availability in the soil

According to Pratiwi et al. (2023) and Walpajri et al. (2023), using bio-activators can speed up the decomposition process of agricultural waste into compost. Compost that has been cooked will look brownish-black, with a temperature of around 30°C and does not emit an odor, it is usually called biocompost.

The use of fermented solid livestock waste has been widely reported to be successful in supporting growth and controlling plant diseases. For example, chicken manure can increase soil fertility and, at the same time, control root rot disease caused by *Phytophthora*. From the research results of Aryantha (2002), chicken and cow manure composted for 5 weeks successfully fertilized *Lupinus albus* plants, while controlling root rot disease by *Phytophthora cinnamomi*. Agricultural waste, such as chicken and cow manure, is positively correlated with microbial activity and the population of antagonistic microbes (actinomycetes and endospore-producing bacteria) in the soil. The diversity of microbial types also appeared to be highest in soil treated with chicken manure.

Livestock waste is a by-product that cannot be released from the agricultural system. Livestock solid waste that is not handled correctly can negatively impact the agricultural land itself and the broader environment, such as global warming and climate change. However, optimal use of livestock waste can contribute to increasing farmers' income and improving the quality of agricultural land so that it can be used sustainably. Animal waste in manure is a valuable source of nutrients and organic material for maintaining soil fertility and crop production. Animal studies have shown that 55–90% of animal feed's nitrogen and phosphorus content is excreted in feces and urine (Tamminga et al., 2000), which is usually used as manure.

According to Cleaner Production (2007), waste management is managing waste from start to finish, including collection, transportation, processing, and application, accompanied by monitoring and waste management regulations. Because waste management affects many different facets of society and the economy, it can be seen as the "entry point" to accomplishing sustainable development goals. According to Wilson and Velis (2015), waste management is related to sustainable production and consumption, poverty reduction, food security, and climate change. However, it is also possible to classify waste management as a "system bottleneck." This is influenced by factors such as population density and distribution, socioeconomic aspects of the surroundings, attitudes and behavior, and society culture (Sahil et al., 2016).

Correctly managing waste will have several detrimental effects. Therefore, sustainable waste management is required to accomplish several sustainable development goals. Responsible consumption and production are demonstrated through sustainable waste management, which will influence the amount of available clean water sources. Management of sustainable waste can also reduce pollution and increase life.

The waste not managed well will produce excess methane and CO_2. This will naturally impact the change of the existing climate, so waste management can be one method for reducing climate change, whereas properly managed and utilized waste can provide several benefits, and one of them is enhancement of economy in society. Recycling and reusing potential waste can boost the economy by providing alternative enhancements and improving overall economic efficiency.

Jarvie (1987) defines sustainable development as adequate development that needs a generation moment. This, without sacrifice, fulfillment needs generations to come. Sustainable development is specified in the Brundtland Report in three aspects: economic, social, and environmental. One way that can be done is with the management of sustainable waste. To achieve sustainable development from an economic perspective, it is necessary to consider methods to increase the economy over a certain period of time, without depleting natural resources.

Quality organic fertilizers from dirt animals allow agriculture sector to reduce its dependency on chemical and organic fertilizers, which can increase land fertility and sustainability. Utilizing dirt animals as input for the conversion process of bioenergy allows farmers to take profit from the market for product waste. Utilizing dirt cows and cow urine in a way appropriate to become fertilizer cages, pesticides, medicines, and products daily can provide and produce opportunities of work in rural areas and possibly protect the land from chemical fertilizers and restore its fertility (Gupta et al., 2016).

In a sense, the term "management waste cattle" refers to the procedures and actions necessary for the collection, storage, processing, and use of waste livestock in the form of animals, whether liquid, porridge, or solid, from the start to the disposal phase. To manage waste livestock, it is critical to provide nutrition to significant plants through the analysis of acceptable waste, allowing farmers to precisely control what they apply to the soil and plant. The dung animal is a source of valuable power that will yield a good harvest when applying excellent practices in practice management. Animal waste from farms is divided into three categories: solid, slurry, and wastewater. Waste congested is processed using a dried or composted method. Examples of waste/dirt livestock/animals include cow pen waste (urine, feces, washing water, milk residue, and waste feed), cow manure, and poultry manure (mixed fertilizer cage, water, feed spills, feathers, and bedding materials), rendering by-products and other waste obtained from cultivation livestock.

Dirt animal/ cattle waste consists of solid, semi-solid, and liquid by-products produced by domesticated animals for meat, milk, eggs, and agriculture products and others for human consumption (Koninger et al., 2021).

One of the main interests of using livestock manure or manure is:

1. Make the best use of the nutrients found in manure while protecting natural resources.
2. Manure application can increase soil organic matter in the medium-/long-term application period.
3. Manure helps reduce soil compaction.
4. Waste cattle increase the stability of aggregate land.
5. Animal waste increases water infiltration and retention.

Approach to reducing pollution from solid waste can be done through feed. Very high levels of efficiency in producing meat, poultry, and animal products are necessary for feeding. Volume and efficiency industry must be expanded more if we want to have sufficient food for future generations and contribute to the alleviation of hunger and shortage of nutrition in all continents.

Several approaches to nutrition can help reduce the effect of pollution from dirt animals. Management of nutrition in a substantial way can reduce the amount of nitrogen and phosphorus excreted by pigs (Nahm, 2002). Manipulation feed is designed to increase digestibility to reduce the amount of nutrition in dirt animals that will be wasted in the environment. Use specific enzymes in feed to solve several problems in nutrition. Nutrient-rich feed often comes out of excreted pollutants through dirt (Haque, 2018) . Using probiotics in feed can also help animals utilize nutritious food more efficiently as well as their potential utility as an alternative to conventional antibiotics (Arsene et al, 2021). The presence of pathogens or pathogen potential (coliform) in the intestinal tract can impair digestion and nutrition absorption, leading to excretion that contains excessive nutrition in feces (Van't Klooster et al., 1998).

The addition of enzymes specific to feed can solve problems in nutrition. Among them, giving enzymes can lower levels of nitrogen and phosphorus released through feces. Phytase changes phosphorus in the form of sour phytate, which, in a way, is biologically available for animals; it has been researched that providing protease in pig feed can increase in vitro nitrogen digestion in weaned pigs. Some of the reasons for using enzymes in animal feed include reducing the impact of livestock waste pollution on the environment (Olivier Munezero and In Ho Kim, 2022).

Specifically in West Java and East Java, small-scale dairy farms without grazing dairy farming are concentrated in Indonesia. Due to inadequate feed, poor reproduction, and issues with animal health, most farms have limited areas available for forage production, and productivity is below potential (De Vries and Wouters, 2017). Due to this circumstance, getting enough high-quality animal feed and recycling animal waste into fertilizer entails much work. In places with high livestock stocking densities, the year-round availability of cattle food presents both quantity and quality difficulty. Only a small amount of manure is used to fertilize crops on many farms, and inadequate waste management and manure quality damage waterways and rivers. These difficulties are typical in many places with high livestock densities (Marion et al., 2020)

The use of dairy cow manure by farmers assumes that one dairy cow produces 20 kg of manure per day, an average of one. Dairy farming households have 94 kg of manure per day. In West Java, dairy cattle manure production reaches around 2.45 thousand tons daily or 73.5 million tons monthly. Only 42.8% of dairy farmers in the research location use dairy cow manure, partially or wholly, as fertilizer, an energy source (biogas), and a medium for growing worms. However, 57.2% of them practically dump it into the environment without processing it. A similar condition was found in Lembang District, where waste (feces, urine, or both) was dumped into the environment. Farmers who employ dairy cow dung do so under the assumption that, on average, each dairy cow generates 20 kg of manure each day. Daily manure production for dairy farming homes is 94 kg. Most farms sell or distribute it to other farms as a soil amendment. Akin to how some farmers must manage animal waste irresponsibly and adequately spread it throughout the pen owing to limited land, doing so compromises the pen's cleanliness and produces an unpleasant odor. Residents will find barns inconvenient, because most are close to farmers' residences and occasionally even in densely populated areas (Ariningsih et al., 2022).

An example of a polluted river is the Citarum River, which stretches for 297 km upstream at Situ Cisanti, which is located at the foot of Mount Wayang, Bandung Regency, and empties into the North Coast of Java Island, Muara Gembong, Bekasi Regency, West Java, which is one of the dirtiest rivers in the world with very high levels of pollution. One source of pollution of the Citarum River is livestock.

The Citarum Harum program aims to reduce environmental pollution by preventing the entry of pollutants from various sources, including livestock along the Citarum River. The Citarum Harum program manifests the government and community's concern about environmental pollution and is an accelerated program to control pollution and damage to the Citarum River basin. Livestock activities along the Citarum River are a source of pollution.

The Citarum Harum program controls the disposal of livestock feces, which contains volatile substances that aesthetically can reduce air and water quality and contain pathogenic microorganisms that can be transmitted to humans and animals. Once it reaches the receiving water body, organic material from livestock feces serves as a substrate for the growth of aerobic microorganisms, which can rapidly deplete available dissolved oxygen, or for anaerobic organisms, which produce a variety of harmful end products. Proper use of livestock waste in biogas, compost, and vermicompost is beneficial for increasing sustainable crop yields. Several approaches, such as physical, chemical, and biological, can be taken, which can help reduce environmental pollution from solid livestock waste.

In all intensive farming locations, livestock waste disposal could be better managed. They are just thrown away. Farmers rarely use manure, because it is a hassle, and they do not have time, so they immediately throw it away. Using manure for fertilizer is the most common practice carried out by farmers. They use it for food crops, horticultural crops, grass and other forages, and other crops. However, only a few farmers process manure for commercial purposes. For practical reasons, food crops and horticultural farmers prefer to use subsidized chemical fertilizers. The land area for dairy farming and forage production still needs to be improved. Because there is no grazing land, dairy cows are kept in pens, tied up, without access to pasture. Limited land is also a challenge in recycling livestock manure as fertilizer. Applying manure to land for grass/forage production can lead to overfertilization, because much of the fertilizer is concentrated on a small part of the land (de Vries et al., 2020).

The lack of incentives and sanctions causes most farmers to throw manure in the surrounding environment. The reasons put forward by dairy farmers illustrate that dairy cattle management is impractical. A study conducted by Zahra et al. (2021) revealed several obstacles to using manure, mainly due to limited land, lack of manure market, and lack of economic profitability of solutions. Regarding potential large-scale users, the main obstacle is the use of manure related to price (cost), quality, and practical aspects, including continuous fertilizer availability and ease of handling, transportation, and application.

The demand for milk will keep rising along with the population, income, and understanding of the value of a balanced diet. Industrial dairy products are anticipated to expand to accommodate the rising demand for milk. However, if handled correctly, the manure created by dairy farms will be well spent, posing a growing threat to the environment due to the increased pollution brought on by the dense dairy cows. Additionally, switching to intensive feeding results in less pollution than smallholder farming, which could result in intense wasting. To lessen the environmental impact, this condition forces the deployment of good fertilizer management. Policymakers and stakeholders should support the reuse of dairy farm manure by offering the proper incentives, technologies, and methods to both boost and decrease the usage of potentially polluting nutrients, according to Ariningsih et al. (2022). Implementing dairy conservation agriculture is one alternative to using manure from dairy livestock. As small-scale dairy farmers only have access to a small amount of agricultural land, they must use manure for agriculture on property other than their own. Solid waste is managed by drying it to produce fertilizer on a plot of land with limited space.

Horticulture, tea and coffee plantations, and coffee agroforestry have enormous potential, and the opportunity to use dairy cow manure is a form of integration that benefits all parties. Horticultural, tea, and coffee farmers are encouraged to use and become a market for organic fertilizer derived from dairy cow dung. Agriculture in these areas can be changed to organic farming if necessary. To connect dairy farmers who produce organic fertilizer with farmer organizations, dairy cooperatives must be supported. To compensate for the lack of land for processing and storing organic fertilizer, cooperatives are also encouraged to hunt for property for processing manure managed by cooperatives. Thus, there needs to be a solution to collect manure from dairy farmers to manure processing units and transport organic fertilizer to end users.

The dairy farming sector in West Java needs a spatial development strategy, and land-based dairy farming is the key to sustainable development (De Vries and Wouters, 2017). Moreover, it is on top of technological fixes for the manure disposal issue that makes manure recycling so tricky.

4.4 RESOURCE RECOVERY

Large-scale animal husbandry has developed due to increasing demand for animal products. As a result of animal husbandry, livestock systems have evolved from traditional, decentralized, and widespread to large-scale, intensive, and specialized systems. In other words, every nation produces a lot of animal waste, and this concentrated animal waste harms the soil, groundwater, and atmosphere by contaminating them severely. Therefore, recycling livestock manure using a commercial model is very important for the environment and the economy. Waste management using the latest technology includes anaerobic digestion and aerobic composting, which uses various types of microorganisms in an aerobic or anaerobic environment to convert them into fertilizer natural gas. Developed countries have tried to apply a circular economy business model to livestock manure recycling with the 3R principle. The rational use of circular economy theory and business models to recycle livestock manure is a practical step in managing livestock manure in various countries for a sustainable livestock future (Liu et al., 2021).

There are numerous methods to recover solid waste materials, including recycling. Materials from solid waste can be used to create new goods. Energy utilization: The waste-to-energy (WtE) process can use solid waste as an energy source. Implementing circular economy strategies is a valuable tool for enhancing sustainability. The three primary approaches are linear, recycling, and circular economy, each of which has its own relationship to waste management (Mingyu Yang et al., 2023).

Another impact of ruminant farming is GHG. Total GHG emissions from dairy farming in Indonesia were 1.4 gigatons of CO_2 in 2000 (including land use changes), with a sizable contribution from deforestation and land use changes (Ministry of Environment, 2010; Widiawati et al., 2018). Agriculture contributes 5% of total GHG emissions in Indonesia. Dairy farming contributes to global warming through emissions of methane (CH_4) from enteric fermentation and manure storage, nitrous oxide (N_2O) from manure storage, fertilizer application, and carbon dioxide from energy use (CO_2).

Only a few studies have estimated GHG emissions from Indonesian dairy production. Permana et al. (2012) in De Vries et al (2017) estimate that GHG emissions from dairy farming in Indonesia are 671 gigagrams of CO_2 per year (CH_4 and N_2 emissions from enteric fermentation).

Dairy cow manure is an essential source of GHG in the atmosphere. The importance of attention focused on forming enteric methane and nitrous oxide in agricultural environments and exposed manure management sites (Richard and Yabar, 2023). Farmers with access to the earthworm market can use vermicomposting to treat waste. Farmers also receive vermicompost, which has significant economic value, and earthworms. Vermicomposting for 30 days is sufficient to provide high-quality organic fertilizer, as claimed by Ramos et al. (2022). For dairy farmers in West Java, more land is needed to manage dairy cow manure.

It is impossible to overstate the effects of agricultural output on the environment, climate change, human health, and animal and plant health. To protect the world, the earth, and its inhabitants from dangers, it has been suggested, for instance, that GHG emissions must be drastically reduced (Mingyu Yang et al., 2023). Ruminant farming produces approximately 37% and 65% of global methane and nitrous oxide emissions, respectively, which is more potent than carbon dioxide. Indiscriminate burning of agricultural solid waste produces climate-relevant emissions. Improper handling of agricultural solid waste affects climate change, and climate change, in turn, hinders food production. It is impossible to overstate the harmful effects of improperly disposing of agricultural solid waste (Hosam and Hassan, 2021).

Reusing solid waste is an effort to reduce the volume of waste and pollutants. Effective management of agricultural solid waste means several options for handling agricultural solid waste. There have been several attempts to use agricultural solid waste as affordable feed for animals, a way to recycle garbage, and a source of animal protein.

Scientific concepts are needed to understand how waste management and processing work, as well as the nature of the impacts resulting from the disposal and distribution of waste in the environment. Manual source separation prior to collection is essential in solid waste management system design. Resources recovery procedures that can be utilized to restore natural resources include the following (Bupe et al., 2018):

- **Environmental engineering:** This includes the processes used to transform a polluted environment into a safe and healthy environment.
- Waste management refers to the procedures to control and prepare waste for resource recovery.
- Procedures for reusing resources that already have been utilized are referred to as resource reuse.
- **Replanting:** This includes the processes used to replant plants cut down or lost in an area.
- **Resource conservation:** This includes processes used to protect natural resources from future destruction or damage.
- **Reclamation:** Reclamation makes previously processed or abandoned regions more usable.
- **Renewable natural resource recovery processes:** These processes are used to recover renewable natural resources, such as water, air, solar energy, wind, and more.

It is crucial to keep in mind that the recovery of natural resources must be done in a sustainable and environmentally responsible manner to avoid causing harm to the current environment and allow for its efficient and effective utilization.

The ground, plants, and livestock must receive nutrients from livestock manure again.

- Reduces environmental pollution and disruption to society
- Saves loss of valuable nutrients
- Maintains healthy soil and temperature

The solution will depend on the specific circumstances of the location, and it is necessary to revitalize the management of livestock waste and its by-products in an environmentally friendly and economically oriented manner with the 3R principle—namely, reduce, reuse, and recycle (repeat). Apart from that, revitalizing the maintenance and handling of livestock waste must also refer to the circular economy or economic cycle as a profitable business practice by utilizing waste and by-products/by-products from livestock activities (Ambar, 2019). To reduce waste and animal by-products or achieve zero waste, the circular economy is described as a system that preserves the value of goods, materials, and resources in the economic cycle for as long as feasible. In this context, the circular economy does not only focus on reducing waste using the 3R principle but also on how to design the use of waste and by-products into products that are economically valuable and have high selling value.

The application of circular economy principles will enable farmers to survive when there are shocks to fluctuations in feed and livestock prices. The mindset of managing livestock operations must be integrated to produce high economic value. Livestock waste, which has been considered waste, with the circular economy concept, makes it a source of income or, in other words, a "gold mine".

Management of livestock waste from upstream to downstream from production to post harvest, such as livestock manure, feed waste, rumen contents, skin, bones, and biogas sludge, can be reused into by-products with economic value. The 3R principles and circular economy can integrate field farms with the nonagricultural sector.

Strengthening the competence of resource managers in the management of waste from livestock and by-products is a comprehensive solution for improving farm management maintaining strong competitiveness. Empowerment of public farmers in rural areas is crucial for increased competence, and reinforcement of institutional models through both formal and informal learning. Additionally, entrepreneurial competence is essential for developing integrated livestock farming as a driver for the local rural economy. The involves converting waste and livestock by-products into valuable economic assets that enhance competitiveness.

4.5   CONCLUSION

Solid waste from livestock can be a problem for the environment if it is not managed correctly. The impact is that it becomes a source of pollution for water, soil, and air. Considering the impact of solid livestock waste on the environment, it is essential to minimize it by proactively managing waste through aerobic fermentation and anaerobic fermentation so that organic fertilizer, biogas, and biofuel can be obtained. The orientation of livestock solid waste management can be carried out based on a circular economy so that livestock waste management activities become appealing to maintain sustainable livestock.

REFERENCES

A. Permana. 2021. Waste management must get priority from various parties. www.itb.ac.id/berita/pengelolaan-limbah-besar-mendapatkan-prioritas-dari-berbagai-party/58244

Ambar. 2019. So that livestock waste management has economic value. UGM Faculty of Animal Husbandry. http://troboslivestock.com/detail-berita/2019/12/26/57/12465/agar-pengelolaan-limbah-peternakan-berbesar- Ekonomis

M. Arsène, M.J. Anyutoulou, K.L. Davares, S.L. Andreevna, F A Vladimirovich, B.Z. Carime, R. Marouf, I. Khelifi, 2021. The use of probiotics in animal feeding for safe production and as potential alternatives to antibiotics. Veterinary World, EISSN: 2231-0916. www.veterinaryworld.org/Vol.14/February-2021/3

I.P. Aryantha, 2002. Development of Sustainable Agricultural System, One Day Discussion on The Minimization of Fertilizer Usage, Menristek-BPPT, 6th May 2002, Jakarta.

B.G. Mwanzaa, C. Mbohwaa, A. Telukdariea, 2018. The influence of waste collection systems on resource recovery. Review. 15th Global Conference on Sustainable Manufacturing. *Procedia Manufacturing* 21, (846–853). www sciencedirect.com

Cleaner Production. 2007. *Pengelolaan Limbah Industri Pangan*. Jakarta, Indonesia: Direktorat Jenderal Industri Kecil Menengah Departemen Perindustrian.

D.C. Wilson, C.A. Velis. 2015. Waste management – still a global challenge in the 21st century: An evidence-based call for action. *Waste Management & Research*, 33(12). 1049–1051. DOI: 10.1177/0734242X15616055

D. Richards, H. Yabar. 2023. Promoting energy and resource recovery from livestock waste: Case study Yuge Farm, Japan. *Case Studies in Chemical and Environmental Engineering* 7, 100299.

E.Y.P. Pratiwi, Suhartini, A.K. Sudrajat. 2023. The effect of different bio-activators on compost quality of agricultural waste. *Indonesian Journal of Bioscience (IJOBI)* 1(1), 37–44. https://journal.uny.ac.id/index.php/ijobi/

E.N. Uchewa, M.O. Otuma, P. Brooks. 2005. Animal waste management strategies, a review. *Animal Research International* 2(1), 267–274.

E. Ening Ariningsih, A.R. Irawan, H.P. Saliem. 2022. Dairy cattle manure utilization by smallholder dairy farmers in West Java, Indonesia. E3S Web of Conferences 361, 03013. https://doi.org/10.1051/e3sconf/202236103013.

E. Maulidya. 2019. Waste management with waste minimization. Nature Lover Student, Faculty of Law, Gajah Mada University. Pengelolaan Sampah dengan Minimasi Limbah – MAJESTIC-55 (ugm.ac.id)

F. Walpajri, F.W. Siregar, A.N. Ilyosa, M. Wiyaga. 2023. Effectiveness of Various Types Bio-Activators to Speed up the Composting Process and Quality of Compost Fertilizer. *International Journal of Progressive Sciences and Technologies (IJPSAT)* 36(2), 630–636.

G.A. Malomo, A.S. Madugu, and S.A. Bolu. 2018. Sustainable animal manure management strategies and practices. IntechOpen. https://doi.org/10.5772/intechopen.74361

G.P. Tangkas, d.Y. Trihadiningrum. 2016. Kajian Pengelolaan Limbah Padat Peternakan Sapi Simantri Berbasis 2R (Reduce dan Recycle) di Kecamatan Seririt, Kabupaten Buleleng. *Jurnal Teknik ITS* 5(2). ISSN: 2337-3539.

Indonesian Ministry of Environment. 2010. Indonesia Second National Communication under the United Nations Framework Convention on Climate Change (UNFCCC). Ministry of Environment, Republic of Indonesia, Jakarta.

K.K. Gupta, K.R. Aneja, D. Rana. 2016. Current status of cow dung as a bioresource for sustainable. *Bioresources and Bioprocessing* 3, 28.

H. Martin. 2005. *Agricultural Composting Basics*. Factsheet. Ontario Ministry of Agriculture, Food, and Rural Affairs.

H.M. Saleh, A.I. Hassan. 2021. Chapter 1: Introductory: Solid waste. In *Strategies of Sustainable Solid Waste Management*, edited by Hosam M. Saleh. Intechopen. DOI: 10.5772/intechopen.95327.

I.S. Anugrah, S. Sarwoprasodjo, K. Suradisastra, d.N. Purnaningsih. 2014. Integrated agriculture system (Simantri): Its concept, implementation, and role in agricultural development in Bali Province. *Forum Penelitian Agro Ekonomi* 32(2), 157–176.

J. Sahil, M.H.I. Al Muhdar, F. Rohman, I. Syamsuri. 2016. Sistem Pengelolaan dan Upaya Penanggulangan Sampah Di Kelurahan Dufa-Dufa KotaTernate. *Jurnal BIOeduKASI* 4(2). ISSN:2301-4678

J. Koninger, E. Lugato, P. Panagos, M. Kochupillai, A. Orgiazzi, M.J.I. Briones. 2021. Review manure management and soil biodiversity: Towards more sustainable food systems in the EU. Agricultural Systems, 194.

Kementerian Lingkungan Hidup dan Kehutanan. 2021. Carbon waste management to reduce green house gas from landfill in industrial activity. pslb3 Menlhk. https://pslb3.menlhk.go.id/portal/read/carbon-waste-management-to-reduce-green-house-gas-from-landfill-in-industrial-activity

E.Y.P. Pratiwi, Suhartini, Ahmad Kamal Sudrajat. 2023. The effect of different bio-activators on compost quality of agricultural waste. *Indonesian Journal of Bioscience* (IJOBI) 1(1), 37-44. https://journal.uny.ac.id/index.php/ijobi/

K.H. Nahm. (2002). Efficient Feed Nutrient Utilization to Reduce Pollutants in Poultry and Swine Manure. *Critical Reviews in Environmental Science and Technology*, 32(1), 1–16 1064-3389/02/$.50 © 2002 by CRC Press LLC

M.A. Herrero, J.C.P. Palhares, F.J. Salazar, V. Charlón, M.P. Tieri, A.M. Pereyra. 2018. Dairy Manure Management Perceptions and Needs in South American Countries. Frontiers in Sustainable Food Systems 2, 22.

M. de Vries A. P. (Bram) Wouters Theun V. Vellinga. 2017. Environmental impacts of dairy farming in Lembang, West Java. Estimation of greenhouse gas emissions and effects of mitigation strategies. Working Paper No. 221 CGIAR Research Program on Climate Change, Agriculture and Food Security (CCAFS).

M. de Vries, B. Wouters, D. Suharyono, A. Sutiarto, S.E. Berasa. 2020. *Effects of Feeding and Manure Management Interventions on Technical and Environmental Performance of Indonesian Dairy Farms (Wageningen)*. Wageningen: Wageningen Livestock Research (Wageningen Livestock Research report 1237).

M.E. Jarvie. 1987. Brundtland Report. Word Commission on Environment and Development (WCED). Fact-checked by The Editors of Encyclopaedia Britannica Article History.

M. Yang, L. Chen, J. Wang, G. Msigwa, A.I. Osman, S. Fawzy, D.W. Rooney, P.-S. Yap. 2023. Circular economy strategies for combating climate change and other environmental issues. *Environmental Chemistry Letters* 21, 55–80. https://doi.org/10.1007/s10311-022-01499-6

O. Munezero, I.H. Kim. 2022. Effects of protease enzyme supplementation in weanling pigs' diet with different crude protein levels on growth performance and nutrient digestibility. *Journal of Animal Science and Technology* 64(5), 854–862. https://doi.org/10.5187/jast.2022.e51.

R.F. Ramos, N.A. Santana, N. de Andrade, I.S. Romagna, B. Tirloni, A. de Oliveira Silveira, J. Domínguez, R.J.S. Jacques. 2022. Vermicomposting of cow manure: Effect of time on earthworm biomass and chemical, physical, and biological properties of vermicompost. Bioresource Technology 126572.

Salundik, 2019. *Animal Waste Management in Indonesia*. Jakarta, Indonesia: ILDEX Indonesia. Indonesia Convention Exhibition.

S.S. Parihar, K.P.S. Saini, G.P. Lakhani, A. Jain, B. Roy, S. Ghosh, B. Aharwal. 2019. Livestock waste management: A review. *Journal of Entomology and Zoology Studies* 7(3), 384–393.

S. Tamminga, A.W. Jongbloed, M.M. Van Eerdt, H.F.M. Aarts, F. Mandersloot, N.J.P. Hoogervorst, H. Westhoek. 2000. The forfaitaire excretie van stikstof door landbouwhuisdieren [Nitrogen excretion standards by animal agriculture]. Lelystad, the Netherlands: Institute handicap Dierhouderij en Diergezondheid, 71.

T. Liu, H. Chen, J. Zhao, P.G. Soundari. 2021. Resource Recovery and Recycling from Livestock Manure: Current Statue, Challenges, and Future Prospects for Sustainable Management. In *Sustainable Resource Management*, Volume I. John Wiley & Sons, Ltd, 137–166. https://doi.org/10.1002/9783527825394.ch6

C.E. Vanklooster, C.M.C. van der Peet-Schwering, A.J.A. Aarnink, N.P. Lenis. 1998. Pollution issues in pig operations and the influence of nutrition, housing and manure handling. In: *Progress in Pig Science*, edited by J. Wiseman, M.A Varley, J.P. Chadwick. Nottingham: Nottingham University Press, 507–518.

W. Al Zahra, M. de Vries, H. de Putter. 2021. *Exploring Barriers and Opportunities for Utilization of Dairy Cattle Manure in Agriculture in West Java.* Indonesia. Wageningen University and Research.

Y. Widiawati, M.I. Shiddieqy, E.S. Rohaeni, Y.N. Anggraeny, Setiasih, Wardi, Firsoni, Antonius, W.T. Sasongko, M.C. Hadiatry, S. Widodo, H. Bansi, A. Herliatika, S. Asmairicen, S. Puspito, W. Widaringsih, N. Miraya E.M.W. Andreas, S. Riyanti. 2023. Sistem Pemeliharaan Ternak Ruminansia yang Adaptif terhadap Perubahan Iklim. Bab 8penerbit.brin.go.id.

I.B. Wisnuardhana. 2009. Membangun Desa Secara Berkelanjutan dengan "Simantri" (Sistem Pertanian Terintegrasi). Dinas Pertanian Tanaman Pangan Provinsi Bali. Denpasar. distanprovinsibali.com/berita/simantri.doc

5 Industrial, Construction, and Demolition Waste

Management and Resource Recovery

Nurhanim Abdul Aziz,
Putri Anis Syahira Mohamad Jamil,
Muhammad Hasnolhadi Samsudin, and
Norli Ismail

5.1 OVERVIEW OF INDUSTRIAL, CONSTRUCTION, AND DEMOLITION WASTE

5.1.1 Industrial Waste Ash and Geopolymers

Industrial waste ash and geopolymer technology have gained considerable attention as a sustainable approach for construction materials. This chapter provides an overview of industrial waste ash and geopolymer, highlighting their significance, properties, and potential applications. The utilization of industrial waste ash in geopolymer production not only addresses the challenges of waste disposal but also offers a sustainable alternative to traditional cement-based materials. The chapter also includes key references that delve deeper into the subject matter.

Industrial waste ash is a residue or by-product generated from various industrial processes, such as coal combustion, metal smelting, and manufacturing (Chindaprasirt et al., 2007). These waste ashes, including fly ash, bottom ash, and slag, have inherent mineral properties and can be utilized as valuable resources in the construction industry. Geopolymer is an innovative cementitious material formed through the reaction of aluminosilicate sources with alkaline activators (Abdullah et al., 2012; Amran et al., 2021). Geopolymers offer numerous advantages, including high compressive strength, excellent chemical resistance, and reduced carbon emissions compared to traditional Portland cement-based materials. The utilization of industrial waste ash as a raw material in geopolymer production enhances its eco-friendliness and sustainability.

Geopolymers exhibit excellent material properties, such as fire resistance, durability, and low shrinkage (Siyal et al., 2018). The utilization of industrial waste ash in geopolymer systems can influence their mechanical properties, workability, and long-term properties (Chindaprasirt et al., 2007). Moreover, geopolymer-based

DOI: 10.1201/9781003543176-5

construction materials find applications in various sectors, including concrete production, infrastructure development, and structural engineering (Liew et al., 2016). Researchers have conducted extensive studies on the properties, synthesis techniques, and performance of geopolymer materials incorporating industrial waste ash. Investigations focus on optimizing the geopolymerization process, understanding the influence of raw materials on the properties, and exploring new applications and mix designs for enhanced performance and durability of the geopolymer (Chindaprasit et al., 2007; Shehata et al., 2021).

The utilization of industrial waste ash in geopolymer production offers a sustainable approach to developing construction materials. Geopolymers demonstrate promising mechanical properties and environmental benefits when compared to traditional cement-based materials. Ongoing research aims to further enhance the understanding of geopolymer technology and its applications in the construction industry.

Industrial waste ash and geopolymer technology have emerged as an environmentally friendly and sustainable approach for the production of sustainable construction materials. This section explores the relationship between industrial waste ash and geopolymer, highlighting their interconnectedness and the benefits they offer in terms of waste management and sustainable material production. The utilization of industrial waste ash as a precursor in geopolymer production not only reduces waste disposal issues but also provides a viable alternative to conventional cement-based materials.

Industrial waste ash refers to the residue or by-product generated from various industrial processes such as coal combustion, metal smelting, and manufacturing. Common types of industrial waste ash include fly ash, bottom ash, and slag (Chindaprasirt et al., 2007). These waste materials possess pozzolanic and cementitious properties, making them suitable for use in geopolymer production. Geopolymer is an innovative cementitious material formed by activating aluminosilicate precursors with alkaline solutions. Geopolymer technology offers several advantages over conventional Portland cement, including reduced carbon emissions, enhanced durability, and improved chemical resistance (Zhou et al., 2021). The incorporation of industrial waste ash in geopolymer formulations provides a resource recovery opportunity that enhances their sustainability and performance.

The relationship between industrial waste ash and geopolymer is symbiotic. Industrial waste ash serves as a valuable source material for geopolymer production, providing the necessary aluminosilicate precursors (Chindaprasirt et al., 2007). The utilization of waste ash in geopolymer formulations not only reduces the consumption of natural resources but also diverts waste from landfills, thereby mitigating environmental concerns associated with waste disposal. Geopolymerization of waste ash creates a durable and eco-friendly construction material, promoting a circular economy approach. The combination of industrial waste ash and geopolymer technology offers numerous benefits. Geopolymers incorporating waste ash exhibit improved mechanical properties, such as compressive strength and flexural strength (Meesala et al., 2020). Additionally, geopolymer-based materials have demonstrated suitability for various applications, including concrete production, infrastructure development, and structural engineering. The use of waste ash in geopolymer systems contributes to sustainable construction practices and reduces the environmental impact of the construction industry.

Researchers have conducted extensive studies on the utilization of industrial waste ash in geopolymer production (Zhou et al., 2021; Xiao et al., 2020). Investigations focus on optimizing the geopolymerization process, studying the influence of different waste ash types on geopolymer properties, and exploring innovative techniques for geopolymer synthesis using waste ash precursors. These studies contribute to advancing the understanding of the relationship between industrial waste ash and geopolymer and facilitate the development of more sustainable construction materials.

The synergistic relationship between industrial waste ash and geopolymer technology provides a sustainable and environmentally friendly approach for the production of construction materials. The utilization of waste ash as a precursor in geopolymer formulations not only addresses waste management challenges but also enhances the performance and sustainability of the resulting materials. Ongoing research and development efforts continue to expand the knowledge and applications of industrial waste ash in geopolymer technology, contributing to a greener and more efficient construction industry.

5.1.2 Construction and Demolition Waste

CDW is a solid waste generated from any construction and demolition activities due to the improvement, preparatory, repair, or alternation work. These activities include civil and building construction, renovation, and demolition such as site clearance, roadwork, land excavation and formation. It is also categorized as an inert solid waste and a complete loss of physical waste from construction sites. Currently, the construction industry plays a vital role in economic growth to improve quality of life by providing necessary infrastructures such as roads, housing, commercial development, and other facilities. In Malaysia, the government has allocated 5,555 projects for the development sector by improving residential, and non-residential building projects, social amenities, and infrastructure projects (CIDB, 2012). Due to the circumstances, the high demand for construction development has generated a high volume of CDW.

European statistics reported a total of 859 million tonnes of CDW which covered 32.9% of the total solid waste (Eurostat, 2011). Many countries recorded CDW has a high number of annual generation (Table 5.1). In the UK and the US were 70 million

TABLE 5.1

Annual generation of CDW in four countries throughout the world (Eurostat, 2011)

Year	Amount	Country
2000	70 million tonnes	UK
2003	323 million tonnes	US
2007	1 million tonnes	China
2012	4 million tonnes	Malaysia

tonnes and 323 million tonnes of CDW, respectively, were reported by Designing Building Wiki (2017). Hong Kong Environment Protection Department reported municipal solid waste landfills had received 3,158 tonnes day^{-1} of CDW (Hong Kong Waste Reduction, 2015). Meanwhile, Malaysia recorded a total of 100,000 to 175,000 tonnes of waste, and the amount is believed to be increasing each year (The Star Online, 2006). This waste was generated from CDW and is relatively high due to the contribution of construction projects. Azis et al. (2012) stated that waste material from CDW was 9% by weight of purchase materials. The generation of CDW is usually due to problems in planning, material, human, and environment such as frequent changes in design, low quality of materials, worker mistakes, weak ability on site management, and bad weather.

High project development on housing, infrastructures, or commercial buildings has generated a high volume of CDW. In Malaysia, Bercham Landfill (Ipoh) and Jelutong Landfill (Penang) are legal dumping sites collecting CDW (Azarmi and Kumar, 2016) . However, illegal dumping is considered a common disposal practice in Malaysia. Approximately, 42% of waste from illegal dumping sites in Johor district is CDW (Rahmat and Ibrahim, 2007). CDW was also illegally dumped along the roadside and also in tropical mangrove swamps (Faridah et al., 2004).

Illegal dumping has governed public concerns owing to the risk to the environment and human health. High transportation costs and distance of location projects are among the key reasons to select illegal dumping as a disposal practice. Illegal dumping occurs when the distance of the project location is far from the gazetted landfill and the contractors refuse to expense additional costs, especially for waste management. According to the Lestari Universiti Kebangsaan Malaysia (UKM) research, only 68% of the contractors agreed to pay for services on collection and disposal (Pereira et al., 2005). Thus, providing a systematic and efficient waste management service such as re-using or re-cycling construction waste materials is required to minimize the amount of CDW being dumped at illegal sites or landfills.

5.2 OVERVIEW OF WASTE AND HEALTH HAZARDS

5.2.1 Industrial Waste: Ash and Its Risks

Ash waste from industrial processes, particularly the burning of coal, oil, or other fossil fuels in power plants, factories, and incinerators, is referred to as industrial ash waste. Depending on the exact origin and content, this waste is sometimes referred to as bottom ash, fly ash, or coal ash (Kumar et al., 2020). With varying elemental compositions (both nutrient and toxic elements) depending on the type and nature of the coal used for combustion and the method of production, fly ash, an amorphous mixture of ferroaluminosilicate minerals, has been produced in enormous quantities, of the order of 600 million tonnes worldwide (Garcia-Sanchez, 2015). Coal ash, the most prevalent type of industrial ash waste, consists of fine particles and solid materials that are left behind after the combustion of coal. It contains a range of substances, including heavy metals such as arsenic, mercury, lead, cadmium, chromium, and selenium, as well as other toxic chemicals (Figure 5.1). These toxins have the potential to contaminate air, groundwater, drinking water, and waterways if they

FIGURE 5.1   Coal ash and fly ash (Singh and Siddique, 2014).

are not managed properly (United States Environmental Protection Agency, 2023). Any type of waste that is dumped improperly constitutes irreversible damage to the environment. Hence, scientific waste management is necessary to lower the risks to the environment.

The exposure of both humans and animals to these harmful compounds after their release into the environment is the main cause of the health risks connected to industrial ash pollution. Industrial waste can significantly worsen airborne pollution, endangering both human and environmental health while also contributing to the deterioration of air quality. When coal ash is not properly managed, it can become airborne as particulate matter and be carried by wind over long distances (Yousuf et al., 2020). Combustion processes in power plants, factories, and incinerators release fine particles known as particulate matter into the air. These particles can vary in size, with smaller ones ($PM_{2.5}$) being more harmful as they can penetrate deep into the respiratory system. Industrial waste, including ash and other combustion by-products, can contribute to the formation of PM and increase its concentration in the air. Inhaling these fine particles can lead to respiratory problems, including bronchitis, asthma, and other lung diseases (Borm, 1997; Zierold and Odoh, 2020).

Industrial emissions can disperse over large areas and contribute to regional and global air pollution. Pollutants released by industries in one region can be transported by wind and affect air quality in distant locations. This transboundary pollution can have widespread environmental and health impacts, affecting both populated areas and remote ecosystems. The adverse effects of industrial waste on airborne pollution underline the importance of adopting cleaner production technologies, implementing effective pollution control measures, and enforcing stringent environmental regulations (Fowler et al., 2020).

Many industrial processes involve the burning of fossil fuels or the use of chemicals that release toxic pollutants into the atmosphere. These emissions can include sulfur dioxide (SO_2), nitrogen oxides (NO_x), volatile organic compounds (VOCs), heavy metals, and other harmful substances. These pollutants can have direct health effects and contribute to the formation of secondary pollutants, such as ground-level ozone and smog (Jambhulkar et al., 2018). Acid rain is created when certain industrial emissions, especially sulfur dioxide and nitrogen oxides, undergo chemical reactions in the atmosphere. Acid rain harms ecosystems by causing water pollution, soil acidification, and harm to infrastructure, plants, and aquatic life (Wang et al., 2020).

Industrial activities are a significant source of greenhouse gas emissions, primarily carbon dioxide (CO_2) but also methane (CH_4) and nitrous oxide (N_2O). These gases trap heat in the atmosphere, causing climate change and global warming. The burning of fossil fuels in industrial settings, as well as the emissions of gases from chemical reactions and waste management, are the main causes of industrial greenhouse gas emissions (Shobeiri et al., 2021).

Heavy metal concentrations found in coal ash are frequently significant and can seep into neighboring water sources or the groundwater. Aquatic habitats and drinking water supplies are both at risk from this contamination (Mokarram et al., 2020). The main causes of water contamination are believed to involve landfill leachate, runoff from urban areas, industrial waste, and agricultural runoff, as well as municipal, industrial, and agricultural wastewater (Samadi et al., 2009). One of the major risks to human health is heavy metal poisoning of water (Saha and Paul, 2019). As very permanent and largely non-degradable pollutants, heavy metals permeate the environment and contaminate both surface and groundwater resources (Burgess, 2015). Consuming or coming into touch with contaminated water can lead to several health concerns, such as cancer, neurological problems, developmental disorders, and organ damage (Mokarram et al., 2020). When ash waste is disposed of or stored inappropriately, the contaminants can seep into the soil, affecting agricultural lands and vegetation. Plants can absorb these pollutants, which may then enter the food chain, potentially exposing humans and animals to harmful substances (Shaheen et al., 2014).

To mitigate the health hazards associated with industrial ash waste, proper management practices are crucial. This includes effective containment, storage, and disposal methods, as well as regulatory oversight, monitoring, and enforcement. Additionally, transitioning to cleaner and more sustainable energy sources can reduce the generation of coal ash and its associated risks.

5.2.2 Construction and Demolition Waste and Its Risks

CDW refers to the waste generated from construction, renovation, and demolition activities, including materials such as concrete, wood, metals, bricks, plastics, and other construction-related debris. This waste can have various health effects, both directly and indirectly (Ali et al., 2019). Construction and demolition activities often generate a significant amount of dust and particulate matter. When inhaled, these fine particles can irritate the respiratory system, leading to respiratory symptoms such as coughing, wheezing, and shortness of breath. Prolonged exposure to high levels of dust can also contribute to the development of chronic respiratory conditions, including bronchitis and asthma (Azarmi and Kumar, 2016).

Asbestos is a fibrous mineral that was once widely utilized in building materials, and older structures may contain it. Asbestos fibers can become airborne during demolition or restoration and breathing them in poses serious health concerns. Asbestos exposure can result in lung conditions such asbestosis, lung cancer, and the uncommon but deadly malignancy mesothelioma (Musk et al., 2020). Construction materials, such as paints, adhesives, solvents, and treated wood, may contain hazardous chemicals and toxic substances. Improper handling, disposal, or burning of these materials can release harmful pollutants into the air, soil, and water. Direct contact, inhalation of toxic fumes, or ingestion of contaminated food or water can result in acute or chronic health effects, including respiratory illnesses, organ damage, developmental issues, and cancer.

Every stage of the building process produces waste and noise, from the creation of the raw materials to the erection and finishing of structures. Construction and demolition activities produce high levels of noise, which can have detrimental effects on human health. The activities that take place at the construction site and the tools that are employed, such as bulldozers, air compressors, pneumatic hammers, loaders, pavement breakers, and dump trucks, are the sources of noise pollution in the construction of buildings (Sholanke et al., 2019). Prolonged exposure to excessive noise levels can lead to hearing loss, increased stress levels, sleep disturbances, cardiovascular disease, voice loss, concentration loss, and other negative impacts on mental and physical well-being. It might also result in less productivity from construction workers (Ning et al., 2019). The quantity of noise and waste created by the construction industry must be significantly reduced in this era of integrating sustainability into the creation of buildings and their settings (Rahman and Ali, 2018).

Construction and demolition sites may harbor settings that encourage the reproduction of pests like rodents and mosquitoes which later lead to vector-borne diseases. Vector-borne diseases are infectious diseases that are transmitted to humans through the bites of vectors, such as mosquitoes, ticks, fleas, or flies. The spread of these disease-carrying pests can be facilitated by accumulated debris, stagnant water, and poor waste management including dengue fever, malaria, and leptospirosis (Krystosik et al., 2020). Although construction waste itself does not directly cause vector-borne diseases, it can create conditions that promote the breeding and proliferation of disease-carrying vectors. There are frequently items or waste at construction sites that can catch rainwater and form stagnant pools (Figure 5.2). Species like Aedes mosquitoes that spread diseases such as dengue fever, chikungunya, and the Zika virus flourish in these stagnant water sources. Standing water is where mosquitoes lay their eggs, and the larvae grow into adult mosquitoes that can infect people with diseases (Ortiz et al., 2022).

Construction waste, including discarded materials, can create hiding places and shelters for rodents, such as rats and mice. These rodents are known to be carriers of diseases like leptospirosis and hantavirus. If construction waste is not effectively managed, it can attract rodents and provide them with suitable habitats, increasing the risk of disease transmission. Construction activities often involve clearing vegetation, excavating land, or altering natural habitats. These disturbances can displace vectors or bring them into closer contact with humans. For example, cutting down trees and clearing bushes can disrupt the habitats of ticks or mosquitoes, causing them to seek alternative hosts, potentially including humans (Oh et al., 2022).

Therefore, it is highly significant to effectively manage industrial waste and CDW to reduce the risks associated with illness and protect both workers and nearby communities.

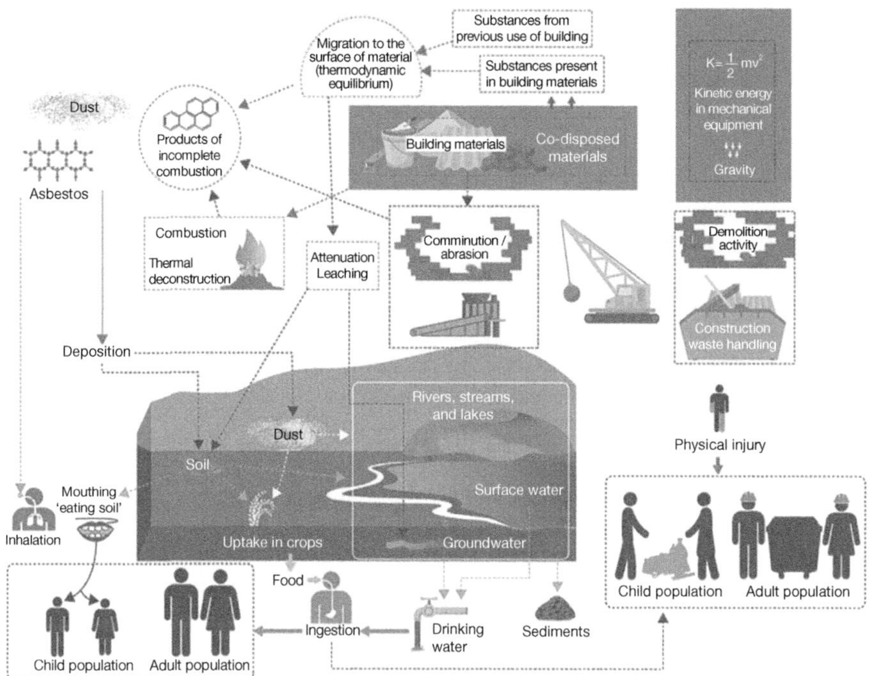

FIGURE 5.2 Overview of main hazards associated with construction and demolition waste (CDW) (Cook et al., 2022).

5.3 IMPORTANCE OF WASTE MANAGEMENT FOR PUBLIC HEALTH

Proper management of industrial waste and CDW, including segregation, containment, recycling, and safe disposal, is essential to mitigate health risks and protect both workers and the surrounding environment. Compliance with relevant regulations, adoption of best practices, and implementation of appropriate safety measures are crucial in minimizing the health hazards associated with these types of waste. Promoting sustainable practices, reducing emissions through cleaner energy sources, and implementing proper waste management strategies can help mitigate the impacts of industrial waste on air quality and protect human health and the environment (Khan et al., 2022).

Effective waste management plays a crucial role in safeguarding public health. Improper handling and disposal of medical, biological, or hazardous waste can result in the transmission of infectious diseases. Pathogens, such as bacteria, viruses, and parasites, present in waste can contaminate water, soil, or air, and potentially infect individuals who encounter the waste. Healthcare settings, if not equipped with proper waste management protocols, can contribute to the spread of healthcare-associated infections. Waste streams may contain hazardous substances such as heavy metals, toxic chemicals, or carcinogens. Improper disposal of such waste can lead to the release of these substances into the environment, increasing the risk of human exposure. Direct contact, inhalation of toxic fumes, or ingestion of contaminated food or water can result in acute or chronic health effects, including respiratory illnesses, organ damage, developmental issues, and cancer (Abdel-Shafy and Mansour, 2018).

Inadequate waste management, such as improper storage of waste or stagnant water in discarded containers, can create breeding grounds for disease-carrying vectors such as mosquitoes and flies. These vectors can transmit diseases such as dengue fever, malaria, or Zika virus. Proper waste management, including proper containment and removal of waste, reduces the potential breeding sites for vectors and helps control the spread of vector-borne diseases (Cruvinel et al., 2020). To mitigate the risk of vector-borne diseases associated with construction waste, it is important to implement proper waste management practices and maintain a clean and organized construction site. This includes the following:

- Promptly removing and properly disposing of construction waste to prevent the accumulation of stagnant water and the creation of harborage sites for vectors.
- Regularly inspecting the site for potential breeding sites and eliminating or treating them to prevent mosquito breeding (Degroote et al., 2018).
- Implementing vector control measures, such as applying larvicides or insecticides, when necessary (Sheela et al., 2017).
- Providing appropriate protective measures, such as insect repellents and protective clothing, to workers to minimize their exposure to vector bites (Degroote et al., 2018).
- Educating workers and site personnel about the risks of vector-borne diseases and promoting personal hygiene practices to reduce the likelihood of infection (Suradi, 2019).

Addressing these health hazards through effective waste management practices is crucial for protecting public health, promoting environmental sustainability, and ensuring the well-being of communities.

5.4 WASTE GENERATION AND COMPOSITION

5.4.1 INDUSTRIAL WASTE: ASH

This section provides a comprehensive summary of waste ash generation and composition for geopolymer production. It explores the types of waste ash commonly used in geopolymer formulations and highlights their potential as sustainable precursors. Waste ash, including fly ash, slag, and glass powder, offers a promising alternative to traditional cementitious materials, reducing environmental impact and promoting waste management.

Waste ash is generated from various industrial and municipal processes, including coal combustion, solid waste incineration, and glass manufacturing. These processes produce fly ash, slag, rice husk ash (RHA), palm oil fuel ash (POFA), and glass powder as common types of waste ash (United States Environmental Protection Agency, 2023). The quantity and composition of waste ash vary depending on the source and the specific industrial or municipal operation. Waste ash exhibits different compositions depending on the source material and the process from which it is derived. Fly ash, for example, is predominantly composed of silicon dioxide (SiO_2), aluminum oxide (Al_2O_3), and calcium oxide (CaO) (Chindaprasirt et al., 2012). Incineration ash from municipal solid waste incinerators may contain a mixture of metals, oxides, and unburned organic matter (Mehdizadeh et al., 2022). RHA, POFA, and glass powder are primarily composed of amorphous silica (SiO_2) and various oxides such as calcium oxide (CaO) and aluminum oxide (Al_2O_3) (Chindaprasirt et al., 2012; Jiang et al., 2022).

5.4.1.1 Types and Sources of Industrial Waste Ash

Industrial waste ash refers to the residues generated during various industrial processes, which can be potentially used as a valuable resource for geopolymer production. Geopolymers are a class of cementitious materials that can be synthesized using industrial waste ash as a substitute for traditional Portland cement. There are different types and sources of industrial waste ash used in geopolymer production, for example, fly ash, slag, RHA, and POFA, with each having different properties depending on its origin and production process.

Fly ash is one of the most widely studied and commonly used industrial waste ashes in geopolymer production. It is a fine particulate residue obtained from coal combustion in thermal power plants. Fly ash contains a significant amount of amorphous aluminosilicate materials, which can react with alkali activators to form geopolymeric gels. This waste ash can be sourced from both coal-fired power plants and industrial boilers. Bottom ash is another type of waste ash that can be utilized in geopolymer production. It is the coarse residue collected at the bottom of furnaces during the combustion of coal or biomass. Bottom ash primarily comprises non-combustible materials, including glassy particles, clinkers, and unburned carbon. Incorporating bottom ash in geopolymer mixes can enhance their mechanical properties and reduce environmental impacts (Sathonsaowaphak et al., 2009).

Utilizing industrial waste materials such as slag (granulated blast furnace slag) and ground granulated blast furnace slag (GGBFS) in geopolymer production not only offers environmental benefits but also enhances the mechanical properties of the resulting materials. Slag is a by-product generated during the iron and steel manufacturing process. It is produced when molten slag is rapidly quenched with water or air, resulting in a granular material with cementitious properties. Slag typically consists of silicates and aluminosilicates and possesses pozzolanic and latent hydraulic characteristics. It consists of various oxides, including calcium, silicon, and aluminum. Slag possesses latent hydraulic properties, making it a suitable ingredient for geopolymer synthesis. Incorporating slag in geopolymer formulations can enhance the long-term strength and durability of the resulting materials (Sathonsaowaphak et al., 2009).

GGBFS is a finely powdered by-product obtained by further processing and grinding granulated blast furnace slag. It is commonly used as a partial replacement for Portland cement in concrete production, offering improved workability, durability, and reduced heat of hydration. Recycling slag and GGBS for geopolymer production helps divert industrial waste from landfills, reducing the strain on natural resources. By utilizing these waste materials, the carbon emissions associated with their disposal are mitigated, resulting in a lower environmental impact. Geopolymers incorporating slag and GGBS exhibit improved mechanical properties, such as increased compressive strength, reduced permeability, and enhanced chemical resistance. Utilizing these materials in geopolymer production promotes sustainable construction practices by offering an alternative to traditional cement-based materials (Wang et al., 2020).

RHA originates from the rice milling process, which occurs in rice-producing countries. When rice is milled to produce white rice, the rice husk, also known as the rice hull, is separated as a by-product. The rice husk is the protective coating on the seed or grain of rice, consisting of hard materials such as silica and lignin to safeguard the seed during the growing season. RHA is abundantly available and renewable, making it an attractive agricultural by-product for various applications. It contains a high proportion of silica compared to other plant residues, which contributes to its pozzolanic properties and suitability for use in geopolymer production. The combustion of rice husk in a separate boiler generates RHA, which can be recycled for different purposes, including energy production and utilization in construction materials (Aghajanian et al., 2022).

POFA is a waste material generated from the combustion of palm oil fuel in power plants or palm oil mills. POFA is primarily sourced from palm oil mills and power plants that utilize palm oil biomass as a renewable energy source. Palm oil mills generate significant quantities of waste biomass, including palm kernel shells, fiber, and empty fruit bunches. The combustion of these biomass residues in power plants produces POFA as a by-product. The availability of POFA is particularly significant in palm oil-producing regions, such as Southeast Asia, where palm oil cultivation is widespread (Aldahdooh et al., 2013). It is increasingly being explored as a sustainable and environmentally friendly source for geopolymer production. POFA offers several advantages, including its abundance as a waste by-product and its pozzolanic properties, which contribute to the strength and durability of geopolymer materials.

The characteristics of POFA can vary depending on factors such as the source of palm oil, combustion conditions, and post-treatment processes. Different types of POFA may exhibit variations in chemical composition, particle size distribution, specific surface area, and pozzolanic reactivity. These variations influence the suitability of POFA for geopolymer production (Ranjbar et al., 2014; Olivia et al., 2016). Figure 5.3 illustrates a simplified representation of the geopolymer process, demonstrating the utilization of potential industrial waste materials derived from metal industries such as aluminum and steel, thermal plants including coal-based power plants, and burned biomass residues such as RHA and POFA. The process involves evaluating minerals based on their classifications, compositions, and reactivity, considering their compatibility with chemical activators.

5.4.1.2 Chemical and Physical Composition Analysis of Waste Ash

The chemical and physical composition of waste ash plays a crucial role in the production of geopolymers. Geopolymers are cementitious materials formed by the reaction of aluminosilicate materials with alkaline activators. Analyzing the chemical composition of waste ash is essential to determine its suitability for geopolymer production. Different types of waste ash, such as fly ash, bottom ash, and boiler ash,

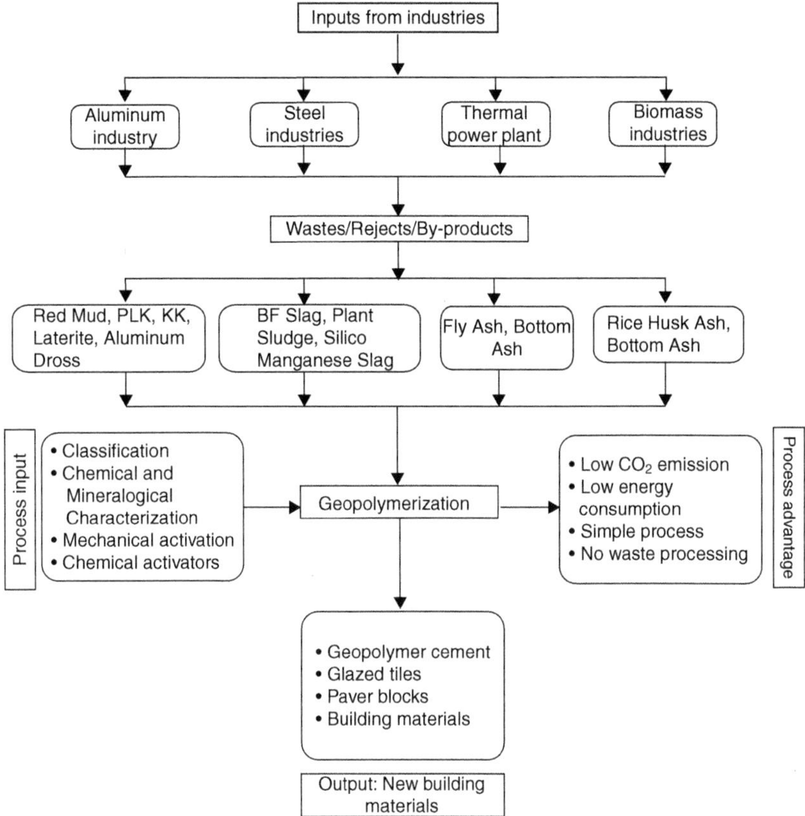

FIGURE 5.3 Process diagram for recycling industrial waste ash by geopolymer production (Hagedorn, 2003).

can have varying chemical compositions, including silica (SiO_2), alumina (Al_2O_3), iron oxide (Fe_2O_3), calcium oxide (CaO), and other trace elements (Sathonsaowaphak et al., 2009; Zarina et al., 2013). The presence and concentration of these elements influence the reactivity and strength development of geopolymers.

Higher silica content in waste ash typically leads to the formation of more geopolymeric gels, resulting in improved mechanical properties of geopolymers. Silica-rich ashes, such as fly ash, are commonly used due to their reactivity and ability to form strong geopolymeric bonds. Alumina acts as an essential component for the formation of geopolymeric gels. It contributes to the mechanical strength and chemical stability of geopolymers. Higher alumina content can enhance the binding capacity and increase the compressive strength of geopolymer products. Iron oxide and calcium oxide present in waste ash can affect the setting time and workability of geopolymer mixes (Zarina et al., 2013). These oxides may have a retarding effect on the geopolymerization process, resulting in longer setting times or decreased workability. A summary of the chemical composition of industrial waste ash is detailed in Table 5.2.

TABLE 5.2
Chemical composition in fly ash, slag, rice husk ash, and palm oil fuel ash

Raw material	Fe_2O_3	Al_2O_3	SiO_2	TiO_2	Na_2O	CaO	LOI	Average PCS	SiO_2/Al_2O_3
Fly ash (Shehata et al., 2021)	0.03	23.57	54.79	ND	0.31	3.17		78.36	2.32
Slag (GGBFS) (Singh and Siddique, 2014)	0.2–1.6	7–12	27–38	ND	ND	34-43	ND	34.2–51.6	2.52-3.42
Rice husk ash (RHA) (Singh and Siddique, 2014)	0.26	0.39	94.95	ND	0.25	0.54	ND	95.6	
Slag (BFS) (Yousuf et al, 2020)	1.2	9.6	32.30	2.2	0.5	38.5	ND	43.10	
GGBS (Borm, 1997)	0.30	8.51	40.30	ND	ND	37.01	ND	86.12	4.70
Palm oil fuel ash (POFA) (Zierold and Odoh, 2020)	4.54	5.66	53.82	ND	0.1	4.24	10.49	59.48	

Apart from chemical composition, the physical characteristics of waste ash, such as particle size distribution and specific surface area, also influence the properties of geopolymers. Finer particles and a higher specific surface area generally promote better reactivity and increased strength development in geopolymer products (Sathonsaowaphak et al., 2009).

5.4.2 Construction and Demolition Waste (CDW)

5.4.2.1 Types of Construction and Demolition Waste

Table 5.3 presents a comparison of waste composition between 2001 and 2018. Waste concrete was the major CDW generated from the site, accounting for 73%, followed by plastic (12%), gypsum (10%), wood (4%), and packaging products (1%). Several authors also reported the same findings where they found concrete was the highest waste generated from CDW (Hammes and Verstraete, 2002). Concrete was generally used as a construction material in structural applications, making this material the main waste in CDW generation. In contrast, Bianchini et al. (2005) reported that wood was the highest amount of waste generated. However, this is not surprising that concrete and wood are the main waste from the construction sector due to the application of these materials in scaffolding, formwork, and as a basic material for building structures (Foo et al., 2013).

Concrete is considered the main waste generated in CDW due to the high demands in major infrastructure projects, housing, and commercial development. Poor management of concrete and aggregates at construction sites by unskilled workers further contributes to the high generation of waste concrete. The amount of waste concrete generated from construction and demolition sites, however, potentially could be reduced as waste concrete could be recycled as a lightweight component such as pedestrian blocks, unbound structure layers of road, cement mortar, and concrete.

5.4.2.2 Characteristics of Construction and Demolition Waste

There are many types of waste generated by CDW, such as concrete, ceramics, metal, plastic, bricks, tiles, asphalt, gypsum, paper, glass, sand, and wood. However, due to the high demand for large infrastructure projects, housing, and commercial development. There are three main materials used in the production of concrete, namely cement, sand, and coarse aggregates. Therefore, crushed concrete waste from municipal waste has a varying proportion of cement attached to the sand and/or coarse aggregates. The proportion of cement paste on the crushed waste concrete influences the degree of deterioration of the concrete.

A high proportion of attached cement paste results in a low density of the concrete aggregate but high water absorption. This is due to the high porosity and low density of attached cement paste, which require high water absorption when mixing the concrete. The higher water consumption affects the workability and hardened process of the fresh concrete. In addition, excessive water demand during the mixing of the concrete leads to the formation of a thin layer of cement slurry on the surface of waste concrete aggregate (WCA), which penetrates into the porous old cement mortar, micro-cracks, and voids. The cement slurry forms a strong interfacial transition zone

TABLE 5.3

Composition of different types of CDW from 2001 until 2018

	Total amount of CDW (%)					Current Study
	Country					
Types of CDW	(Florida) (Jang and Townsend, 2001)	(Malaysia) (Faridah et al., 2004)	(Italy) (Bianchini et al., 2005)	(Malaysia) (Lachimpadi et al., 2012)	(Malaysia) (Foo et al., 2013)	(Malaysia)
	2001	2004	2005	2012	2013	2018
Concrete	56.3	12.3	33.8	60.0	60.0	73.0±1.2
Plastic	0.7	0.4	1.0	1.0	1.0	12.0±1.8
Gypsum	4.4	—	—	—	—	10.0±0.8
Wood	8.0	69.1	1.8	17.0	17.0	4.0±0.4
Packaging product	5.8	2.0	—	1.0	1.0	1.0±0.2
Brick	—	6.6	39.3	3.0	3.0	-
Metal	1.8	9.6	2.8	2.0	2.0	-
Soil and sand	—	—	19.0	15.0	15.0	-
Others (glass, asphalt, tile, ceramic)	23.0	—	2.3	1.0	1.0	-

(ITZ) between the cement paste and WCA, but also forms an ettringite, portlandite, and calcium silicate hydrate (C–S–H), which promotes the deterioration of the fresh concrete.

WCA showed that the density of the oven dry density (ODD) and surface dry density (SDD) of WCA were 1.64 ± 0.11 g cm^{-3} and 1.77 ± 0.19 g cm^{-3}, respectively. The density of WCA in this study was in accordance with the requirement of British Standard (BS) EN 12620:2002 (2006) aggregate was more than 2 g cm^{-3} (BS EN 12620-100:2002-2, 2006; British Standard Institution, 2006). The water absorption of WCA was $10.67 \pm 0.13\%$ of dry mass, slightly exceeding the standard requirement of BS EN 12626:2002 (2002) where the value should be less than 10% of dry mass. However, the percentage obtained in this study was within the common range for waste concrete. De Juan and Gutierrez (2009) also found the percentages of water absorption were in the range of 8% to 13% due to the deterioration of attached cement mortar.

Crushed WCA consisted of approximately half percent of attached cement paste, which was $47.76 \pm 8.43\%$, while another half percent was from the concrete aggregate. The percentage of attached cement mortar in WCA was slightly akin to the virgin concrete aggregate, but the difference was in the attached cement paste. The attached cement paste in WCA experienced deterioration due to high porosity and low density (de Juan & Gutierrez (2009).

The density (ODD and SDD) of WCA was lower than virgin material (coarse concrete aggregate), but higher in water absorption. Low density and high water absorption indicated that the attached cement paste consisted of more void spaces with pore water in WCA, further indicating that WCA has deteriorated. The presence of void spaces encouraged the pore water to fill the spaces by penetrating the water surface. Furthermore, the attached cement paste in WCA faced a longer neutralization rather than in virgin material. The lesser density in WCA was influenced by a greater volume of void spaces between particles. This is part of the carbonation process. The deteriorated attached cement paste in WCA had more ettringite originating from the early ettringite formation (EEF) and delayed ettringite formation (DEF) (Tam et al., 2009), resulting in the increment of calcium and sulfate in WCA that led to the expansion of ITZ. Therefore, the high volume of void spaces and ITZ between particles affected the density and water absorption of WCA.

Low density and high water absorption of WCA influence the concrete mixing process. It causes more water demand and delays the hydration process during concrete hardening (Padmini et al., 2009). Excessive water leads to the cement slurry formation in void spaces and old cracks, which enhances the formation of ettringite and C–S–H in ITZ and subsequently expands the micro-crack. The expansion of microcracks in concrete definitely reduces the compressive strength and affects its durability. Low density and high water absorption of WCA potentially influence the compressive strength of the concrete. Thus, several drawbacks of WCA such as high-water absorption and low density should be resolved beforehand in order not to negatively affect the compressive strength of the recycled products.

Major minerals, i.e., ettringite, quartz, calcite, minor of albite, and portlandite are found in WCA. The presence of ettringite as a major mineral in WCA indicated that the concrete sample was rich in EEF and DEF which further showed the increment of crystal in void spaces. Massive distribution of ettringite in void spaces triggered the swell pressure and consequently, enhanced the length of micro-crack between particles, which finally reduced the durability and strength. Therefore, it can be concluded that the presence of ettringite was a signal to WCA deterioration.

A low amount of portlandite in WCA showed the pozzolanic effect of attached cement paste had totally dissolved in pore water as another mechanism of the carbonation process in concrete for chemical stability (Jagad et al., 2023). Dissolution of portlandite in the carbonation process caused solubility of inorganic elements, and precipitation of calcite ($CaCO_3$) and C–S–H in pore water. The richness of calcite and C–S–H in WCA explained the reason why CaO and SiO were highly detected in X-ray fluorescence (XRF) analysis. Moreover, the dissolution of portlandite caused the formation of ettringite which potentially affected the compressive strength. As a result, the massive distribution needle shape of ettringite in recycled concrete caused the enlargement of micro-crack at ITZ that slowly led to the loss of concrete mass and shortened the service life of the recycled product.

The presence of quartz and calcite in WCA indicated the test samples did not only consist of broken attached cement paste (after the crushing process) but also fractions of concrete aggregate. Angulo et al. (2010) reported that quartz originated from rock, while calcite originated from cement. This finding was similar to the result obtained in the analysis of attached cement paste. The content of attached cement paste and concrete aggregate was relatively low, approximately 50% of the total WCA. Therefore, utilization of WCA as a part of the recycled materials enables the production of great compressive strength of recycled product due to the less amount of deteriorated attached cement paste.

Chemical analysis indicated that the major oxides that existed in the WCA were CaO and SiO_2 (Figure 5.4). The obtained results showed that the carbonation that

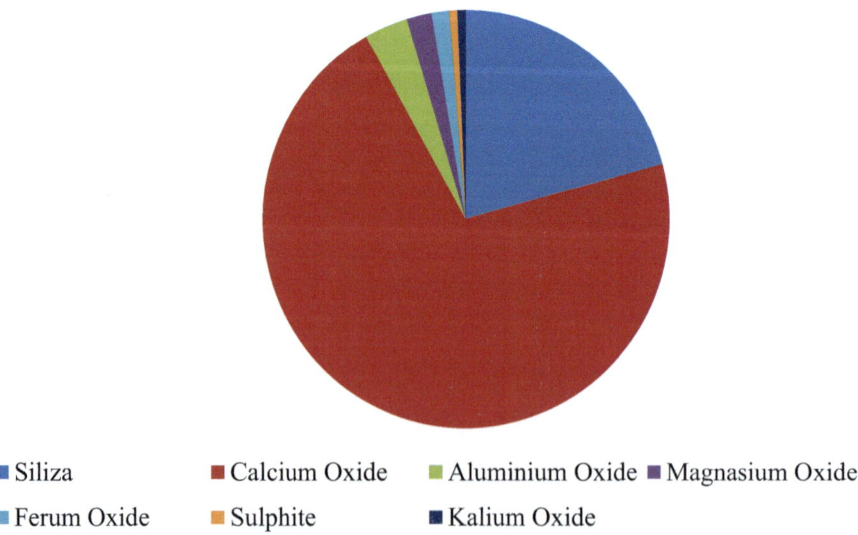

FIGURE 5.4   Major chemical elements found in WCA.

occurred in WCA was high due to the dissolution of portlandite to calcite and C–S–H. Besides portlandite, decalcification and polymerization of C–S–H contributed to the increment of CaO. Therefore, it was not surprising that CaO was highly found in WCA via X-ray diffraction (XRD) analysis.

The dissolution of portlandite initiated the leaching of inorganic elements into the pore water, while, ettringite tended to incorporate these leached inorganic elements particularly As, Cr, Mo, and Se by substituting sulfate with its structure (Cornelis et al., 2006). Thus, the high alkalinity of the cement matrix not only dissolves portlandite but influences the solubility of sulfate within pore water as well. Briefly, the presence of three main minerals in WCA, i.e., ettringite, calcite, and portlandite are indicators of deterioration of attached cement paste in waste concrete. Deterioration of attached cement paste from the dissolution of portlandite in WCA contributed to the changes in the porosity which finally affected the solubility of inorganic elements such as sulfate. The sulfate then dissolved in pore water and leached out from WCA due to the chemical stability of the attached cement paste and the surrounding environment.

Two major inorganic elements, i.e., Ca and SO_4^{2-} leached out from WCA (Figure 5.5), while 12 minor inorganic elements leached out from the same concrete. This result was similar to XRD and XRF analyses where Ca and SO_4^{2-} were the two main elements in calcite and ettringite (Nurhanim, 2016). Ca was highly concentrated in the leachate compared to SO_4^{2-}. The leaching trend of Ca significantly showed two different released patterns in carbonated aged concrete (WCA) and fresh concrete (control) ($p < 0.05$). High concentrations of Ca leaching out in aged WCA than control, ranging from 10.06 mg L^{-1} to 149.15 mg L^{-1} and 8.45 mg L^{-1} to 83.98 mg L^{-1}, respectively.

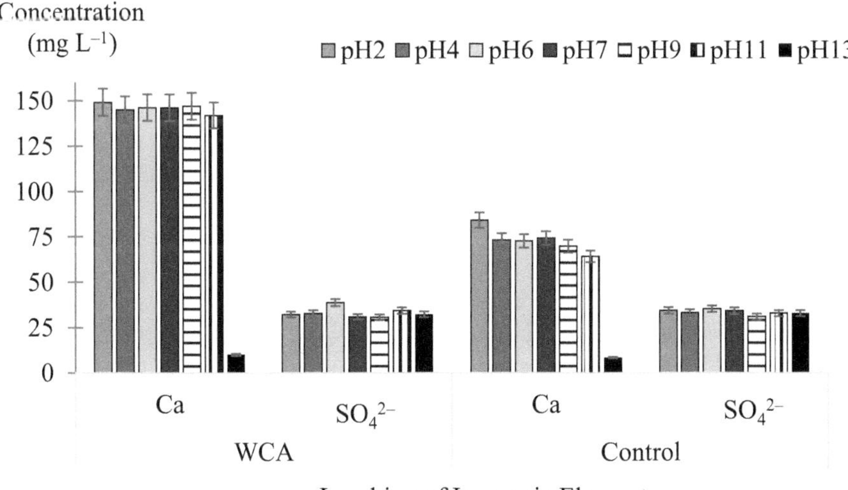

FIGURE 5.5 Major inorganic elements leaching out from WCA and control at different pH values.

Concentrations of SO_4^{2-} leaching out from WCA were in the range of 30.78 mg L^{-1} to 38.90 mg L^{-1} and control were 31.14 mg L^{-1} to 35.26 mg L^{-1}. The leachability of SO_4^{2-} from both test samples was likely unaffected by deteriorated attached cement paste, although the presence of ettringite as a source of SO_4^{2-} was high in WCA. Therefore, the concentration of the pollutants that can be leached out was not similar to the total content of the pollutants. This might be due to the stability of ettringite that prevented SO_4^{2-} solubilized from its structure.

The leaching trend of the cement-based material was related to the degree of carbonation. Aged concrete (WCA) had a higher degree of carbonation compared to fresh concrete (control). The carbonation process produces calcite ($CaCO_3$) from the reaction of decalcification of C–S–H, portlandite, and ettringite with CO_2 and OH$^-$ from pore water. The carbonation process in cement-based material caused changes in the pore structure due to the increase of calcite after being converted from portlandite. Besides, solubilized inorganic elements from the carbonation decreased the water permeability in the pore structure. Therefore, carbonated waste concrete was found to leach more inorganic elements, particularly Ca.

Most of the minor inorganic elements (Mg, Fe, Zn, Mn, Pb, Cu, Cd, As, Cr, Se, and Ni) in aged WCA were low in their concentrations (<0.1 mg L^{-1}) and reported in Figure 5.5. Each element showed different leaching trends at different pH values. The Leachability of Ca in WCA was high in acidic regions (pH 2) with a concentration of 149.15 mg L^{-1}. High Ca leached out at low pH due to decalcification of C–S–H, dissolution of portlandite, and unstable state of calcite and ettringite.

Generally, the leaching behavior of these inorganic elements demonstrated a similar trend as previous studies. Leached elements are mainly originated from the deteriorated attached cement paste (Solpuker et al., 2014). Deterioration of attached cement paste causes inorganic elements to dissolve in pore water, enhances the pore size, and decreases the water permeability (Mullauer et al., 2012). This condition causes a soluble element in pore water to have surface complexation with minerals. The surface complexation between the soluble elements in pore water and minerals potentially increases the leaching concentration if comes in contact with an acid solution.

The high concentration of SO_4^{2-} in the cement-based materials (WCA and control) was probably due to the presence of As, Cr, and Se in pore water that caused a retention mechanism. These three elements temporarily occupied in the pore water before being substituted with SO_4^{2-} in the ettringite structure. This reaction caused more SO_4^{2-} to solubilize in pore water, subsequently reacting with attached cement paste. Exposure of SO_4^{2-} to the attached cement paste in WCA extensively formed ettringite in void spaces and generated stress between particles responsible for the occurrence of micro-crack. Massive distribution needle shape of ettringite further led to an expansion of micro-crack, which, consequently resulted in the degradation of compressive strength. According to Galvin et al. (2013), soluble SO_4^{2-} in pore water could penetrate the micro-crack in concrete and affect the material structure and strength.

Although none of the regulations proposed an acceptance limit for CDW leachate, WCA definitely required proper disposal management. More attention should be given to the possible harmful effects of waste concrete from CDW on the environment.

Although the acceptable limit of SO_4^{2-} concentration leached out from WCA is unable to be determined, the leaching of SO_4^{2-} from the bulky form of waste concrete to the water surface could lead to aesthetic effects, i.e., taste, odor, or color. Leached SO_4^{2-} could potentially form volatile organic compounds such as hydrogen sulfide. Acidic hydrogen sulfide is volatile from aqueous solution into gas phases and could cause air pollution and water by triggering bad odor and taste problems. Furthermore, the common disposal practice of CDW in our country is dumping CDW together with domestic and municipal waste at municipal waste landfills, which subsequently causes intolerable odor and taste to humans and the environment.

In conclusion, the presence of SO_4^{2-} in WCA needs to be lessened to avoid re-crystallization of DEF. It is essential to control the sulfate content and water absorption of WCA to avoid an advancement of ettringite (DEF) formation which could negatively affect the strength and quality of recycled products.

5.5 RESOURCE RECOVERY TECHNOLOGY

Many researchers have tried to use different types of CDW as they can be used as substitute materials in the construction industry. Bricks, plastics, ceramics, and glass are the most common types of construction waste used in resource recovery in the construction industry. The innovative approach aims to overcome some challenges in the production of cement and concrete aggregates. Carbon dioxide emissions and costly energy supply, including transport activities, water and explosives consumption, the casting and laying process, and electricity, have a major impact on the carbon footprint of the construction industry. Therefore, resource recovery technology offers the industry the opportunity to reduce its dependence on natural resources and lower the cost of raw materials. In addition, the excessive energy consumption for the production and transport of aggregates and cement is another reason why a recycling approach is widely used.

5.5.1 CHEMICAL TREATMENT

Previous research was focused on the production of great compressive strength of recycled products when utilizing WCA as a part of the material. Admixture of superplasticizer and dissolution by acid have enhanced the compressive strength and gotten rid of the attached cement paste onto the WCA surface, but indirectly have degraded the recycled product. Admixture superplasticizer during cement mortar mix has caused corrosion with embedded steel reinforcement, while, dissolution by hydrochloric acid solution has degraded the composition of aggregate (Binici, 2007). Therefore, chemical treatment is an inappropriate solution for WCA.

5.5.2 PHYSICAL TREATMENT

Similar to chemical treatment, physical treatment is also having its own drawbacks. The addition of fly ash and blast furnace slag in concrete and cement mortar mix as well as a replacement of raw concrete aggregate with WCA at an appropriate ratio are examples of the physical treatment for WCA. The physical treatment methods had overcome the low density of WCA by producing a great compressive strength of recycled product. However, these physical treatment methods failed to reduce the water absorption and sulfate content in WCA. Additional fly ash and blast furnace slag have caused an increment in the retardation rate of cement hydration due to high water demand (Cheah and Ramli, 2012). Furthermore, sulfate content in fly ash is high with the range of 2.20% to 12.50%, which is not suitable to be used together with WCA because it has a high possibility to expand DEF. Briefly, chemical and physical treatments conducted by previous studies have proven that both treatments are unsuitable to be applied on WCA because these treatments have a high possibility to trigger other problems such as corrosion, expanded DEF, high water demand, and degradation of aggregate compositions.

Recycling waste ash as a raw material for geopolymer production presents many environmental advantages. It helps in reducing waste disposal and landfilling, thus minimizing the environmental impact associated with waste accumulation. By diverting waste ash from landfills, geopolymer production contributes to sustainable waste management practices. Geopolymers have lower carbon dioxide (CO_2) emissions compared to traditional cement-based materials. The production of cement, a major component of concrete, is a significant source of CO_2 emissions. By utilizing waste ash in geopolymer production, CO_2 emissions can be reduced due to lower energy requirements and a decreased reliance on non-renewable resources (Tchakoute et al., 2015).

Waste ash, such as fly ash and bottom ash, is a by-product of various industrial processes like coal combustion. Recycling waste ash for geopolymer production offers a viable solution for utilizing this abundant waste material. Geopolymerization transforms waste ash into a durable and resource-efficient material suitable for construction applications (Sathonsaowaphak et al., 2009). Geopolymer materials derived from waste ash exhibit promising mechanical and durability properties. Studies have shown that geopolymer composites made from waste ash possess satisfactory strength and durability characteristics, making them suitable for structural and non-structural applications (Abdulkareem et al., 2014). The incorporation of waste ash enhances the performance and extends the service life of geopolymer-based materials. Recycling waste ash for geopolymer production is an essential practice that offers environmental benefits, reduces CO_2 emissions, and utilizes waste materials effectively. Geopolymers derived from waste ash demonstrate favorable mechanical and durability properties, making them a viable alternative to traditional cement-based materials.

5.5.2.1 Geopolymers: Fundamentals and Applications

Geopolymers are inorganic polymers formed by chemically activating aluminosilicate materials, such as fly ash or slag, with alkaline solutions. The process of geopolymerization involves the dissolution of the precursor materials, followed by polymerization and solidification to create a three-dimensional (3D) network structure (Duxson et al., 2007). An understanding of the chemical reactions, curing processes, and material characteristics is essential in harnessing the full potential of geopolymers (Figure 5.6).

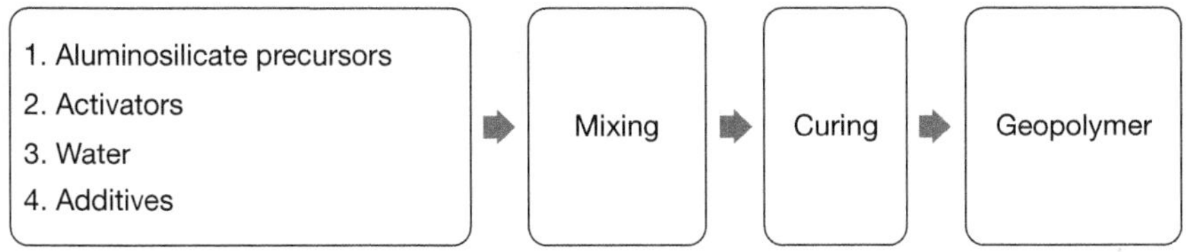

FIGURE 5.6 The process of geopolymer production (Achal et al., 2011).

Geopolymers demonstrate excellent mechanical properties, including high compressive strength, tensile strength, and flexural strength. Their performance can be tailored by adjusting the composition and curing conditions. Geopolymers also demonstrate enhanced durability, such as resistance to chemical attack, fire, and abrasion. These properties make geopolymers suitable for various applications in construction, infrastructure, and advanced materials.

The application examples of geopolymer are as folows:

a. **Construction Materials:** Geopolymers have been extensively investigated as alternatives to traditional cement-based materials. They can be used as binders in concrete, mortar, and grout formulations, offering improved strength, durability, and reduced environmental impact.
b. **Waste Stabilization and Remediation:** Geopolymers have shown immense potential in stabilizing and immobilizing hazardous waste materials. The solidification of waste with geopolymers helps in reducing leaching and minimizing the risk of environmental contamination.
c. **High-Temperature Applications:** Geopolymers exhibit excellent thermal stability, making them suitable for high-temperature applications. They find use in thermal insulating materials, refractory composites, and fire-resistant coatings.
d. **Advanced Functional Materials:** Geopolymers can be engineered to possess specific properties for various advanced applications. Examples include geopolymer-based composites for structural reinforcement, geopolymer foams for lightweight insulation, and geopolymer-based catalyst supports for chemical reactions.

    Geopolymers offer a sustainable and versatile alternative to traditional cement-based materials, with a broad range of applications across different industries. Understanding the fundamentals of geopolymerization, along with the mechanical and durability properties, is crucial in harnessing the full potential of geopolymers.

5.5.2.2 Geopolymer Chemistry and Reaction Mechanisms

Understanding the chemistry and reaction mechanisms involved in geopolymer formation is essential for optimizing their properties and expanding their utilization. Geopolymers are formed through the geopolymerization process, which involves the chemical activation of aluminosilicate materials, such as fly ash or slag, with alkaline solutions. The key chemical reactions in geopolymerization include the dissolution of precursor materials, hydrolysis and condensation of silicon (Si) and aluminum (Al) species, and the formation of a 3D geopolymeric network (Duxson et al., 2007). The choice of precursor materials, alkaline activators, and curing conditions significantly influences the geopolymer chemistry and resulting properties.

As for geopolymer reaction mechanism, it involves three stages. It starts with dissolution and ionization, followed by polymerization and solidification. The first step in geopolymerization is the dissolution of aluminosilicate materials, leading to the release of silicon and aluminum ions. This process involves the hydrolysis of complex minerals and the release of Si and Al species in solution. After dissolution, the silicon and aluminum species undergo hydrolysis and condensation reactions, leading to the formation of polymeric chains. Crosslinking occurs through the linking of Si and Al species with hydroxyl groups, resulting in the formation of a 3D geopolymeric network (Fowler et al., 2020). As the geopolymer network forms, the material solidifies and gains strength. The curing process involves the transformation of the gel-like precursor into a hardened geopolymer matrix, which is influenced by factors such as curing temperature, time, and alkaline activator composition (Shehata et al., 2021).

Water plays a crucial role in geopolymerization reactions. It acts as a solvent for the dissolution of precursor materials, provides hydroxyl groups for condensation reactions, and influences the viscosity and workability of the geopolymer paste (Tchakoute et al., 2015). Alkaline activators, such as sodium hydroxide (NaOH) or potassium hydroxide (KOH), provide the necessary alkalinity to initiate and facilitate the geopolymerization process (Shehata et al., 2021). The concentration and type of alkaline activators can affect the kinetics, degree of polymerization, and resulting properties of geopolymers. The dissolution, polymerization, crosslinking, and curing processes play significant roles in the formation and properties of geopolymers. Water and alkaline activators have a profound impact on the reaction kinetics and resulting geopolymer characteristics.

5.5.3 BIOLOGICAL TREATMENT

Biological methods have been introduced to overcome many environmental issues due to the low demand for energy, materials, gas emissions, and waste generation (Chen and Lin, 2004). Bacterial concrete technology has received tremendous concern in the field of civil engineering to heal micro-cracks on concrete and mortar (Mohannadoss et al., 2015). This technology uses microorganisms to precipitate calcium carbonate ($CaCO_3$) due to their metabolic activity (Ghosh et al., 2005) and is the same concept as microbial mineral precipitation (Mohannadoss et al., 2015), biomineralization (Hamilton, 2003) and microbial-induced carbonate precipitation (MICP) De Muynck et al., 2010). These methods utilized ureolytic bacteria to hydrolyze urea into ammonium and carbonate. Released ammonium is then spread to surrounding concrete and mortar and subsequently increases pH value and enriches the carbonate ion on concrete or mortar due to the accumulation of insoluble $CaCO_3$ which fills up the micro-crack of the concrete surface and improves the impermeability (Hammes and Verstraete, 2002).

Previous studies have demonstrated that the surface treatment has improved many properties of concrete and mortar, particularly in mechanical strength, water absorption, and sulfate content equivalent to fresh concrete. Furthermore, the use of a suitable microbe in concrete technology could enhance the durability of concrete as well as improve its resistance toward the sulfate attack. In an experiment demonstrated by Achal et al. (2011), the compressive strength of bacteria-treated mortar increased up to 36% and the water was absorbed six times less than that of the control sample. Mohannadoss et al. (2015) showed that the compressive strength of mortar was increased up to 25% after 28 days of curing which further verified the benefit of microbial activities within the void spaces between cement-aggregate matrices.

5.6 RESOURCE RECOVERY TECHNOLOGY BY PHYSICAL TREATMENT FOR INDUSTRIAL WASTE ASH AS A GEOPOLYMER PRECURSOR

Geopolymer precursors play a crucial role in resource recovery from industrial waste ash through physical treatment. By employing physical methods, valuable resources can be extracted from industrial waste ash and utilized as a sustainable material for geopolymer production. Extant research suggests that industrial waste ash, such as wastewater sludge incineration ash (WSIA), can be effectively utilized as a sustainable material in the 3D printing industry (Ki et al., 2021). Geopolymers, which are environmentally friendly cementitious materials, can be developed using industrial waste ash (Hoang et al., 2020). These geopolymer materials help reduce carbon dioxide emissions associated with the cement industry. Industrial waste-based geopolymers also have potential applications in soil stabilization, ground improvement, and pavement construction (Sharma and Kumar, 2020).

Physical methods of metal recovery from industrial waste are considered superior due to low chemical use and sludge production (Hoang et al., 2020). This shows that physical treatment techniques can be employed to recover valuable metals from industrial waste ash, which can then serve as a precursor for geopolymer production. The mix proportions of various industrial waste ash, such as fly ash and GGBFS, can be adjusted based on the desired strength grade of concrete required for geopolymer application (Reddy et al., 2018). Geopolymer synthesis using fly ash as a precursor has been explored for immobilizing hazardous waste, such as lead–zinc slag (Li et al., 2018). Physical treatment methods can be employed to recover resources from industrial waste ash, which can then be utilized as a geopolymer precursor. By employing these resource recovery techniques, industries can promote sustainability and reduce waste while contributing to the circular economy.

5.6.1 SUITABILITY OF WASTE ASH FOR GEOPOLYMER PRODUCTION

The suitability of waste ash for geopolymer production has been extensively studied, and diverse types of waste ash have been investigated for their potential use in geopolymer synthesis. Several waste ashes, such as fly ash, red mud, RHA, waste glass, slag, and mine tailings, have been explored as potential raw materials for geopolymer production (Hossain et al., 2021) These waste ashes contain components like silicon and aluminum, which are crucial for geopolymerization reactions (Milad et al., 2021).

The mechanical and durability properties of geopolymer concrete (GPC) containing various waste materials have been analyzed. Studies have shown that the absence of proper guidelines for mix design and sodium silicate content can affect the strength and durability of geopolymer concrete. However, the utilization of waste materials in geopolymer concrete has the potential to enhance its mechanical and durability properties (Podolsky et al., 2021). Geopolymers developed using industrial waste, such as geopolymer slag, red mud, and fly ash, have shown better durability performance compared to other industrial waste (Azad et al., 2021). Furthermore, the utilization of industrial waste in geopolymers has a positive impact on the environment, society, and economy.

Various studies have investigated the suitability of specific waste ashes for geopolymer applications. For example, fly ash-based geopolymers have been found to be suitable as landfill geoliner due to their low permeability and adsorption capability for inorganic pollutants. Incinerated sewage sludge ash (ISSA) has also been studied for its suitability in producing geopolymer concrete (Chowdhury et al., 2021; Istuque et al., 2019). Overall, the research indicates that waste ashes from different industrial processes can be suitable for geopolymer production. The utilization of waste ashes in geopolymers not only provides a sustainable solution for waste management but also offers the potential for improving the mechanical, durability, and environmental performance of geopolymer-based materials.

5.6.2 Geopolymerization Techniques

Geopolymerization is a process that involves the conversion of aluminosilicate materials in the form of waste ash into geopolymers, which are amorphous inorganic materials with covalent bonds. It is a complex physicochemical process that plays a crucial role in the development of geopolymeric materials with desirable properties. The geopolymerization process typically starts with oligomers, such as dimers, trimers, tetramers, or pentamers, which provide the structural units for the 3D macromolecular structure. These oligomers undergo chemical reactions and polymerization to form the geopolymer matrix (Davidovits, 2020). The geopolymerization technique requires aluminosilicate precursors, which can include materials like metakaolin, fly ash, slag, clays, volcanic ashes, or other sources rich in aluminum or silicon (Jia et al., 2020; Giacobello et al., 2022). These precursors are mixed with an alkaline solution, often containing sodium silicate, and subjected to specific curing conditions, such as elevated temperatures or ambient conditions, to initiate the geopolymerization process (Salunkhe et al., 2022).

Various experimental techniques are employed to gain a detailed understanding of the mechanistic steps involved in geopolymerization. These techniques can include XRD, Fourier-transform infrared spectroscopy (FTIR), nuclear magnetic resonance (NMR) spectroscopy, scanning electron microscopy (SEM), transmission electron microscopy (TEM), and other analytical methods. The specific conditions and parameters, such as the type and concentration of alkaline activators, curing temperature, duration, and the addition of other additives, can significantly influence the geopolymerization process and the resulting properties of the geopolymeric materials (Provis et al., 2009; Singh and Singh, 2019; Zhu et al., 2022).

Geopolymerization has gained attention as an environmentally friendly and sustainable alternative to traditional concrete based on ordinary Portland cement (OPC). Geopolymer concrete offers higher strength and deformation properties while reducing the carbon footprint associated with OPC production. Geopolymerization is a complex process that involves the transformation of aluminosilicate materials into geopolymers through chemical reactions and polymerization. It requires aluminosilicate precursors, alkaline activators, and specific curing conditions. Understanding the geopolymerization process requires employing various experimental techniques to analyze the structural and chemical changes that occur during the process. Geopolymerization has emerged as an environmentally friendly alternative to OPC-based concrete.

5.6.3 Performance and Properties of Geopolymer Materials

Geopolymers are known for their excellent mechanical strength and durability, making them a promising alternative to traditional construction materials like Portland cement concrete. Geopolymers offer several advantages over OPC, including increased mechanical strength, enhanced durability, resistance to fire and chemical corrosion, and stability in acidic and alkaline environments (Cong and Cheng, 2021; Castillo et al., 2021). The mechanical strength of geopolymers, particularly their compressive strength, is a crucial property for their analysis and application. Geopolymers can demonstrate compression strengths higher than 100 MPa, and their strength is influenced by factors such as alkali content, alkali cation, and Si/Ai ratio. The higher the compression strength of geopolymers, the better their resistance to carbonation and mechanical loads (Lingyu et al., 2021). In terms of durability, geopolymers demonstrate excellent performance. They possess high resistance to fire, chemical corrosion, and thermal fluctuations. Geopolymers also exhibit improved stability in acidic and alkaline environments, making them suitable for various applications (Castillo et al., 2021).

Numerous studies have explored the mechanical properties and durability of geopolymers. These studies have investigated parameters such as compressive strength, flexural strength, modulus of elasticity, ultrasonic pulse velocity, shrinkage, expansion, creep, weight loss, carbonation resistance, sulfate resistance, and corrosion resistance (Lingyu et al., 2021). Experimental work has been conducted to identify optimal mix proportions for high-strength geopolymer concrete using various binders, aggregates, and supplementary cementitious materials. It is important to note that the mechanical strength and durability of geopolymers can vary depending on factors such as the specific geopolymer formulation, curing conditions, activators, and raw materials used. Therefore, further research and development are ongoing to optimize the performance and explore the potential applications of geopolymers in civil engineering and construction (Jagad et al., 2023).

In summary, geopolymers have excellent mechanical strength, including high compressive strength, and offer enhanced durability compared to OPC-based materials.

They exhibit resistance to fire, chemical corrosion, and thermal fluctuations while maintaining stability in various environments. Ongoing research aims to understand and optimize the mechanical properties and durability of geopolymers for diverse applications in the construction industry.

5.6.4 SUCCESSFUL GEOPOLYMER PROJECTS UTILIZING WASTE ASH

There have been several successful geopolymer projects launched that have utilized waste ash as a key ingredient. Geopolymers offer a sustainable and environmentally friendly solution by utilizing industrial and municipal waste materials in construction applications. Some of the examples of successful geopolymer projects that have utilized waste ash are listed below:

i. **Utilization of hospital or bio-medical waste ash in geopolymer concrete:** This research investigated the impact of incinerated bio-medical waste (IBWA) ash as a partial substitute for the aluminosilicate source material in geopolymer concrete (Kumare et al., 2022). This project successfully converted hazardous IBWA as a source material for geopolymer production.

ii. **Utilization of waste RHA for sustainable geopolymer:** In many rice-producing countries, RHA is a common waste by-product. This review article discusses the utilization of waste RHA in sustainable geopolymer production (Hossain et al., 2021).

iii. **Municipal solid waste incinerated ash in geopolymer concrete:** The "WasteToWealth" project aims to optimize the treatment method of municipal solid waste incinerated (MSWI) bottom ash and manufacture artificial aggregates as a secondary raw material in geopolymer concrete applications (Adesanya et al., 2021).

iv. **Industrial by-products in geopolymer production:** This review article addresses various industrial by-products such as FGD residue, fly ash, blast furnace slag, and glass waste that can be used in the production of geopolymers (De Oliveira et al., 2021)

These projects demonstrate the potential of utilizing waste ash materials in geopolymer applications, providing sustainable alternatives to traditional construction materials. Geopolymers offer benefits such as improved mechanical performance, durability, and environmental impact compared to conventional materials. By effectively utilizing waste ash, these projects contribute to waste management and resource conservation efforts.

5.7 RESOURCE RECOVERY BY BIOLOGICAL TREATMENT FOR CONSTRUCTION AND DEMOLITION WASTE

5.7.1 *BACILLUS SUBTILIS* AS SULFATE REDUCTION BACTERIA

Table 5.4 shows the effect characteristics of *B. subtilis* on sulfate reaction in WCA. The ability of *B. subtilis* to simplify the complex ion of sulfate into sulfur elements for

TABLE 5.4
Effect characteristics of *B. subtilis* on sulfate reduction in waste concrete aggregate

Characteristics of *B. subtilis*	Advantage
Rod shape	Easier transport inorganic element from substrate *via* EPS
Reduce sulfate by dissimilatory pathway	Consumed sulfate although under depletion oxygen for energy conservation
Adapted to high sulfate content	Efficiently consume maximum concentration of sulfate for life process of *B. subtilis*
High negative charge on cell surface	High affinity to aggregate surface with strong electrostatic force for maximized sulfate reduction

metabolite reaction in gaining energy has categorized this gram-positive bacterium as sulphate-reducing bacteria. This genus of *Bacillus* cell is facultative anaerobe bacteria that is able to survive under depletion of oxygen to generate more energy in the form of adenosine triphosphate (ATP). *B. subtilis* uses sulfate as an alternative to oxygen as an electron acceptor. This reaction is a dissimilatory pathway to reduce sulfate and involves the consumption of hydrogen ions as electron acceptors for energy conservation as Equation 5.1 (Miao et al., 2012):

$$SO_4^{2-} + 9H^+ + 8e^- \rightarrow HS^- + 4H_2O \tag{5.1}$$

High-alkaline environment of concrete would not restrain the *B. subtilis* cell growth due to the high adaptation of this cell (Miao et al., 2012). Furthermore, the rod shape of *B. subtilis* cell was easier to transport inorganic elements from the substrate than the spherical shape (Gupta et al., 2013). This shows the cell shape of *B. subtilis* cell also has a significant role in reducing the sulfate content in aggregate.

Moreover, growing bacterial cells release greater hydrogen ions compared to dead cells due to the deprotonation of functional groups from the extracellular polymeric substance (EPS) in membrane cells. This character caused the zeta potential value of *B. subtilis* to reach the range of −35 to −55 mV due to the high negative charge on its surface due to the release of hydrogen ions. Therefore, a rich negative charge on *B. subtilis* surface potentially has high affinity adhered to a rich positive charge of WCA due to the strong electrostatic force.

The morphological characteristic of *B. subtilis* analyzed by SEM micrographs is depicted in Figure 5.7. The bacteria were rod-shaped with a length of approximately 2.084 μm. This length demonstrated that *B. subtilis* had a high surface area, thus, allowing a maximum adhesion on the surface of WCA to absorb as much as possible inorganic elements from WCA.

The high surface area and the rod shape of *B. subtilis* potentially required a minimum retention time to maximize the number of adhered *B. subtilis* onto the WCA surface. The cell shape was one of the crucial factors controlling the adhesion between bacteria cells and minerals. Previous research has reported that rod-shaped bacteria

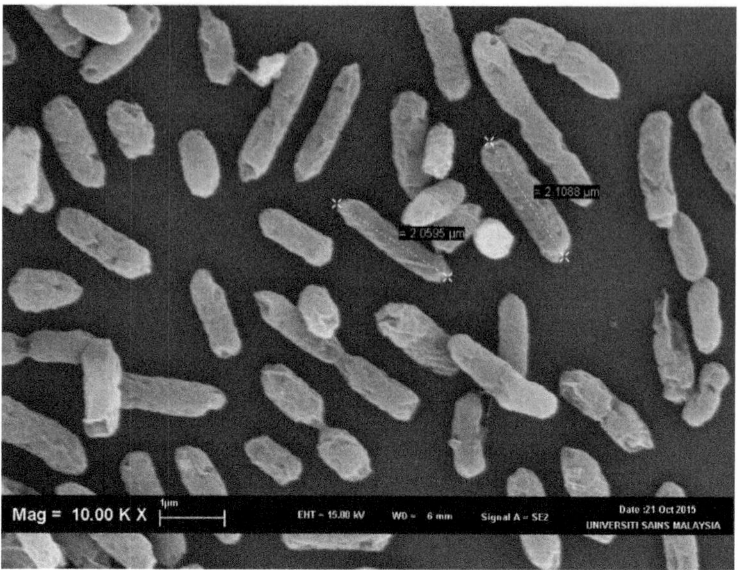

FIGURE 5.7 SEM micrograph of *B. subtilis* with length measurement.

were easier to transport inorganic elements from minerals *via* EPS and required less retention time than spherical shapes (Gupta et al., 2013). The lesser the retention time for bacteria cells to adhere onto the mineral surface, the more effective the bacteria cells transporting out the inorganic elements in the minerals. The surface of WCA with adhered *B. subtilis* is clearly shown in Figure 5.8(B). The surface was rough as compared to the surface of WCA without *B. subtilis* (Figure 5.8(A)). The rough surface with numerous holes showed high adhesion of *B. subtilis* onto WCA which was similar to the previous study of He et al. (2012).

5.7.2 Capability of *B. subtilis* in Concrete and Mineral Technologies

Numerous microbial species have been tested and further utilized in bacterial concrete technology either aerobic, anaerobic, or facultative bacteria. Most of the species are from aerobic bacteria in the genus *Bacillus* such as *Bacillus pasteurii*, *Bacillus amyloliquefaciens*, *Bacillus lentus*, *Bacillus sphaericus*, and *Bacillus megaterium* (Kumar et al., 2020). Species other than *Bacillus* have also been utilized in bacterial concrete technology such as *Pseudomonas aeruginosa*, *Escherichia coli*, *Acinetobacter* sp., and *Myxococcus xanthus*. Hence, considering all benefits attained by using *B. subtilis*, interdisciplinary approaches by combining biological knowledge in developing new, low cost and safe building structures are in favor and are extensively studied by many researchers nowadays (He et al., 2012).

Recent scientific studies on *B. subtilis* have proven its ability to treat micro-crack similar to that of other *Bacillus* sp. in concrete treatment (Knipper and Speck, 2012). Research has demonstrated that *B. subtilis* could increase the compressive strength of

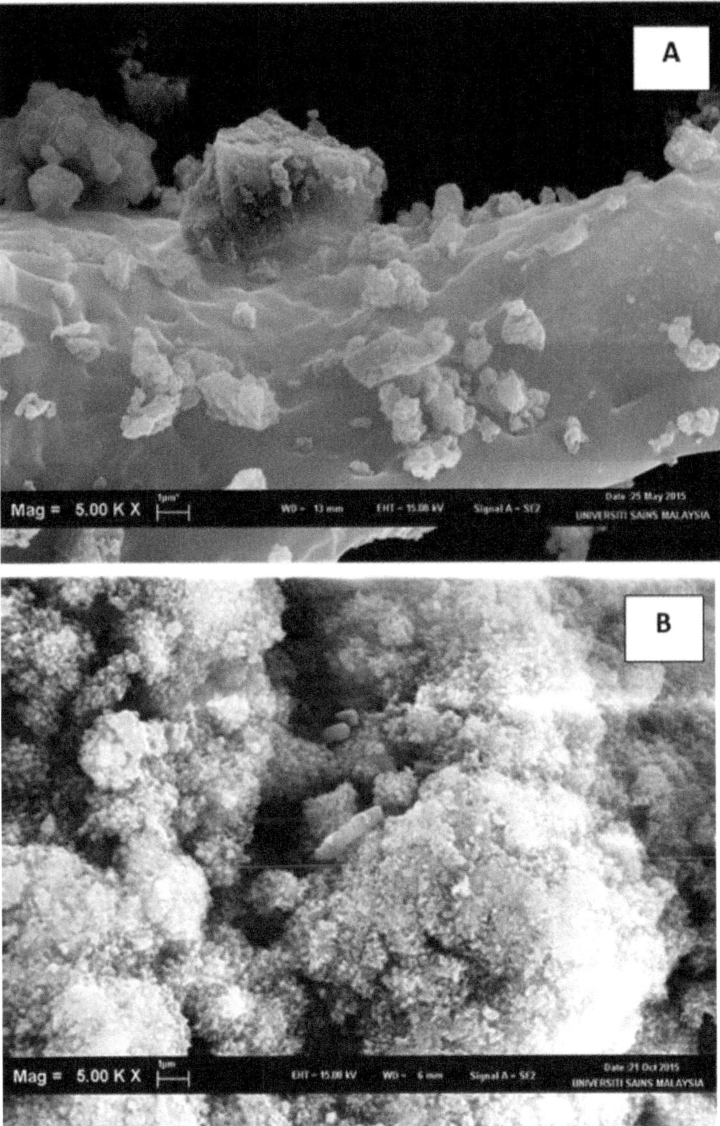

FIGURE 5.8 SEM micrographs. **(A)** Surface of WCA without *B. subtilis*. **(B)** Surface of WCA with adhesion of *B. subtilis*.

recycled concrete by up to 20 % (Bai et al., 2016). Besides, *B. subtilis* also reduces the water absorption of concrete aggregate (Sahoo et al., 2016). Another example is a study by Pei et al. (2013) in which, *B. subtilis* has increased the compressive strength of mortar by 15.8%. Besides improvements related to compressive strength, pore size, and water absorption, concrete treated with *B. subtilis* is also highly resistant to sulfuric acid attack (Van et al., 2010). Besides that, *B. subtilis* heals the micro-crack deeper and could produce harder concrete (Miao et al., 2012). Achal et al. (2011) reported that *B. subtilis* successfully healed the crack of mortar up to 27.2 mm in depth and subsequently increased 40% of compressive strength as compared to control.

It further shows that treating the micro-crack deeper into the concrete using bacteria has a great benefit rather than just treating the crack on the surface (Reddy et al., 2010). Achal et al. (2011) claimed that the efficiency of bacterial concrete technology occurred when bacteria penetrated deeper into the micro-crack. However, this approach is merely useful in micro-crack remediation treatment on concrete surfaces and does not cover the whole surface area of WCA. A treatment on the whole surface of WCA could potentially reduce the sulfate content and water absorption, avoiding the DEF formation, which subsequently limits the crack formation.

Previous research has revealed that all compounds found in concrete (either from attached cement paste or aggregate) are rich in sulfate-based compounds such as ettringite, gypsum, monosulfate, and metal-sulfate ($NaSO_4$) (Okabe et al., 2007). Bacteria from the genus *Bacillus* such as *B. subtilis* are highly efficient in consuming the sulfate under aerobic conditions and are also able to survive in high-alkaline environments (pH > 12). Besides, the dissimulation of sulfate by *B. subtilis* occurs in low oxygen availability, of which, sulfate acts as an oxidizing agent (Hagedorn et al., 2003). The adhesion of *B. subtilis* on the micro-crack surface would serve as a nucleation site, where the bacteria would reduce the sulfate in ettringite located in the micro-crack.

B. subtilis cells have been widely utilized in industrial fields typically for treating various heavy metals in wastewater (Gruyer et al., 2013). Apart from that, several studies have found that *B. subtilis* has a high affinity for various mineral surfaces intermediated by EPSs in cell membranes to remove inorganic elements (Adel, 2014). Quartz, corundum, silicate, and sulfide mineral are among the minerals treated with *B. subtilis*.

5.7.3 Metabolite of *B. subtilis*

Differences inorganic elements have varied in their function for the cell's life. For instance, essential elements of Ca, K, Cr, Na, N, S, PO_4^{3-}, and Zn would act as catalysts for biochemical reactions; stabilizers of protein structure and maintain osmotic balance of membrane cell structure. Meanwhile, the elements of Fe, Co, and Ni are important for redox reactions. However, Ag, Al, Cd, Hg, and Pb have been categorized as non-essential elements due to their non-biological role in cell growth. Moreover, excessive exposure to inorganic elements could be detrimental to the cell. Therefore, metabolite reaction in the anabolic pathway of *B. subtilis* responsible for controlling the movement of inorganic elements through membrane cells. Simple inorganic elements would be consumed by the cell to gain energy for cell growth and reproduce and repair their membrane cell structure. The consumption of inorganic elements from the surrounding cell surface involves various mechanisms (Dong et al., 2014).

The movement of the inorganic element from the surrounding cell surface is regulated by various mechanisms such as efflux and influx mechanisms, impermeability, precipitation within membrane cell structure, and release of the metal-complexing agent. Typically, these mechanisms sometimes act as a resistance system for the bacterial cells. Excessive exposure to high concentrations of certain inorganic elements could be detrimental to their cell growth. Naturally, the gradient concentration of elements between the cell and the surrounding cell surface could trigger the diffusion process. Active transport of protein and cytoplasm were responsible for export rapidly and indirectly caused permeability barrier at membrane cell structure. Besides that, the presence of protein in membrane cells acts to regulate the excessive diffusion of inorganic elements by binding together at the cell surface to prevent detrimental on sensitive cellular components. Moreover, the bacterial cell is able to export out non-essential elements by releasing of metal-complexing agent (glutathione) on the surrounding cell surface with the aim to absorb these elements before transporting them out from the cell.

The catabolic pathway is another metabolite of *B. subtilis* under depletion of oxygen purposely for energy conservation and life process (cell movement and synthesize more EPS on membrane cells. This process involved complex molecules of functional groups in EPS as an electron acceptor to synthesize sulfate into simple sulfur ions (Equation 5.2) (Achal et al., 2013). Therefore, the rich functional group of organic molecules in EPS could enhance sulfate reduction in aggregate.

$$SO_4^{2-} + CH_3COO \text{ (complex molecule)} \rightarrow HS^- + 2HCO_3 \tag{5.2}$$

5.7.4 Mechanism Adhered to *B. subtilis*

EPS consists of various organic macromolecules such as polysaccharides, proteins, nucleic acids, glycoproteins, and phospholipids which link up by ionized functional groups (carboxyl, hydroxyl, carbonyl, amino, and phosphate) into long chains located on membrane cell (Figure 5.9). The EPS matrix serves many functions such as capturing nutrients; controlling cell attachment; enhanced resistance to environmental

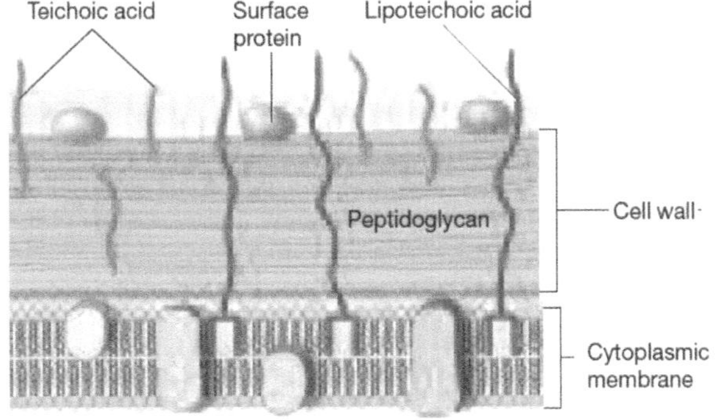

FIGURE 5.9 The membrane cell structure of gram positive bacteria cell (Geoffrey, 2010).

stress and disinfectants. Development of EPS matrix on membrane cell depending on factors of growth phase and species of bacteria (Adel, 2014). Thick peptidoglycan of gram-positive bacteria allocated more adhesion sites for inorganic elements compared with gram-negative bacteria. Extensive developing extracellular molecules on EPS structure by bacterial cells at lag and exponential phase cause provision more surface charges and chemical interactions.

The ionized function group as intermediate for *B. subtilis* adhered with opposite charge inorganic elements due to the strong electrostatic force. Besides electrostatic force, 40% to 60% of the hydrophobic functional groups from amino acids caused hydrophobic behavior of *B. subtilis* cells to have repulsion force to water molecules. Therefore, the *B. subtilis* cells have efficiently mobilized and deposited onto the mineral surface before absorbing the inorganic element. This is due to the deposition of *B. subtilis* cells onto mineral surfaces aided by hydrophobicity and electrostatic force between WCA and *B. subtilis*, which is controlled by numerous factors either physical, chemical, or biological properties (Adel, 2014).

Furthermore, the allocation adhesion site by the ionized functional group of bacterial cells could likely extend linkage patterns via cross-bridging and phosphodiester bonds. For example, the binding of SO_4^{2-} through cationic bridging by pre-existing membrane cell-bound metallic cation (Equation 5.3).

$$\left(Al,Fe\right)_n\left(OH\right)_{3n-x}{}^{x+} + xSi\left(OH\right)_4 \rightarrow \left(Al,Fe\right)_n\left(OH\right)_{3n-x}$$
$$\left[OSi\left(OH\right)_3\right]_x + xH^+ \tag{5.3}$$

Highly negative charge in EPS of *B. subtilis* potentially increases the number of adhered cells onto WCA. The high affinity of *B. subtilis* cells adhered onto the WCA surface reduces the presence of inorganic elements in WCA and subsequently leads to dissolution of ettringite due to the metabolic activities to reduce sulfate into simpler sulfur for the life process.

To enhance the adhesion of *B. subtilis* onto WCA, several factors such as bacterial cell concentration, ratio between bacterial and substrate, point zero charges (PZC) of zeta potential, retention time, bacterial growth phase, and bacterial cell shape are crucial and exhibit a tremendous effect in reducing a significant number of inorganic elements in WCA. Many researchers have investigated and demonstrated that these factors aided *B. subtilis* in multiplying its number to efficiently adhere to the mineral surface (Hong et al., 2013).

5.7.5 Utilization of *B. subtilis* in Waste Concrete Aggregate Treatment for Sulfate Reduction

Physical and chemical treatment on CDW was focused on improving low-density and high-water absorption of WCA to enhance the mechanical strength of recycled products. However, both treatments depended on super-plasticizers that trigger problems with corrosion and degrade the aggregate. Furthermore, direct use of WCA with high-DEF content could deteriorate the aggregate due to the recrystallization of DEF in ITZ. Recently, the adaptability of *B. subtilis* in highly alkaline environments of concrete and the capability consume sulfate as part of metabolite activities possibly to utilize in WCA treatment. Sulfate reduction bacteria from the genus of *Bacillus*, not only reduce sulfate content and inorganic elements in WCA but are able to precipitate $CaCO_3$ in existing micro-crack on the whole surface of aggregate as part of ureolytic activities. Thus, treated WCA could prevent from deterioration process (carbonation and ettringite formation) and redistribution contaminant in the pore solution of recycled product. Utilizing *B. subtilis* in WCA treatment not only reduces the amount of CDW dumped at the landfills or along the roadsides but also saves the environment by reducing the natural resource dependency and energy consumption, as well as overcoming the drawback characteristic of sulfate content.

The limiting factors of waste concrete to be recycled are high sulfate content and high-water absorption in attached cement paste. High sulfate content and water absorption in WCA potentially enhance DEF and expand micro-crack formation. This micro-crack formation should be minimized to improve concrete durability and workability. Most of the applications focus on the precipitation of $CaCO_3$ by bacterial cells filling in the micro-crack. Although bacterially reduced sulfate and water absorption in waste concrete by adhesion are considered new technologies, they are potentially feasible to treat and recycle the waste concrete. However, this concept needs further analysis and some areas need to be clarified before the recycled WCA can be used. For example, ensuring the produced recycled WCA possesses two main features, i.e., low water absorption and low sulfate content in ettringite structure. Both features are crucial to prevent any conducive environment from facilitating DEF formation which finally prevents the formation of micro-cracks that affect the mechanical strength of the recycled product.

Furthermore, the most effective condition for WCA treatment by *B. subtilis* needs to be examined to identify the super-saturated level of *B. subtilis* to highly reduce sulfate and water absorption. By far, no major breakthrough has been achieved in the field of interdisciplinary microbiology in waste concrete; nonetheless, a treatment of concrete using microorganisms is definitely a promising technology and the potential gains from research are also enormous.

5.7.6 Waste Concrete Aggregate Treated with *B. subtilis* as Potential Recycled Aggregate

B. subtilis adhesion test was applied in WCA treatment particularly to reduce the sulfate content before the expansion of ettringite which could trigger the micro-crack formation as well as reduce the compressive strength of the recycled product. Four main factors were examined to observe the effectiveness of *B. subtilis* in treating WCA, i.e., the effect of *B. subtilis* concentration, *B. subtilis*: WCA ratio, zeta potential, and retention time.

The highest sulfate reduction (22.2%) was achieved at the concentrations of 0.2×10^8 CFU mL^{-1} and 12×10^8 CFU mL^{-1}. At this stage, *B. subtilis* adhesion reached maximum colony formation at about 5.59×10^8 CFU mL^{-1} and 4.69×10^8 CFU mL^{-1}. Increasing the *B. subtilis* concentration did not increase the number of *B. subtilis* adhering to WCA. A high concentration of *B. subtilis* indicates the stationary phase of these bacteria. When cultured at the highest concentration, the cell already reached a mature cell with larger size and surface properties compared to the lag phase. The high electrostatic and hydrophobic forces of *B. subtilis* and WCA therefore resulted in a minimal number of adherent cells (cells in the stationary phase) due to the limited space provided by WCA. In addition to the influence of cell size, the lag phase of *B. subtilis* cells tended to have a high affinity for WCA to absorb inorganic elements because of the active metabolic process in lag phase cells. Therefore, a high *B. subtilis* concentration (stationary phase) did not reduce the sulfate in WCA as much as possible.

The presence of functional groups in the EPS of *B. subtilis* allocated many adhesion sites for metal and sulfate binding *via* cross-bridging and phosphodiester bond by electrostatic force between bacterial cells and the mineral (Sheng et al., 2008). The electropositive charge from diaminopimelic acid, amine group, and phosphodiester bond tended to attract the sulfate ion in WCA. Therefore, the adhesion site allocated by such function groups acts as an intermediate for *B. subtilis* absorbing sulfate for metabolite reaction on sulfate reduction. Reduction of sulfate content in rich functional groups of bacterial cells occurred at a high rate. The WCA were rich functional groups in the lag and initial exponential phases. This showed that *B. subtilis* relied on WCA for essential nutrient up-take and high possibly occurred chemotaxis process. As a result, high *B. subtilis* adhered onto the WCA surface in the lag and initial exponential phase. Bacterial cells at the exponential phase actively enhanced the content of functional groups on the EPS by demonstrating high ionic strength. EPS of *B. subtilis* in the lag phase and initial exponential phase showed several functional groups from different chemical bonds (C=O, C–O–C, C–OH, C=N, CH, N–H, P–OH) compared to the stationary phase. This finding showed that the lag and initial exponential phases of *B. subtilis* consisted of more functional groups, i.e., carboxylic (C=O, C–O–H, C–O–C), amine (C=N, N–H), and phosphoryl (P–OH) groups compared to the stationary phase. Rich functional groups at lag and initial exponential phase showed *B. subtilis* cells keep developing EPS components for growth.

Theoretically, the overgrowth of *B. subtilis* causes the nutrients and the adhesion sites to become limited. This situation further leads to unbalanced net electrostatic and hydrophobic force between the bacteria cells and WCA and affects the adhesion of *B. subtilis* cells on the surface material (Kjervik et al., 2018). The bacteria then change the protein expression as a mechanism of protection to survive under these extreme conditions (limited nutrients and unbalanced net electrostatic and hydrophobic force) (Celik et al., 2019). This is done by altering the physical and/ or chemical properties of *B. subtilis* membranes, for instance, by increasing the surface area of hydrophobic water-insoluble growth substances and the bioavailability of hydrophobic substances (Bahera and Mulaba-Bafubiandi, 2017). The remaining bacteria that do not adhere to the surface material are unable to gain the nutrients; hence, they could not survive and causing the growth of the bacteria become retarded.

The best *B. subtilis* concentration for WCA treatment was selected at 0.2×10^8 CFU mL^{-1} due to the ability of *B. subtilis* to reduce maximum sulfate content with the highest number of adhesions of *B. subtilis*, even at low concentrations. This showed that *B. subtilis* was able to survive and was well-adapted to high alkaline WCA.

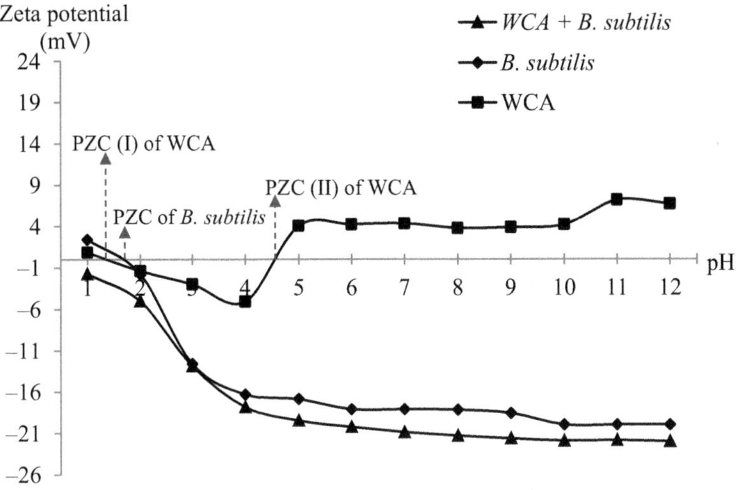

FIGURE 5.10 Zeta potentials of WCA, *B. subtilis,* and adhesion between WCA and *B. subtilis.*

The zeta potential values of WCA were in the range of −6 mV to 3 mV at below pH 4 (Figure 5.10). Zeta potential analysis is important to analyze the point zero charge (PZC) value. The PZC for WCA was noted at a strong acidic region with different pH values, namely pH 1.8 and pH 4.5. Fowle and colleagues claimed that quartz is responsible for the PZC at low pH due to the dissolution of quartz likely occurring around pH 2 (Caulier et al., 2019). Meanwhile, C–S–H in a deteriorated aggregate tends to be dissolved at a pH below 5. Therefore, the dissolution of different minerals (quartz and C–S–H) caused two PZC to occur for WCA.

The dissimilation of sulfate ion occurred when WCA was introduced to acid due to the conversion of sulfate to hydrogen sulfide (H_2S) when metal had solubilized in solution (Fowle et al., 2004). Hydrogen ions from acid could act as an electron acceptor when exposed to the sulfate ion as a redox reaction (Equation 5.4). As shown in Figure 5.11, the highest sulfate reduction (84.91%) was at pH 4.

$$SO_4^{2-} + 9H^+ + 8e^- \rightarrow HS^- + 4\,H_2O \tag{5.4}$$

According to the leaching behavior of inorganic elements, metals such as Ca, Fe, Zn, Mg, Pb, Mn, Cu, Cd, and Cr were present in leachate at acidic regions (pH 2 and pH 4). Furthermore, the dissimilation of sulfate occurred at low pH values and led to the leaching of various metals in WCA. Thus, the sulfate reduction was high (73.58% to 84.91%) in the acidic region (pH 3 and pH 4) compared to the alkaline region (pH 9 to pH 12) which showed a lower reduction of sulfate content (41.51% to 69.81%) (F (p > 0.05) = 32.92) (Figure 5.11). Therefore, leached metals in WCA caused a reduction of the zeta potential value of WCA due to the reduction of positive charges below pH 4, ranging between −6 mV and 3 mV. As a result, PZC of WCA mostly occurred at acidic regions between pH 1 and pH 5.

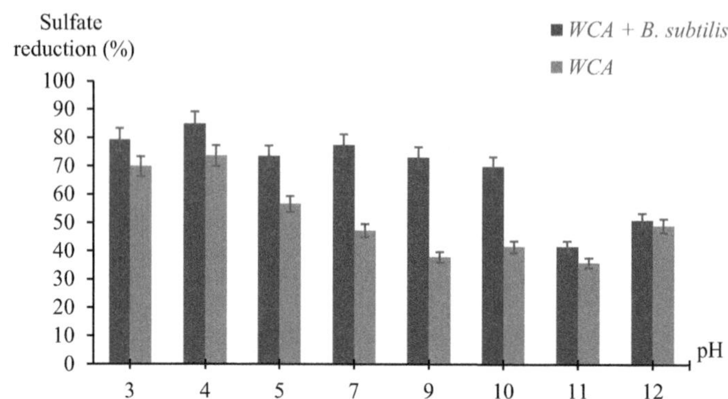

FIGURE 5.11 Amount of sulfate reduction in WCA treated with *B. subtilis* and WCA (control).

Generally, maximum adhesion occurs at PZC due to the high affinity of opposite electric charges (Hong et al., 2013). The electrostatic double layer becomes thinner with a balanced net charge, which is easier for *B. subtilis* to mobilize towards WCA (Hong et al., 2013). The electric charge on WCA and *B. subtilis* surfaces posed a balanced net electric charge at low pH values. Based on Figure 5.12, the PZC for *B. subtilis* was at pH 1.8. In a study by Hong et.al. (2013), the PZC value for *B. subtilis* was pH 2.1. El-Midany and Abdel-khalek also showed a low PZC value at pH < 2.2 (El-Midany and Abdel-khalek, 2014a). The low pH value of PZC for *B. subtilis* was due to the changes in the ionizable state of functional groups (carboxyl, amino, and hydroxyl group) and the high amount of gluronic acid in EPS (El-Midany and Abdel-Khalek, 2014b).

Moreover, the solubility of H_2S affected the stability of DEF by releasing more sulfate ions from DEF structure and subsequently replaced it with inorganic elements of As, Se, Cu, and Cr (Vegas et al., 2011). The leaching test in this study showed that the concentrations of leached sulfate ion at pH 2 to pH 6 were 30 mg L^{-1} to 40 mg L^{-1}. Therefore, the substitution of sulfate by four main inorganic elements (As, Se, Cu, and Cr) on DEF structure increased the net positive charge of zeta potential and assisted the sulfate reduction in WCA in the acidic region. Besides, As, Se, Cu, and Cr, most of inorganic elements such as Fe, Mg, Zn, and Mn display low leachability at pH 6 to pH 11. As a result, the zeta potential value in WCA gradually increased from 0.86 mV to 7.18 mV after pH 4.5.

Although PZC of *B. subtilis* potentially had a strong electrostatic force between *B. subtilis* and WCA, high acidity of *B. subtilis* PZC was not suitable to be applied because *B. subtilis* was unable to survive longer in acidic regions (pH 3 to pH 5). Figure 5.12 shows the lowest *B. subtilis* colony (between 7.15 × 10^8 CFU mL^{-1} and 8.6 × 10^8 CFU mL^{-1}) adhered onto WCA at low pH values (pH 3 to pH 5). Furthermore, the addition of acid could destroy the *B. subtilis* structure and removed all metals from the WCA which subsequently reduced the electrostatic force. In this study, inorganic elements, mainly Mg, Zn, Fe, and Pb, were highly leached in acidic regions. Consequently, only a small amount of *B. subtilis* adhered onto WCA due to the lack of inorganic elements as nutrient supplies. In addition, changing the pH to

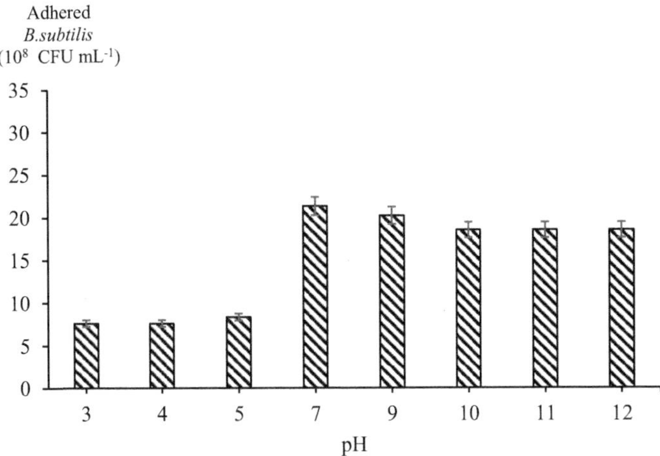

FIGURE 5.12   Effect of pH on adhered *B. subtilis*.

acidic regions could degrade the concrete aggregate. The dilute acid solution could dissolve the cement paste and affect the remaining aggregate fraction.

Therefore, the adjustment of WCA pH (pH 11 to pH 12) to the PZC pH (pH 1.8 and pH 4.5) was not suitable to be applied in WCA treatment due to the difficulty in changing the high alkalinity of WCA and the effect of acid solution to membrane structure of *B. subtilis*. The threshold levels of *B. subtilis* adhered onto WCA were from pH 10 to pH 12 with a colony number of 18.5×10^8 CFU mL^{-1}. These threshold levels showed that high number of *B. subtilis* were able to exploit the alkaline environment for growth.

The maximum values of zeta potential of *B. subtilis* were in the range of −21.93 mV to −22.03 mV at pH 10, pH 11, and pH 12 (most alkaline). This data showed that *B. subtilis* carried high negative charge on the surface. Hong et al. reported the maximum zeta potential of *B. subtilis* was from −27.3 mV to −31.7 mV (Geoffrey, 2010). High negative charge carried by *B. subtilis* was due to the deprotonating of amino group (El-Midany and Abdel-Khalek, 2014b). Moreover, ionized functional groups of carboxylic, amine and phosphoric were highly detected by FTIR analysis, of which these functional groups contributed to high negative charge on *B. subtilis* surface.

While the maximum values of zeta potential of WCA were from +6.71 mV to + 7.18 mV at pH 11 and pH 12 (most alkaline region), high positive charge carried by WCA was due to the presence of various metals and minerals. WCA consisted of several metals such as Ca, Al, Mg, Fe, K, Na, and Cr. Numerous metals and minerals present in WCA contributed to high positive electric charge. Therefore, high negative electric charge of *B. subtilis* and high positive electric charge of WCA tended to have strong electrostatic force between them. This strong electrostatic force led to the strong adhesion of *B. subtilis* onto WCA surface. This strong adhesion between *B. subtilis* and WCA is shown in Figure 5.7.

The maximum zeta potential values of WCA with adhered *B. subtilis* ranged between −21.87 mV and −22.03 mV at WCA with pH 11 and pH 12, respectively.

This obtained results demonstrated a strong adhesion between *B. subtilis* and WCA due to the electric surface charge of both *B. subtilis* and WCA were at an unstable stage which attracted each other. However, the zeta potential values of WCA with adhered *B. subtilis* did not reduce the net negative electric charge carried by *B. subtilis*. This showed that the metals in WCA had reduced by *B. subtilis*. Sheng et al. reported that the presence of metal from the metal sulfate and metal oxide in WCA deprotonated the functional group of carboxyl for metabolic activities, but leaving sulfate and oxide on the surface. Although the adhesion between *B. subtilis* and WCA was strong, the zeta potential remained high in negative electric charge due to the rich deprotonated functional groups (Hong et al., 2013). Besides, the leaching behavior of inorganic elements in WCA resulted in the decrease of positive electric charge from WCA. In addition, the up-take of inorganic elements in WCA by *B. subtilis* as essential nutrients for the production of more EPS contributed to the reduction of positive electric charge of WCA. Therefore, rich deprotonated functional groups of *B. subtilis* and solubility of metal oxides (MgO, Al_2O_3, Fe_2O_3) in WCA as well as highly leaching of inorganic element potentially alter the zeta potential values of WCA at pH 11 from 7.18 mV to the −21.87 mV when WCA adhered with *B. subtilis*.

The sulfate reduction was high at pH 3 to pH 5 in WCA, ranging from 56.60% to 73.58%. However, the reduction was enhanced, ranging from 73.58% to 84.91% when WCA was treated with *B. subtilis*. Although adhered *B. subtilis* on WCA at acidic region (pH 3 to pH 5) showed a weak adhesion compared to pH 7 to pH 12, the sulfate reduction seemed high at acidic region. Acidic region caused the occurrence of the PZC for *B. subtilis* and WCA, where *B. subtilis* surface and WCA at PZC pH had a net electric charge with a strong electrostatic force between *B. subtilis* and WCA. Therefore, sulfate highly mobilized and transported out from WCA.

Changing the pH of WCA by adding an acid or alkali was highly detrimental to the environment as inorganic elements tended to leach out from the WCA and destroyed the *B. subtilis* structure as well degrade concrete aggregate. Although the pH did not change, WCA treated with *B. subtilis* tremendously reduced the sulfate content from 25.8% to 41.51% at pH 11. In this study, the material pH at pH 11–12 was used in the WCA treatment by *B. subtilis* due to the environmental concern.

The maximum reduction was reached at 4 day^{-1} with 81.5%. The sulfate content did not show any reduction when the retention time was increased to 5 day^{-1}. The retention time significantly differed on the sulfate reduction by *B. subtilis*, indicating that *B. subtilis* required 4 days to consume all essential nutrients from inorganic elements (metal and anion). These elements were used for biochemical reaction typically for stabilizing the protein structure, osmotic balance, metabolic activities, enzyme, and DNA through electrostatic force and osmotic pressure (Abdel-Khalek and El-Midany, 2013).

Treated WCA (TWCA) by *B. subtilis* was utilized to produce recycled cement mortar (RCM). The physical characteristics of TWCA were initially evaluated to observe the efficacy of *B. subtilis* in treating WCA using the adhesion test (Table 5.5). The value of water absorption of TWCA decreased by 3.23 % of dry mass, while the densities of ODD and SDD increased up to 1.84 ± 0.11 g cm^{-3} and 2.07 ± 0.19 g cm^{-3}, respectively. The density and water absorption shows TWCA were complied

TABLE 5.5
Physical characteristics of WCA before and after WCA treatment

Physical Characteristics	Results		BS EN 12620: 2002/PRAI (2006) Standard
	Before Treatment (WCA)	After Treatment (TWCA)	
Oven dry density (ODD)	1.64 ± 0.11 g cm^{-3}	1.84 ± 0.11 g cm^{-3}	2 g cm^{-1}
Saturated surface dry density (SDD)	1.77 ± 0.19 g cm^{-3}	2.07 ± 0.19 g cm^{-3}	2 g cm^{-1}
Water absorption	10.67 ± 0.13 % of dry mass	6.84 ± 0.13% of dry mass	10% of dry mass

with British Standard BS EN 12620:2002 (Nain et al., 2019). Evidently, the WCA treatment by *B. subtilis* did not only reduce the sulfate content but also remediated the density and water absorption of WCA.

Physical characteristics of WCA were improved due to the soaking and air-dried procedures. Soaked WCA in suspension solution followed by air-dried procedures aimed to prevent the suction of excessive water during concrete mixing (British Standard Institution, 2006). Soaked WCA with water during WCA treatment indirectly activated the unhydrated residual attached cement paste which led to the formation of cement slurry on the WCA surface with the self-cementing feature. This self- cementing feature in residual attached cement paste of WCA caused the cement slurry filled in the old cracks and void spaces and further resulted in the reduction in ITZ length due to the ettringite formation. Tam et al. claimed that the formation of cement slurry was able to enhance the compressive strength of recycled concrete but required a large amount of water when waste concrete was mixed with cement and sand (de Juan and Gutierrez, 2009). Besides, the formation of cement slurry indirectly contributed to the increment of WCA density. Therefore, WCA treatment by conducting soaking and air-dried procedures have improved three main drawback characteristics of WCA, namely sulfate content, water absorption value and density.

The production of RCM showed a great compressive strength, i.e., 26 MPa under the compressive test (Figure 5.13). The compressive strength of RCM from TWCA complied with the standard limit requirement for cement mortar production (Zaharieva et al., 2003), where the strength for RCM was higher than the control (20 MPa). The high compressive strength of the RCM confirmed the ability of *B. subtilis* cells to treat WCA. Reducing the sulfate content and water uptake in WCA could prevent the spread of the massive distribution of DEF and subsequently prevent the development of internal microcracks, which could lead to mass loss and reduce the compressive strength. In this study, mixing a new cement paste with TWCA allowed the new cement paste to penetrate into the voids at the ITZ between the adherent cement paste and the concrete aggregate, healing the old microcrack without increasing the formation of ettringite.

Moreover, the precipitation of $CaCO_3$ by the metabolic activity of the remaining *B. subtilis* from the WCA treatment in the microcrack caused the membrane of

FIGURE 5.13   RCM cube produced by TWCA.

B. subtilis to act as a nucleation center for the formation of C–S–H (British Standard Institution, 2009). The massive formation of C–S–H on the WCA surface invades the cavity of the old microcrack and ITZ. Therefore, the RCM had a high compressive strength. The complete packing of the particles by a new cement grain and the formation of C–S–H could prevent continuous mass loss when TWCA was hardened by a new cement paste. The porosity of WCA was then reduced and mass loss could be prevented (Thomas et al., 2009). In other words, the reduction of water absorption of TWCA was supported by the precipitation of $CaCO_3$ in voids and ITZ. C–S–H formation caused crystallization of mineral, which resulted in high dense-RCM with great compressive strength. Besides, deposited of dead *B. subtilis* cells in WCA have plugged the micro-crack and ITZ. High compressive strength of RCM treated with *B. subtilis* was able to overcome the micro-crack problem and potentially prevented excessive DEF formation due to the carbonated, low-density, high-water absorption as well as sulfate content attached cement paste of WCA.

5.8   CONCLUSION AND RECOMMENDATIONS

5.8.1   Summary of Key Findings

Geopolymerization is a process that aims to turn industrial solid waste into a chemically durable cement binder that is both thermally stable and non-combustible. It offers numerous advantages over OPC, such as reducing CO_2 emissions and utilizing waste ingredients. Recycling industrial waste ash through geopolymerization can contribute significantly to promoting sustainability and circular economy principles by converting waste materials into valuable resources.

Based on physical characteristics, WCA was physically deteriorated due to the low-density, high-water absorption, carbonated by attached cement paste and rich in ettringite. Leaching behavior of inorganic elements demonstrated that Ca and SO_4^{2-} were the major leached elements, High concentration of leached Ca and SO_4^{2-} exhibited that WCA was a carbonated concrete and that the attached cement paste has deteriorated. Therefore, biological treatment on WCA treated with *B. subtilis* successfully reduced the sulfate content using three different parameters, i.e., *B. subtilis* concentrations, zeta potential, and retention times. The highest reduction of sulfate content was 81.50% when bacterial concentration was at 0.2×10^8 CFU mL^{-1} and the retention time as increased up to 4 day^{-1} without changing the zeta potential of *B. subtilis* or WCA. RCM using a TWCA could replace a sand portion due to higher compressive strength (26 MPa).

5.8.2 Recommendations for Waste Minimization by Promoting Resource Recovery Technologies

To promote industrial waste ash recycling through geopolymerization, the following recommendations can be considered. Invest in research and development to explore the potential of different types of industrial waste ash, such as fly ash, bottom ash, slags, and red mud, for geopolymerization. Conduct more studies to determine the optimal composition and processing conditions for achieving desired geopolymer properties. Foster collaboration and partnerships between industries, research institutions, and government agencies to facilitate the exchange of knowledge, resources, and expertise in the field of geopolymerization. This can help accelerate the development and implementation of geopolymer technologies for industrial waste ash recycling.

Governments can play a crucial role by establishing supportive regulations and policies that encourage the recycling of industrial waste ash through geopolymerization. This can include incentives, tax benefits, or regulatory frameworks that promote the use of geopolymer-based materials in construction and infrastructure projects. Raise awareness among industries, professionals, and the general public about the benefits of geopolymerization for industrial waste ash recycling. Promote educational programs and training initiatives to build capacity and knowledge in the field of geopolymer technology. Support and promote pilot and demonstration projects that highlight successful applications of geopolymer-based materials using industrial waste ash. These projects can serve as examples of the economic and environmental viability of geopolymerization for recycling industrial waste. Lastly, develop standardized testing methods and certification systems for geopolymer-based materials to ensure their quality, durability, and compliance with relevant industry standards. This can enhance the confidence of stakeholders and facilitate the wider adoption of geopolymer products in various construction applications.

While recycling of CDW needs to have further analysis on the filtrate quality from treated WCA should have an additional treatment before it can be released into the watercourse due to environmental concern. The efficiency of anaerobic bacteria and aerobic bacteria such as *Shewanella* species and *B. subtilis* to reduce the sulfate content in WCA should be conducted as a comparison. The compressive strength of self-healing concrete and recycled concrete from WCA treated with *B. subtilis* should be tested thoroughly and compared to determine the best method to overcome the micro-crack problem and analysis the ettringite formation in recycled aggregate from the WCA treatment using *B. subtilis* so as to promote resource recovery technologies through biological treatment.

REFERENCES

Abdel-khalek, M. A., & El-Midany, A. A. (2013). Application of *Bacillus subtilis* for reducing ash and sulfur in coal. *Environmental Earth Science, 70*(2), 753–760.

Abdel-Shafy, H. I., & Mansour, M. S. (2018). Solid waste issue: Sources, composition, disposal, recycling, and valorization. *Egyptian Journal of Petroleum, 27*(4), 1275–1290.

Abdulkareem, O. A., Bakri, A. M. A., Kamarudin, H., Nizar, I. K., & Saif, A. A. (2014). Effects of elevated temperatures on the thermal behavior and mechanical performance of fly ash geopolymer paste, mortar and lightweight concrete. *Construction and Building Materials, 50*, 377–387.

Abdullah, M. M. A. B., Hussin, K., Bnhussain, M., Ismail, K. N., Yahya, Z., & Abdul Razak, R. (2012). Fly ash-based geopolymer lightweight concrete using foaming agent. *International Journal of Molecular Sciences, 13*(6), 7186–7198.

Achal, V., Mukerjee, A., & Sudhakara, R. M. (2013). Biogenic treatment improves the durability and remediates the cracks of concrete structures. *Construction and Building Materials, 48*, 1–5.

Achal, V., Mukherjee, A., & Reddy, M. S. (2011). Microbial concrete: Way to enhance durability of building structure. *Journal of Materials in Civil Engineering, 23*, 730–734.

Adel, A. S. A. (2014). Screening of bacterial isolates from sewage treated effluent with potential to remove heavy metals and B-lactam antibiotics. PhD Thesis, Universiti Sains Malaysia.

Adesanya, E., Perumal, P., Liimatainen, H., Yliniemi, J., Ohenoja, K., Kinnunen, P., & Illikainen, M. (2021). Opportunities to improve sustainability of alkali-activated materials: A review of side-stream based activators. *Journal of Cleaner Production, 286*, 125558.

Aghajanian, A., Thomas, C., & Sainz-Aja, J. A. (2022). The use of rice husk ash in eco-concrete. In *The Structural Integrity of Recycled Aggregate Concrete Produced with Filler and Pozzolanans.* Elsevier eBooks (pp. 171–197).

Aldahdooh, M., Bunnori, N. M., & Johari, M. a. M. (2013). Development of green ultra-high performance fiber reinforced concrete containing ultrafine palm oil fuel ash. *Construction and Building Materials, 48*, 379–389.

Ali, M. U., Liu, G., Yousaf, B., Ullah, H., Abbas, Q., & Munir, M. A. M. (2019). A systematic review on global pollution status of particulate matter-associated potential toxic elements and health perspectives in urban environment. *Environmental Geochemistry and Health, 41*, 1131–1162.

Amran, M., Debbarma, S., & Ozbakkaloglu, T. (2021). Fly ash-based eco-friendly geopolymer concrete: A critical review of the long-term durability properties. *Construction and Building Materials, 270*, 121857.

Angulo, S. C., Carijo, P., Figueiredo, A. D., Chaves, A. P., & John, V. M. (2010). On the classification of mixed construction and demolition waste aggregate by porosity and its impact on the mechanical performance of concrete. *Materials and Structures, 43*, 519–528.

Azad, N. M., & Samarakoon, S. M. K. (2021). Utilization of industrial by-products/waste to manufacture geopolymer cement/concrete. *Sustainability, 13*(2), 873.

Azarmi, F., & Kumar, P. (2016). Ambient exposure to coarse and fine particle emissions from building demolition. *Atmospheric Environment, 137*, 62–79.

Azis, A. A. A., Memon, A. H.., Rahman, I.A., Nagapan, S., & Imran, Q. B. A. (2012). Challenges fares by construction industry in accomplishing sustainability goals. *IEEE Symposium on Business, Engineering and Industrial Applications.* 23–26 September, Bandung, Indonesia (pp. 630–634).

Bai, H., Nelly, C., Andre, P., & Edvina, L. (2016). Bacteria cell properties and grain size impact on bacteria transport and deposition in porous media. *Colloids and Surfaces B: Biointerfaces, 139,* 148–155.

Behera, S. K., & Mulaba-Bafubiandi, A. F. (2017). Microbes assisted mineral flotation a future prospective for mineral processing industries: A review. *Mineral Processing and Extractive Metallurgy Review, 38*(2), 96–105.

Bianchini, G., Marrochino, E., Tassinari, R., & Vaccaro, C. (2005). Recycling of construction and demolition waste materials: A chemical-mineralogical appraisal. *Waste Management, 25,* 149–159.

Binici, H. (2007). Effect of crushed ceramic and basaltic pumice as fine aggregates on concrete mortar properties. *Construction and Building Materials, 21,* 1191–1197.

Borm, P. J. (1997). Toxicity and occupational health hazards of coal fly ash (CFA). A review of data and comparison to coal mine dust. *Annals of Occupational Hygiene, 41*(6), 659–676.

British Standard Institution. (2006). BS EN 12620:2002 Aggregate for concrete (Amendment to EN 12620:2002), Brussels, Belgium.

British Standard Institution. (2009). BS EN 12390-3: 2009: Testing hardened concrete. Compressive strength of test specimens.

Burgess, G. L. (2015). Effects of heavy metals on benthic macroinvertebrates in the Cordillera Blanca, Peru. Maser Thesis, Western Washington University.

Cao, J., Zhang, G., Mao, Z., Fang, Z., & Yang, C. (2009). Precipitation of valuable metals from bioleaching solution by biogenic sulphides. *Minerals Engineering, 22*(3), 289–295.

Castillo, H., Collado, H., Droguett, T., Sánchez, S. F., Vesely, M., Garrido, P., & Palma, S. (2021). Factors affecting the compressive strength of geopolymers: A review. *Minerals, 11*(12), 1317.

Caulier, S., Nannan, C., Gillis, A., Licciardi, F., Bragard, C., & Mahillon, J. (2019). Overview of the antimicrobial compound produced by members of the *Bacillus subtilis* group. *Frontier in Microbiotechnology, 26,* 1–19.

Celik, P. A., Aksoy, D. O., Koca, S., & Koca, H. (2019). The approach of biodesulfurization for clean coal technologies: A review. *International Journal of Environmental Science and Technology, 16,* 2115–2132.

Cheah, C. B., & Ramli, M. (2012). Mechanical strength, durability and drying shrinkage of structural mortar containing HCWA as partial replacement of cement. *Construction and Building Materials, 30,* 320–329.

Chen, S. Y., & Lin, J. G. (2004). Bioleaching of heavy metals from livestock sludge by indigenous sulfur—oxidizing bacteria: Effects of sludge solids concentration. *Chemosphere, 54,* 283–289.

Chindaprasirt, P., Chareerat, T., & Sirivivatnanon, V. (2007). Workability and strength of coarse high calcium fly ash geopolymer. *Cement and Concrete Composites, 29*(3), 224–229.

Chindaprasirt, P., De Silva, P., Sagoe-Crentsil, K., & Hanjitsuwan, S. (2012). Effect of SiO_2 and Al_2O_3 on the setting and hardening of high calcium fly ash-based geopolymer systems. *Journal of Materials Science, 47*(12), 4876–4883.

Chowdhury, A., Naz, A., & Chowdhury, A. (2021). Waste to resource: Applicablity of fly ash as landfill geoliner to control ground water pollution. *Materials Today: Proceedings, 60,* 8–13.

Cong, P., & Cheng, Y. (2021). Advances in geopolymer materials: A comprehensive review. *Journal of Traffic and Transportation Engineering, 8*(3), 283–314.

Cook, E., Velis, C. A., & Black, L. (2022). Construction and demolition waste management: A systematic scoping review of risks to occupational and public health. *Frontiers in Sustainability, 3*, 43.

Cornelis, G., Gerven, V. T., & Vandecasteale, C. (2006). Antimony leaching from uncarbonated and carbonated MSWI bottom ash. *Journal of Hazardous Materials, 137*(3), 1284–1292.

Cruvinel, V. R. N., Zolnikov, T. R., Obara, M. T., de Oliveira, V. T. L., Vianna, E. N., do Santos, F. S. G., ... & Scott, J. A. (2020). Vector-borne diseases in waste pickers in Brasilia, Brazil. *Waste Management, 105*, 223–232.

Davidovits, J. (2020). *Geopolymer Chemistry and Applications* (5th Ed). France: Institut Geopolymer.

de Juan, M. S., & Gutierrez, P. A. (2009). Study on the influence of attached mortar content on the properties of recycled concrete aggregates. *Construction and Building Materials, 23*, 872–877.

De Muynck, W., De Belie, N., & Verstraete, W. (2010). Microbial carbonate precipitation in construction materials: a review. *Ecological Engineering, 36*(2), 118–136.

De Oliveira, L. L., De Azevedo, A. R. G., Marvila, M. T., Pereira, E. C., Fediuk, R., & Vieira, C. M. F. (2021). Durability of geopolymers with industrial waste. *Case Studies in Construction Materials, 16*, e00839.

de-Brito, J., & Saikia, N. (2013). *Recycled Aggregate in Concrete: Use of Industrial Construction and Demolition Waste*. London: Springer-Verlag.

Degroote, S., Zinszer, K., & Ridde, V. (2018). Interventions for vector-borne diseases focused on housing and hygiene in urban areas: A scoping review. *Infectious Diseases of Poverty, 7*, 1–27.

Designing Building Wiki. (2017). Building a better quality of life: A strategy for more sustainable construction [Online] [Accessed 11 December 2017]. Available from www.designingbuildings.co.uk/wiki/Building_abetter_quality_of_life: A_strategiy_for_more_sustainable_construction

Dong, Z., Yang, H., Wu, D., Ni, J., Kim, H., & Tong, M. (2014). Influence of silicate on the transport of bacteria in quartz sand and iron-coated sand. *Colloids and Surfaces B: Biointerfaces, 123*, 995–1002.

Duxson, P., Fernández-Jiménez, A., Provis, J. L., Lukey, G. C., Palomo, Á., & Van Deventer, J. (2007). Geopolymer technology: The current state of the art. *Journal of Materials Science, 42*(9), 2917–2933.

El-Midany, A. A., & Abdel-khalek, M. A. (2014a). Influence of bacteria–coal electrostatic interaction on coal cleaning. *International Journal of Mineral Processing, 126*, 30–34.

El-Midany, A. A., & Abdel-Khalek, M. A. (2014b). Reducing sulfur and ash from coal using *Bacillus subtilis* and *Paenibacillus polymyxa*. *Fuel, 115*, 589–595.

Eurostat. (2011). Waste statistics, European Commission. [Online] [Accessed 1 May 2012]. Available from https://ec.europa.eu/eurostat/statistics-explained/index.php?title=Waste_statistics#Total_waste_generation

Faridah, A. H. A., Hasmanie, A. H., & Hasnain, M. I. (2004). A study on construction and demolition waste from buildings in Seberang Perai. In 3rd National Conference in Civil Engineering. 20–22 July, Copthorne Orchid, Tanjung Bungah, Malaysia (pp. 1–5).

Foo, L. C., Ismail, A. A., Ade, A., Nagapan, S., & Khairul, I. K. (2013). Classification and quantification of construction waste at housing project site. *International Journal of Zero Waste Generation, 1*(1), 1–4.

Fowle, D. A., Kulezyeki, E., & Robert, J. A. (2004). Linking bacteria–metal interaction to mineral attachment: A role for outer sphere complexation of cation. In *Eleventh International Symposium on Water-Rock Interaction WRI-11, 27 June–2 July*, New York (pp. 1113–1117).

Fowler, D., Brimblecombe, P., Burrows, J., Heal, M. R., Grennfelt, P., Stevenson, D. S., ... & Vieno, M. (2020). A chronology of global air quality. *Philosophical Transactions of the Royal Society A, 378*(2183), 20190314.

Galvin, A. P., Ayuso, J., Barbudo, A., & Jimenez, J. R. (2013). Analysis of leaching procedures for environmental risk assessment of recycled aggregate use in unaved roads. *Construction and Building Materials, 40*, 1207–1214.

Garcia-Sanchez, M., Siles, J. A., Cajthaml, T., Garcia-Romera, I., Tlustoš, P., & Száková, J. (2015). Effect of digestate and fly ash applications on soil functional properties and microbial communities. *European Journal of Soil Biology, 71*, 1–12.

Geoffrey, M. G. (2010). Metals, minerals and microbes: Geomicrobiology and bioremediation. *Microbiology, 156*, 609–643.

Ghosh, P., Mandal, S., Chattopadhyay, B. D., & Pal, S. (2005). Use of microorganism to improve the strength of cement mortar. *Cement and Concrete Research, 35*, 1980–1983.

Giacobello, F., Ielo, I., Belhamdi, H., & Plutino, M. R. (2022). Geopolymers and functionalization strategies for the development of sustainable materials in construction industry and cultural heritage applications: A review. *Materials, 15*(5), 1725.

Gruyer, N., Dorais, M., Alsanius, B. W., & Zagury, G. J. (2013). Simultaneous removal of nitrate and sulfate from greenhouse wastewater by constructed wetlands. *Journal of Environmental Quality, 42*, 1256–1266.

Gupta, S., Rathi, C., & Kapur, S. (2013). Biological induced self-healing concrete: a futuristic solution for crack repair. *International Journal of Applied Sciences and Biotechnology, 1*(3), 85–89.

Hagedorn, C., Blanch, A. R., & Valerie, J. H. (2003) Microbial source tracking: methods, applications and case studies: Chemical – based fecal source tracking methods. Virginia Polytechnic, Institute and State University, Blacksburg, USA.

Hamid, Z. A., & Kamar, K. A. M. (2010). Modernising the Malaysian Construction Industry in Special Track 18th CIB Workld Building Congress, United Kingdom.

Hamilton, W. A. (2003). Microbially influenced corrosion as a model system for the study of metal–microbe interactions: A unifying electron transfer hypothesis. *Biofouling, 19*, 65–76.

Hammes, F., & Verstraete, W. (2002). Key roles of pH and calcium metabolism in microbial carbonate precipitation. *Environmental Science Biotechnology, 36*, 118–136.

He, H., Hong, F. F., Tao, X. X., Ma, C. Y., & Zhao, Y. D. (2012). Biodesulfurization of coal with *Acidithiobacillus caldus* and analysis of the interfacial interaction between cells and pyrite. *Fuel Processing Technology, 101*, 73–77.

Hoang, N. H., Ishigaki, T., Kubota, R., Tong, T. K., Nguyen, T. T., Nguyen, H., Yamada, M., & Kawamoto, K. (2020). Waste generation, composition, and handling in building-related construction and demolition in Hanoi, Vietnam. *Waste Management, 117*, 32–41.

Hong Kong Waste Reduction. (2015). Monitoring of solid waste in Hong Kong [Online] [Accessed 1 January 2017]. Available from www.wastereduction.gov.hk/en/assitancewizard/waste_red_sat.htm

Hong, Z., Chen, W., Rong, X., Cai, P., Dai, K., & Huang, Q. (2013). The effect of extracellular polymeric substances on the adhesion of bacteria to clay minerals and geothite. *Chemical Geology, 360–361*, 118–125.

Hossain, S. S., Roy, P., & Bae, C. (2021). Utilization of waste rice husk ash for sustainable geopolymer: A review. *Construction and Building Materials, 310*, 125218.

Hossain, S. S., Roy, P., & Bae, C. (2021b). Utilization of waste rice husk ash for sustainable geopolymer: A review. *Construction and Building Materials*, 310, 125218.

Istuque, D. B., Soriano, L., Akasaki, J. L., Melges, J. L. P., Borrachero, M., Monzó, J., Payá, J., & Tashima, M. M. (2019). Effect of sewage sludge ash on mechanical and microstructural properties of geopolymers based on metakaolin. *Construction and Building Materials, 203*, 95–103.

Jagad, G., Modhera, C. D., Patel, D., & Patel, V. (2023). Mechanical and microstructural behavior of high strength geopolymer concrete inclusion of various industrial wastes. *Innovative Infrastructure Solutions, 8*(6).

Jambhulkar, H. P., Shaikh, S. M. S., & Kumar, M. S. (2018). Fly ash toxicity, emerging issues and possible implications for its exploitation in agriculture; Indian scenario: A review. *Chemosphere, 213*, 333–344.

Jang, Y. C., & Townsend, T. (2001). Sulfate leaching from recovered construction and demolition debris fines. *Advances in Environmental Research, 5*, 203–217.

Jia, D., He, P., Wang, M., & Yan, S. (2020). *Geopolymerization mechanism of geopolymers.* Springer Series in Materials Science. Springer Science+Business Media.

Jiang, X., Xiao, R., Bai, Y., Huang, B., & Ma, Y. (2022). Influence of waste glass powder as a supplementary cementitious material (SCM) on physical and mechanical properties of cement paste under high temperatures. *Journal of Cleaner Production, 340*, 130778.

Kan, L., Shi, R., Zhao, Y., Duan, X., & Wu, M. (2020). Feasibility study on using incineration fly ash from municipal solid waste to develop high ductile alkali-activated composites. *Journal of Cleaner Production, 254*, 120168.

Khan, S., Anjum, R., Raza, S. T., Bazai, N. A., & Ihtisham, M. (2022). Technologies for municipal solid waste management: Current status, challenges, and future perspectives. *Chemosphere, 288*, 132403.

Ki, D., Kang, S. M., & Park, K. (2021). Upcycling of wastewater sludge incineration ash as a 3D printing technology resource. *Frontiers in Sustainability, 2*.

Kjervik, M., Schwibbert, K., Dietrich, P., Thissen, A., & Unger, W. E. S. (2018). Surface characterization of *Escherichia coli* under various condition by near ambient pressure XPS. *Surface and Interface Analysis, 50*(11), 996–1000.

Knipper, J., & Speck, T. (2012). Design and construction principles in nature and architecture. *Bioinspiration & Biomimetics, 7*, 2–15.

Krystosik, A., Njoroge, G., Odhiambo, L., Forsyth, J. E., Mutuku, F., & LaBeaud, A. D. (2020). Solid wastes provide breeding sites, burrows, and food for biological disease vectors, and urban zoonotic reservoirs: A call to action for solutions-based research. *Frontiers in Public Health, 7*, 405.

Kumar, A., Muthukannan, M., Arunkumar, K., Sriram, M., Vigneshwar, R., & Sikkandar, A. G. (2022). Development of eco-friendly geopolymer concrete by utilizing hazardous industrial waste materials. *Materials Today: Proceedings, 66*, 2215–2225.

Kumar, S. S., Kumar, A., Singh, S., Malyan, S. K., Baram, S., Sharma, J., ... & Pugazhendhi, A. (2020). Industrial wastes: Fly ash, steel slag and phosphogypsum-potential candidates to mitigate greenhouse gas emissions from paddy fields. *Chemosphere, 241*, 124824.

Lachimpadi, S. K., Pereira, J. J., Taha, M. R., & Mokhtar, M. (2012). Construction waste minimisation comparing conventional and precast construction (Mixed System and IBS) methods in high-rise buildings: A Malaysia case study. *Resources, Convention and Recycling, 68*, 96–103.

Li, S., Huang, X., Muhammad, F., Yu, L., Xia, M., Zhao, J., ..., & Li, D. (2018). Waste solidification/stabilization of lead–zinc slag by utilizing fly ash based geopolymers. *RSC Advances, 8*(57), 32956–32965.

Liew, Y.-M., Heah, C.-Y., Mohd Mustafa, A. B., & Kamarudin, H. (2016). Structure and properties of clay-based geopolymer cements: A review. *Progress in Materials Science, 83*, 595–629.

Lingyu, T., Dongpo, H., Jia-Ning, Z., & Hongguang, W. (2021). Durability of geopolymers and geopolymer concretes: A review. *Reviews on Advanced Materials Science, 60*(1), 1–14.

Singh, M., & Siddique, R., (2014). Compressive strength, drying shrinkage and chemical resistance of concrete incorporating coal bottom ash as partial or total replacement of sand. *Journal of Cleaner Production, 112*, 620–630.

Meesala, C. R., Verma, N. K., & Kumar, S. (2020). Critical review on fly-ash based geopolymer concrete. *Structural Concrete, 21*(3), 1013–1028.

Mehdizadeh, H., Cheng, X., Mo, K. H., & Ling, T.-C. (2022). Upcycling of waste hydrated cement paste containing high-volume supplementary cementitious materials via CO_2 pre-treatment. *Journal of Building Engineering, 52*, 104396.

Miao, I., Brusseau, M. L., Caroll, K. C., Carreon-Diazconti, C., & Johnson, B., (2012). Sulphate reduction in groundwater: Characterization and application for remediation. *Environmental Giochemical Health, 34* (4), 539–550.

Milad, A., Ali, A., Babalghaith, A. M., Memon, Z. A., Mashaan, N. S., Arafa, S., & Yusoff, N. I. M. (2021). Utilisation of waste-based geopolymer in asphalt pavement modification and construction—a review. *Sustainability, 13*(6), 3330.

Mohannadoss, P., Amirreza, T., Rosli, M. Z., Mohammas, I., Muhd Zaini, A. M., Ali, K., & Hesam, K. (2015). Bioconcrete strength, durability, permeability, recycling and effects on human health: A review. Third International Conference Advances in Civil, Structural and Mechanical Engineering, 26–27 April, USA (pp. 1–9).

Mokarram, M., Saber, A., & Sheykhi, V. (2020). Effects of heavy metal contamination on river water quality due to release of industrial effluents. *Journal of Cleaner Production, 277*, 123380.

Mullauer, W., Beddoe, R. E., & Heiz, D. (2012). Effect of carbonation, chloride and external sulphates on the leaching behaviour of major and trace elements from concrete. *Cement Concrete Composites, 34*, 618–626.

Musk, A. W., de Klerk, N., Reid, A., Hui, J., Franklin, P., & Brims, F. (2020). Asbestos-related diseases. *International Journal of Tuberculosis and Lung Disease, 24*(6), 562–567.

Nagapan, S., Ismail, A. R., & Ade, A. (2012). Factors contributing to physical and non-physical waste generation in construction industry. *International Journal of Advances in Applied Sciences, 1*(1), 1–10.

Nain, N., Surabhi, R., Yathish, N. V., Krishnamurthy, V., Deepa, T., & Tharannum, S. (2019). Enhancement in strength parameters of concrete by application of Bacillus bacteria. *Construction and Building Materials, 202*, 904–908.

Ning, X., Qi, J., Wu, C., & Wang, W. (2019). Reducing noise pollution by planning construction site layout via a multi-objective optimization model. *Journal of Cleaner Production, 222*, 218–230.

Nurhanim, A. A. (2016). Leaching behaviour of construction and demolition waste (concrete and gypsum). *Iranica Journal of Energy & Environment, 7*(2), 203–211.

Oh, H. Y. P., Humaidi, M., Chan, Q. Y., Yap, G., Ang, K. Y., Tan, J., ... & Mailepessov, D. (2022). Association of rodents with man-made infrastructures and food waste in Urban Singapore. *Infection Ecology & Epidemiology, 12*(1), 2016560.

Okabe, S., Odagiri, M., Ito, T., & Satoh, H. (2007). Succession of sulphur-oxidizing bacteria in the microbial community on corroding concrete in sewer systems. *Applied and Environmental Microbiology, 73*(3), 971–980.

Olivia, M., Wulandari, C., Sitompul, I. R., Darmayanti, L., & Djauhari, Z. (2016). Study of fly ash (FA) and palm oil fuel ash (POFA) geopolymer mortar resistance in acidic peat environment. *Materials Science Forum, 841*, 126–132.

Ortiz, D. I., Piche-Ovares, M., Romero-Vega, L. M., Wagman, J., & Troyo, A. (2022). The impact of deforestation, urbanization, and changing land use patterns on the ecology of mosquito and tick-borne diseases in Central America. *Insects, 13*(1), 20.

Padmini, A. K., Ramamurthy, K., & Mathews, M. S. (2009). Influence of parent concrete on the properties of recycled aggregate concrete. *Construction and Building Materials, 23,* 829–836.

Pei, R., Liu, J., Wang, S., & Yang, M. (2013). Use of bacterial cell walls to improve the mechanical performance of concrete. *Cement and Concrete Research, 39,* 122–130.

Pereira, J. J., Siwar, C., & Inginieur. (2005). Construction waste management: Are contractor unaware or just recalcitrant. *Ingenieur, 27,* 47–49.

Podolsky, Z., Liu, J. B., Dinh, H., Doh, J., Guerrieri, M., & Fragomeni, S. (2021). State of the art on the application of waste materials in geopolymer concrete. *Case Studies in Construction Materials, 15,* e00637.

Provis, J. L., & Van Deventer, J. S. J. (2009). *Geopolymers: Structures, Processing, Properties and Industrial Applications.* Woodhead Publishing.

Rahman, M. M., & Ali, N. O. (2018). An overview of construction related pollution. In 7th Brunei International Conference on Engineering and Technology (BICET), 12–14 November, Brunei.

Rahmat, N. S., & Ibrahim, A. H. (2007). Illegal dumping site: Case study in the district of Johor Bahru Tengah. In 1st International Conference on Sustainable Materials, 9–11 June, Penang (pp. 89–91).

Ranjbar, N., Dolatshahi-Pirouz, A., Alengaram, U. J., Metselaar, H. S. C., & Jumaat, M. Z. (2014). Compressive strength and microstructural analysis of fly ash/palm oil fuel ash based geopolymer mortar under elevated temperatures. *Construction and Building Materials, 65,* 114–121.

Reddy, M. V., Dinakar, P., & Rao, B. D. (2018). Mix design development of fly ash and ground granulated blast furnace slag based geopolymer concrete. *Journal of Building Engineering, 20,* 712–722.

Reddy, S., Seshagiri, M., Apama, P., & Sasikala, C. (2010). Performance of standard grade bacterial (*Bacillus subtilis*) concrete. *Asian Journal of Civil Engineering (Building and Housing), 1,* 43–55.

Saha, P., & Paul, B. (2019). Assessment of heavy metal toxicity related with human health risk in the surface water of an industrialized area by a novel technique. *Human and Ecological Risk Assessment: An International Journal, 25*(4), 966–987.

Sahoo, K. K., Manoranjan, A., Pradip, S., Robin, D. P., & Suman, J. (2016). Enhancement of properties of recycled coarse aggregate concrete using bacteria. *International Journal of Smart and Nano Materials, 7*(1), 22–38.

Salunkhe, P., Balip, P., Deshmukh, T., Gadge, S., & Vhanmane, P. B. (2022). Geopolymer a sustainable material: A review. In *Smart Technologies for Energy, Environment and Sustainable Development,* Vol 1, Springer eBooks (pp. 29–41).

Samadi, M., Saghi, H., Rahmani, A., & Mirzaee, S. (2009). Zoning of water quality of Valley Mradbeak Hamedan River based on NSFWQI qualitative index by using Geographic Information System (GIS). *Scientific Journal of Hamadan University of Medical Science, 16*(3), 53.

Sathonsaowaphak, A., Chindaprasirt, P., & Pimraksa, K. (2009). Workability and strength of lignite bottom ash geopolymer mortar. *Journal of Hazardous Materials, 168*(1), 44–50.

Shaheen, S. M., Hooda, P. S., & Tsadilas, C. D. (2014). Opportunities and challenges in the use of coal fly ash for soil improvements—a review. *Journal of Environmental Management, 145,* 249–267.

Shahmansouri, A. A., Bengar, H. A., & Ghanbari, S. (2020). Compressive strength prediction of eco-efficient GGBS-based geopolymer concrete using GEP method. *Journal of Building Engineering, 31,* 101326.

Sharma, K., & Kumar, A. (2020). Utilization of industrial waste-based geopolymers as a soil stabilizer—a review. *Innovative Infrastructure Solutions*, 5(3), 97.

Sheela, A. M., Ghermandi, A., Vineetha, P., Sheeja, R. V., Justus, J., & Ajayakrishna, K. (2017). Assessment of relation of land use characteristics with vector-borne diseases in tropical areas. *Land Use Policy*, *63*, 369–380.

Shehata, N., Sayed, E. T., & Abdelkareem, M. A. (2021). Recent progress in environmentally friendly geopolymers: A review. *Science of the Total Environment*, *762*, 143166.

Sheng, X., Ting, Y. P., & Pehkonen, S. O. (2008). The influence of ionic strength, nutrient and pH on bacterial adhesion to metals. *Journal of Colloid and Interface Science*, *321*, 256–264.

Shobeiri, V., Bennett, B., Xie, T., & Visintin, P. (2021). A comprehensive assessment of the global warming potential of geopolymer concrete. *Journal of Cleaner Production*, 297, 126669.

Sholanke, A., Aina-Badejo, T., Aina-Babajide, A., & Jacob, A. N. (2019, September). Noise pollution and waste control techniques in building construction in Nigeria: A literature review. In IOP Conference Series: Earth and Environmental Science (Vol. 331, No. 1, p. 012016). IOP Publishing.

Singh, J. P., & Singh, S. K. (2019). Geopolymerization of solid waste of non-ferrous metallurgy—a review. *Journal of Environmental Management*, *251*, 109571.

Siyal, A. A., Shamsuddin, M. R., Khan, M. I., Rabat, N. E., Zulfiqar, M., Man, Z., Siame, J., & Azizli, K. A. (2018). A review on geopolymers as emerging materials for the adsorption of heavy metals and dyes. *Journal of Environmental Management*, *224*, 327–339.

Solpuker, U., Sheets, J., Kim, Y., & Schwartz, F. W. (2014). Leaching potential of previous concrete and immobilization of Cu, Pb and Zn using previous concrete. *Journal of Contaminants Hydrology*, *161*, 35–48.

Suradi, G. D. (2019). The tourism industry, waste management, and vector-borne diseases: A case study of Curaçao. Master's thesis.

Tam, V. W. Y., Gao, X. F., Tam, C. M., & Ng, K. M. (2009). Physio-chemical reactions in recycle aggregate concrete. *Journal of Hazardous Materials*, *163*, 823–828.

Tchakouté, H. K., Kong, S., Stephan, D., Tchadjié, L., & Njopwouo, D. (2015). A comparative study of two methods to produce geopolymer composites from volcanic scoria and the role of structural water contained in the volcanic scoria on its reactivity. *Ceramics International*, *41*(10), 12568–12577.

The Star Online. (2006). Recycle wood & construction waste. [Online] [Accessed 24 July 2012]. Available from http://thestar.com.my/news/story.asp?file_/2006/10/3southeneast/15505 469&sec=southeneast

Thomas, J. J., Jennings, H. M., & Chen, J. J. (2009). Influence of nucleation seeding on the hydration mechanisms of tricalcium silicate and cement. *Journal of Physical Chemistry C*, *113*(11), 4327–4334.

United States Environmental Protection Agency. (2023, May 23). *Coal Ash (Coal Combustion Residuals, or CCR)*. Available from www.epa.gov/coalash

Van Tittelboom, K., De Belie, N., De Muynck, W., & Verstraete, W. (2010). Use of bacteria to repair cracks in concrete. *Cement and Concrete Research*, *40*, 157–166.

Vegas, I., Ibanez, J. A., Lisbano, A., Saez de Cortazar, A., & Fras, M. (2011). Pre-normative research on the use of mixed recycled aggregates in unbound road sections. *Construction and Building Materials*, *25*, 2674–2682.

Wang, J., Xie, J., Wang, C., Zhao, J., Liu, F., & Fang, C. (2020). Study on the optimum initial curing condition for fly ash and GGBS based geopolymer recycled aggregate concrete. *Construction and Building Materials*, *247*, 118540.

Wang, M. H. S., Wang, L. K., & Shammas, N. K. (2020). Glossary of acid rain management and environmental protection. In *Handbook of Environment and Waste Management: Acid Rain and Greenhouse Gas Pollution Control* (pp. 719–749).

Xiao, R., Ma, Y., Jiang, X., Zhang, M., Zhang, Y., Wang, Y., ... & He, Q. (2020). Strength, microstructure, efflorescence behavior and environmental impacts of waste glass geopolymers cured at ambient temperature. *Journal of Cleaner Production, 252*, 19610.

Yousuf, A., Manzoor, S. O., Youssouf, M., Malik, Z. A., & Khawaja, K. S. (2020). Fly ash: production and utilization in India—an overview. *Journal of Materials and Environmental Science, 11*(6), 911–921.

Zarina, Y., Mustafa, A. B. a. M., Kamarudin, H., Nizar, I. K., Victor, S. A., Petrică, V., & Rafiza, A. (2013). Chemical and physical characterization of boiler ash from palm oil industry waste for geopolymer composite. *Revista De Chimie, 64*(12), 1408–1412.

Zaharieva, R., Buyle-Bodin, F., Skoczylas, F., & Wirquin, E. (2003). Assessment of the surface permeation properties of recycled aggregate concrete. *Cement Concrete Composites, 25*(2), 223–232.

Zhou, S., Lu, C., Zhu, X., & Li, F. (2021). Preparation and characterization of high-strength geopolymer based on BH-1 lunar soil simulant with low alkali content. *Engineering, 7*(11), 1631–1645.

Zhu, X., Qian, H., Wu, H., Zhou, Q., Feng, H., Zeng, Q., ... & Yan, D. (2022). Early-stage geopolymerization process of metakaolin-based geopolymer. *Materials, 15*(17), 6125.

Zierold, K. M., & Odoh, C. (2020). A review on fly ash from coal-fired power plants: chemical composition, regulations, and health evidence. *Reviews on Environmental Health, 35*(4), 401–418.

6 Health Hazards and Waste Management

*Siti Norabiatulaiffa Mohd Yamen,
Mohd Saiful Samsudin, and
Muhammad Izzul Fahmi Mohd Rosli*

6.1 INTRODUCTION

The presence of hazardous waste poses significant risks to humans, animals, and the environment on a large scale. The escalation in waste production is closely linked to the growing global population, resulting in a rapid increase in industrial activities that serve as major sources of hazardous pollutants worldwide. Insufficient management of waste represents a critical problem that can bring about various negative consequences for society and ecosystems. Improper waste management, storage, and disposal can negatively impact human health through inhalation of polluted air, ingestion of polluted water, and consumption of plants grown in contaminated soil that are used as food crops (Awasthi et al., 2018) as shown in Figure 6.1. Several studies have emphasized the health dangers associated with inadequate waste management, offering valuable insights into the detrimental impacts on human welfare. Study by Rushton (2003) also stated that several substances, including cadmium, arsenic, chromium, nickel, dioxins, and polycyclic aromatic hydrocarbons (PAHs), have been identified as potentially carcinogenic based on studies conducted on animals or individuals exposed to high concentrations of these substances. Nonetheless, the presence of conclusive evidence linking these substances to cancer at environmental levels is frequently inconclusive. Besides their potential carcinogenic properties, many of these substances can also induce various toxic effects on the central nervous system, liver, kidneys, heart, lungs, skin, and reproductive system, depending on the level and duration of exposure (Chandra et al., 2006; Karim et al., 2018; Singh, 2015).

The insufficient disposal of solid waste has been underscored by the World Health Organization (WHO) as a significant concern due to its potential to contaminate soil, water, and air, thereby posing health risks to populations residing in the affected areas (WHO, 2015). Insufficient handling of waste can lead to the release of toxic pollutants into the atmosphere, resulting in air pollution and its associated health hazards. When waste is burned openly or incinerated improperly in uncontrolled settings, it can emit particulate matter, heavy metals, organic compounds, and hazardous gases, which have been linked to respiratory problems (Mahajan, 2023; Uzunlulu et al., 2022). A major air pollutant is an airborne particulate matter caused by the incineration of waste and plastic. Air pollution has serious effects for the ecosystem, the greenhouse gas, and

DOI: 10.1201/9781003543176-6

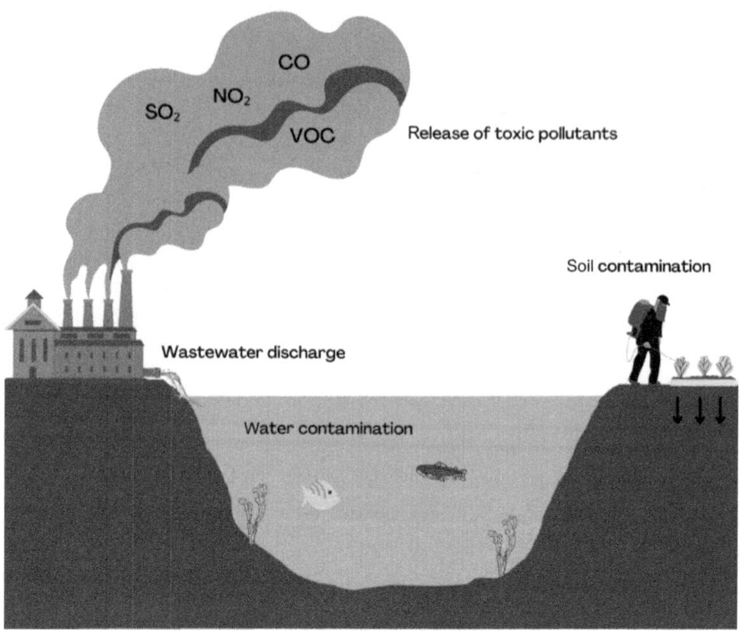

FIGURE 6.1 Improper waste management on environment.

global warming. It can cause respiratory and psychological problems (Awasthi et al., 2018). A comprehensive waste management system is necessary to promote environmental well-being, enhance quality of life, ensure economic viability, and achieve overall sustainability within a society (Gentil et al., 2009).

Harmful substances from waste disposal leach into groundwater and surface water sources, contaminate drinking water supplies, thus increasing the likelihood of waterborne diseases. Contaminants such as heavy metals, pathogens, and organic pollutants can induce gastrointestinal illnesses, hepatitis, and other waterborne infections (Lad et al., 2020). Fowle et al. (1996) also mentioned that drinking contaminated water can cause acute or chronic poisoning, depending on the precise chemicals implicated. Furthermore, the growth of industry and urbanization has presented significant obstacles to handling the disposal of industrial and domestic pollutants, resulting in the contamination of water sources with various forms of inorganic, organic, and biological contaminants. Thus, these pollutants can persist in the water system for a long period heading to long-term contamination in environment (Li et al., 2019; Sarker et al., 2021).

Improper waste management further contributes to soil deterioration, which negatively affects human health and agricultural productivity. During the last century, industrial activity has made the most major contribution to the issue of soil contamination (Kristanti et al., 2023). The waste such as heavy metals (Wu et al., 2015), solvents (Anju et al., 2010), and pesticides (Anju et al., 2010; Odukkathil & Vasudevan, 2013; Rodrigo et al., 2014) contains toxic substances that can seep into the soil. Soil contamination influences and modifies the soil's chemical and biological properties, and it is frequently linked with improper refuse disposal, industrial activities, and nonorganic farming methods (Alaswad, 2019; Verkhovets et al., 2020). When hazardous substances are disposed of incorrectly or untreated waste is dumped on land, it can pollute the soil with heavy metals, pesticides, and other pollutants (Hasan et al., 2021). This contamination disrupts soil fertility, nutrient cycling, and soil microbial activity (Xu et al., 2019). The accumulation of toxic substances in crops grown on contaminated soil poses health risks to humans when consumed, potentially leading to chronic exposure to harmful substances (Cameselle et al., 2013; Mester et al., 2022).

Innovative waste management practices are warranted to address waste management issues. These methods provide sustainable and efficient solutions for waste management, thereby minimizing the negative effects of waste on public health and the environment. Failure to implement innovative waste management methods can have severe consequences. Improper waste disposal practices, such as indiscriminate dumping, can result in the accumulation of waste in our surroundings, attracting pests and disease-carrying vectors (Qasim et al., 2020). Thus, it can lead to the spread a disease. The land and aquatic environments and atmosphere can become polluted with hazardous materials, posing a threat to the health and safety of both human beings and wildlife (Gusat et al., 2019). Therefore, by adopting innovative approaches to waste management, a circular economy that promotes sustainability, resource conservation, and healthier living conditions can cultivate for communities. These practices not only contribute to a more sustainable future by reducing refuse generation, but also by protecting ecosystems.

6.2   HEALTH HAZARDS OF DIFFERENT TYPES OF SOLID WASTE

The improper handling and disposal of various types of solid waste can lead to significant health hazards. Solid waste includes an extensive range of materials generated by human activities, and incorrect waste management can have detrimental effects on health. Understanding the health risks connected with various types of solid waste is essential for dealing with and managing these problems (Figure 6.2).

6.3   MUNICIPAL WASTE

Municipal waste, also known as solid waste or garbage, refers to the discarded materials generated by households, commercial establishments, and institutions (Vinti et al., 2021). Household cleaning products, insecticides, and medications are among the compounds found in municipal waste. These chemicals can seriously endanger the health of communities nearby when they leach into the soil or water sources. Long-term exposure to these pollutants can cause respiratory issues, hormonal abnormalities, and even cancer (Li et al., 2015; Nadal et al., 2020; Liu et al., 2023). A study by García-Pérez et al. (2015) found excess ovarian cancer mortality among women residing near industries, including waste treatment plants and pharmaceutical industry. The translocation of heavy metals from landfills to groundwater, soil, and plants contributes to the contamination of the environment with municipal waste. The concentration of heavy metals in leachate from municipal solid waste

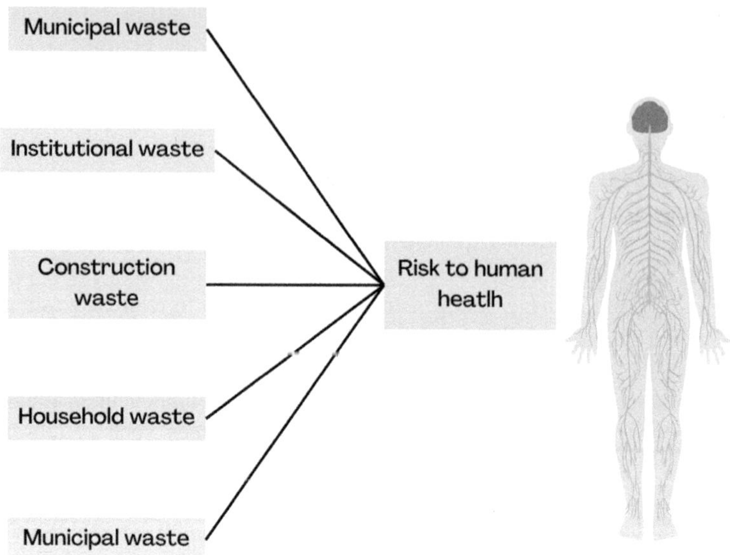

FIGURE 6.2   Different types of solid waste.

(MSW) is influenced by the duration of landfill operation (Olagunju et al., 2018; Onyekwelu & Aghamelu, 2019). In addition, atmospheric contamination is another source of environmental pollution and health risk resulting from landfill fires. Several studies have located that hazardous compounds spread to the environment as a result of municipal waste landfill fires. They contribute significantly to air pollution by emitting carbon monoxide (CO), nitric oxide (NOx), particles of primary organic aerosols (POA), and volatile organic compounds (North & Halden, 2013; Posnack, 2014; Witt et al., 2017).

6.4   INDUSTRIAL WASTE

Industrial waste often contains toxic chemicals, including solvents, acids, pesticides, and heavy metals (lead, mercury, and cadmium). Acute or long-term health effects may result from coming into contact with, swallowing, or inhaling these chemicals. Depending on the precise compounds used, the health hazards can include respiratory problems neurological disorders, organ damage, reproductive issues, and even cancer (Erickson et al., 2019). Tiembre et al. (2009) conducted a study that examined the impact of toxic waste disposal to health risks in Abidjan City. According to the study findings, the waste analysis demonstrated the presence of detrimental substances like mercaptans and hydrogen sulfide. This waste dumping has a detrimental impact, leading to poisoning among the residents in the city. Another study by Leem et al. (2003) stated that workers and people living nearby to municipal waste incinerators and industrial waste incinerators were exposed to hazardous compounds such polychlorinated dibenzo-p-dioxins (PCDDs) and polychlorinated dibenzofurans (PCDFs), as well as the impact these pollutants had on their health. Inhabitants of the region near industrial waste incinerators had greater PCDD/PCDF concentrations than workers and inhabitants of the area near MSW incinerators. According to the study, people who live and work close to MSW and industrial waste incinerators are exposed to dangerous compounds that could have a negative impact on their health. Therefore, workers involved in handling industrial waste are at a higher risk of exposure to hazardous substances, which may cause various occupational illnesses.

6.5 CONSTRUCTION/DEMOLITION WASTE

Most of the health risks associated with construction and demolition waste come from exposure to dust, asbestos, and other hazardous substances commonly found in these waste types. Inhalation of high levels of dust particles can lead to respiratory issues such as asthma, bronchitis, and chronic obstructive pulmonary disease (Critchley, 2013). Construction industry is a major contributor to air pollution in the form of dust. The particulate matter generated by construction equipment and transportation activities can have detrimental health effects (Luo et al., 2021). A study conducted by Xiaodong et al. (2015) in Beijing developed a health damage assessment model to evaluate the effects of dust pollution from various construction activities at different stages and to monetize the evaluation results. The health effect analysis among worker groups shows obvious differences in the dust pollution among various construction activities. The study showed carpenters suffering the most health damage from dust during the main structure construction stage. Cook et al. (2022) published a systematic review of risks to occupational and public health associated with construction and demolition waste. They discovered that the combustible fraction of CDW is mismanaged and disposed of by open burning in many low- and middle-income countries, including an increase in plastics used in the sector. The residues of these materials are likely to be burned, and the high chloride content of PVC poses a serious threat when burnt in open, uncontrolled fires due to the release of dioxins and related substances. Yang et al. (2018) examined the association between increased small airway obstruction and asbestos exposure in individuals with asbestosis. According to the findings, asbestos inhalation not only causes asbestosis and pleural abnormalities but also causes tiny airway anomalies. In addition to cancer and chronic respiratory diseases, asbestos exposure can also potentially impact the heart, underscoring the importance of implementing comprehensive monitoring systems to safeguard workers (Moitra et al., 2022).

6.6 MEDICAL WASTE

Exposure to medical waste, both direct and indirect, can have serious health repercussions, including a wide range of negative impacts. These include the potential for cancer, harm to the reproductive system, respiratory conditions, effects on the nervous system, and other health problems (Alagöz & Kocasoy, 2008; Oli et al., 2016). For example, Barayang et al. (2022) and Exposto et al. (2022) highlighted that improper disposal of medical waste can result in the transmission of diseases. They emphasize the importance of waste management to protect the surrounding community from the spread of disease and injury. According to Cesaro and Belgiorno (2015), most medical waste is potentially hazardous as it may contain pathogens or hazardous substances that necessitate special treatment protocols. Medical waste mishandling or improper disposal can result in the spread of diseases such as human immuno-deficiency virus (HIV), hepatitis B and C, and other bloodborne infections (Rao & Wankhede, 2011; Rogers et al., 2002). Proper management of medical waste is essential for preventing environmental contamination and mitigating the risk of infectious disease transmission.

6.7   HOUSEHOLD WASTE

Numerous household products contain hazardous substances. When these products are disposed, they form household hazardous waste (HHW), posing a risk to human health and the environment (Conn, 1989). For instance, the improper disposal of electronic waste, such as defunct computers, televisions, and batteries, can cause the emission of toxic substances such as lead, mercury, and cadmium. Toxic and hazardous substances contained in e-waste pose hazards to human health and the environment via air, water, and soil (Havinal, 2022). Chen et al. (2018) conducted a study in a huge e-waste recycling industrial park in South China and discovered that all e-waste-related samples contained extremely high concentrations of short-chain chlorinated paraffins (SCCPs) and medium-chain chlorinated paraffins (MCCPs). Chlorinated paraffins (CPs) were abundant in foods of indigenous plant and animal origin, including fish, vegetables, and rice. According to the study, CP exposure may pose significant dangers to e-waste workers and local communities. Numerous articles stated that SCCPs were found in the hair, nails, blood samples, human breast milk samples, placenta tissues, and umbilical cord blood (Aamir et al., 2019; Cao et al., 2017; Han et al., 2021; Qiao et al., 2018; Tomasko et al., 2021; van Mourik et al., 2020).

Additionally, Gutberlet and Uddin (2017) stated that household waste also containing organic substances lures insects, rats, and pigeons, which are the source of many diseases. For instance, pigeons can carry and spread various diseases to humans. Candidiasis, a fungal infection, tuberculosis, giardiasis caused by a parasite found in contaminated food, histoplasmosis, a serious respiratory disease that can be fatal, and salmonellosis, which comes from pigeon droppings, are some examples. Another disease called leptospirosis can be easily transmitted through inhaling or touching tissues of infected animals or rat urine. Household waste often includes various chemicals, such as cleaning products, pesticides, and paints. If these chemicals are not properly disposed of, they can contaminate soil and water sources, posing risks to human health.

6.8   INNOVATIVE WASTE MANAGEMENT METHODS FOR REDUCING HEALTH HAZARDS

Waste management is now tightly regulated in most developed countries and includes the generation, collection, processing, transport, and disposal of waste. In addition, the remediation of waste sites is an important issue, both to reduce hazards while operational and to prepare the site for a change of use (Rushton, 2003).

6.9 SOURCE REDUCTION AND WASTE MINIMIZATION

Waste minimization is an idea that discusses techniques for reducing the generation of hazardous waste. Waste minimization is considered an approach to waste handling that focuses on minimizing waste production as opposed to promoting post-production waste treatment (Cheremisinoff, 2003). This strategy involves both source reduction and on-site recycling. Source reduction involves modifying or enhancing existing processes and implementing more efficient process control measures. Source reduction or avoiding waste generation is a most desirable goal and should be explored first. In the waste management hierarchy (Figure 6.3), source reduction is followed in order by recycling, treatment and disposal. The hierarchy is a valuable policy tool for promoting resource conservation, addressing landfill capacity constraints, mitigating air and water pollution, and protecting public health and safety. Elements of this hierarchy are already in place in many developing countries, where traditional waste prevention, reuse, and recycling practices are prevalent (Jibril et al., 2012).

Waste minimization at industrial plants can be implement through the hierarchical order of waste prevention, reduction, recycling, reuse, and treatment. Source reduction can be achieved significantly by using alternatives and optimization to existing raw materials, processes, and the operational procedures. Because industries are the primary producers of hazardous waste, source reduction at the industrial level is more urgent and complex than waste reduction at the household level. To reduce the production of hazardous waste from industrial sources and the wide range of waste materials that industries produce, modifying their processes has become an essential requirement (Hussain, 2021).

Waste minimization plays a vital role in mitigating present and future risks to human health and the environment, as it provides professionals and organizations with strategies to decrease both the volume and potential harm of waste produced (Noh et al., 2018).

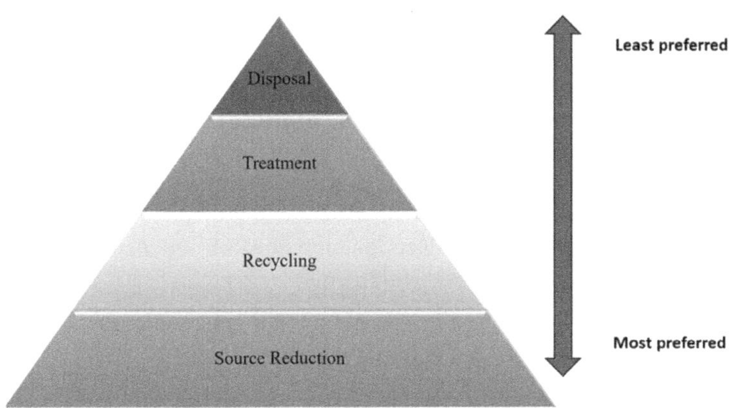

FIGURE 6.3 Hierarchy of waste management.

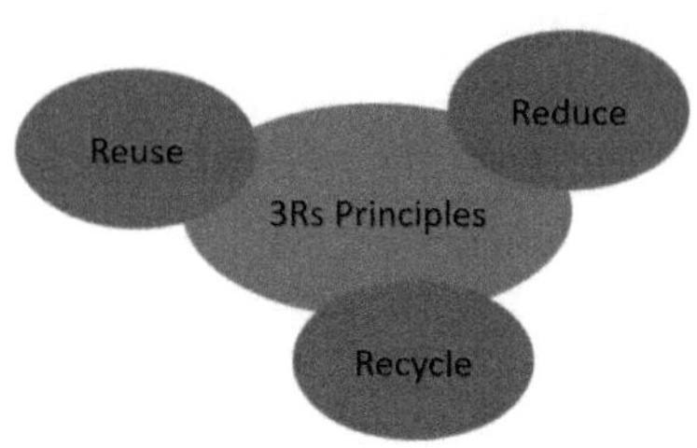

FIGURE 6.4 Three 3Rs principles.

6.10 RECYCLING AND COMPOSTING

Recycling is a process that involves converting waste materials into new products or materials to prevent them from being discarded as waste. It follows the principles of reducing, reusing, and recycling to minimize the impact of waste on the environment. In short, 3Rs techniques include methods and skilled approaches designed to reduce the volume of unused waste materials generated for disposal (Jibril et al., 2012). The concept of recycling revolves around the idea that waste materials can be valuable resources that can be recovered, processed, and transformed into new products, thus conserving natural resources and reducing the need for raw materials extraction (Al-Masry, 2019). Reducing, reusing, and particularly recycling have become the only permissible waste disposal methods. The "zero-waste" policy, which incorporates waste reduction, reuse, recycling, and decomposition, entails zero disposals and zero warming from waste (Figure 6.4).

Composting involves the biological breakdown of organic materials in the presence of oxygen, resulting in the production of stable and sanitized compost. Among various recycling approaches, composting is highly recommended due to its positive impact on the environment and economy (Hoang & Fogarassy, 2020). This method offers numerous environmental advantages, such as the reduction of greenhouse gas emissions, the minimization of leachate quantities when waste is sent to landfills, and the improvement of soil quality when compost is used as a soil amendment (Adhikari et al., 2009; Maulini-Duran et al., 2014; Wei et al., 2017). This technique is considered to be the most efficient, environmentally safe, and agronomically sound, as the compost may be used as a soil conditioner and organic fertilizer, because it contains high levels of nutrients for the soil (Gonawala & Jardosh, 2018). Improper waste disposal poses various health hazards to both environment and human populations. Understanding the waste management hierarchy is essential in comprehending how recycling and composting can effectively reduce health hazards. The waste management hierarchy prioritizes waste reduction and resource recovery over disposal. Recycling and composting help reduce the emissions of pollutants and greenhouse gases associated with waste disposal (Figure 6.5).

6.11 WASTE-TO-ENERGY TECHNOLOGIES

Waste-to-energy technologies is an advanced technologies that turn waste into a viable and usable energy source (Adhikari et al., 2009; Wei et al., 2017). Generally,

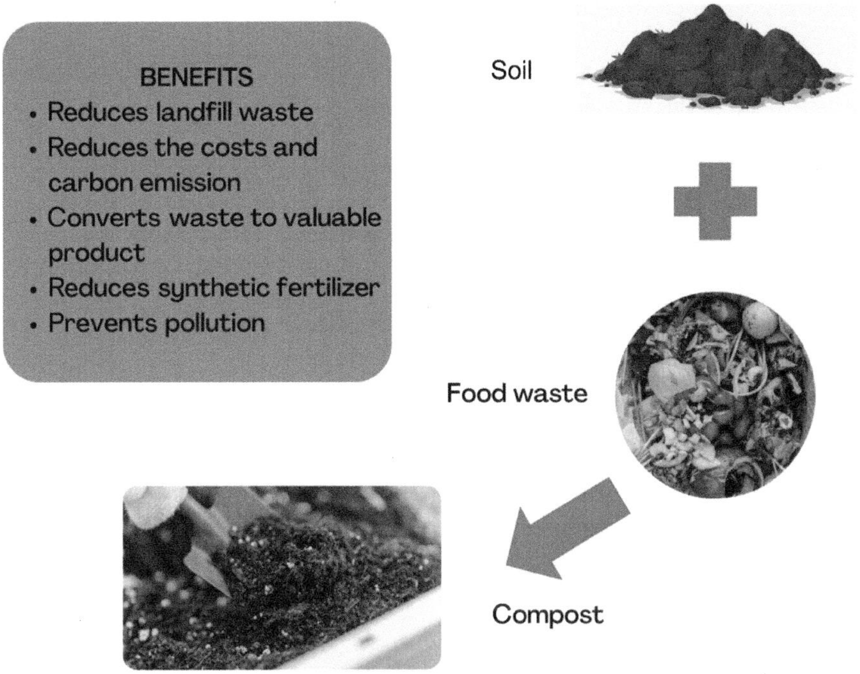

FIGURE 6.5 Food waste composting.

these technologies classified into two main categories: biological treatment technologies, encompassing biochemical processes like anaerobic digestion (Anyaoku & Baroutian, 2018; Fan et al., 2018), and thermal treatment technologies, which involve thermochemical processes such as pyrolysis, gasification, and incineration (Dong et al., 2018; Moya et al., 2017; Rajasekhar et al., 2015; Thakare & Nandi, 2016). The facilities employ advanced emission control technologies to minimize the release of pollutants into the air. These technologies, such as scrubbers (Haji Nawawi & Jaafar, 2021; Streeter, 2016), bag filters, and electrostatic precipitators, capture and remove particulate matter, heavy metals, and other harmful pollutants (Labrada et al., 2019; Wang et al., 2022). The process also involves high-temperature processes that effectively destroy harmful organic compounds and pathogens present in waste. Gasification, for instance, operates at high temperatures, ensuring the complete combustion of organic materials. This helps eliminate the potential for the release of toxic gases (Yujun et al., 2020).

A study by Pourali (2010) stated that the implementation of plasma gasification technology in waste to energy minimizes demand on distressed landfills and provides an environmentally friendly method of disposing of MSW. Most of hazardous waste can be disposed and burned the waste permanently into plasma gasification chamber without generating ash to bury in the landfills. Additionally, the study also mentions that enormous amount of energy from solid waste can be fully processed through plasma gasification technology into electricity, liquid fuels, or polymers. Pyrolysis is a thermal process that occurs at temperatures above 400°C without the presence of oxygen and breaks long chains of polymer molecules into shorter, simpler molecules (Mazalan et al., 2018). It is an innovative waste-to-energy technology that could aid in the development of renewable energy sources and environmental sustainability (Amenaghawon et al., 2021). According to the study, pyrolysis has lower environmental implications than incineration and landfilling. By enabling efficient waste management and decreasing greenhouse gas emissions, it facilitates the utilization of pyrolysis by-products like pyrolysis gas and biochar in the energy and building materials industries, thereby amplifying its positive environmental impacts (Ławińska et al., 2022). Reducing the greenhouse impact as well as air pollution may result in improved air quality and fewer respiratory symptoms among individuals living in nearby regions. Waste-to-energy technologies also can contribute to improved air quality by reducing the release of harmful pollutants compared to traditional waste disposal methods. According to Vehlow (2015), the advanced emission control systems in modern waste-to-energy facilities effectively capture and treat pollutants, reducing their impact on air quality. This leads to improved respiratory health outcomes for nearby communities. The utilization of waste-to-energy technologies allows for the efficient and controlled conversion of waste materials, reducing the volume of waste, preventing the release of harmful substances, and minimizing the risk of pollution (Shah et al., 2021).

6.12   LANDFILL DESIGN AND MANAGEMENT STRATEGIES

Landfill design and management play a crucial role in waste management practices, ensuring the protection of public health and the environment. Proper landfill design is essential to mitigate health hazards associated with waste disposal. It involves implementing innovative approaches that prevent the migration of pollutants and contaminants into the surrounding environment. The design incorporates features such as liner systems, leachate collection systems, and landfill gas management systems, which serve to contain and control the waste's impact on the environment (Mattiello et al., 2013). Furthermore, waste segregation and pretreatment before landfill disposal are vital steps in the management process (Visvanathan et al., 2002). By separating hazardous or toxic waste from the general waste stream, these strategies help prevent the introduction of harmful substances into the landfill.

Implementing gas collection systems and gas-to-energy conversion methods not only mitigate environmental impacts but also reduce potential health risks associated with landfill gas emissions. Landfills produce a significant amount of methane, which is a potent greenhouse gas that contributes to climate change (Aghdam, 2018; Jain et al., 2021). For instance, methane is produced by the anaerobic digestion of biodegradable waste deposited in landfills. The produced methane in landfills can be recovered and utilized for the production of electricity or heat. Higher recovery of methane could result in lower methane emissions into the atmosphere, and therefore lower contribution of landfills to climate change and higher methane recovery can result in higher production of heat and electricity (Aghdam, 2018). Additionally, self-control systems can be developed to monitor and control the quality of leachate and gas emissions from landfills. A hazardous waste landfill in northern Italy developed self-control systems and promoted monitoring and research activity on the field, which made it a model of technological efficiency and management (Marchese et al., 2013).

Monitoring and control systems are essential for identifying and resolving any emerging problems in landfill management. Early detection of potential health hazards can be achieved through advanced technologies such as landfill gas and groundwater monitoring systems, allowing proactive measures to protect public health.

6.13 ADVANCED TREATMENT TECHNOLOGIES FOR HAZARDOUS WASTE

Advanced treatment technologies encompass a range of approaches that leverage scientific advancements and engineering principles to tackle hazardous waste. These technologies aim to mitigate risks by reducing the toxicity and volume of hazardous waste.

Thermal desorption is an effective technology commonly used in engineering practice (Sang et al., 2021) for managing hazardous waste that can remove a wide spectrum of volatile and semi-volatile pollutants such as, total petroleum hydrocarbons (TPH), polychlorinated biphenyls (PCBs), dichlorodiphenyltrichloroethane (DDT), PAHs, and mercury (Hg) found in soils (Zhao et al., 2019). Additionally, the use of low-temperature indirect thermal desorption might lead to sustainable remediation, because it conserves energy (operates at low temperature), possesses advantageous physicochemical characteristics, and allows for quick recolonization of microbial communities in heated soils (Sang et al., 2021; Zhao et al., 2019). According to (Chapman et al., 2011), advanced thermal treatment (ATT) technologies such as gas plasma can transform solid waste into ash type material to environmentally stable products that can be reused in a variety of applications, effectively complete the recycling cycle. Additional uses for gas plasma include the thermochemical treatment of hazardous byproducts of air pollution control (APC) processes. These residues derive from the thermal treatments used to produce ceramic goods from MSW gaseous mitigation systems. Plasma arc also an environmentally friendly method of managing hazardous waste and the process of plasma arc gasification presents a mutually beneficial approach to extract energy from diverse solid and hazardous waste (Deegan et al., 2007). The plasma arc turns the carbon-based components of numerous waste products into a synthetic gas, including MSW, sludge, hazardous waste, utilized tires, and agricultural waste. Then, this gas can be used in boilers, gas turbines, and reciprocating engine generators to produce energy (Pourali, 2010). These technologies can aid in lowering the amounts of harmful substances to harmless levels, hence reducing the possible health risks linked to exposure to these contaminants.

Bioremediation also is an innovative treatment technology that utilizes biological processes to degrade, transform, or remove hazardous compounds from waste streams. It harnesses the natural abilities of microorganisms, such as bacteria and fungi, to break down organic pollutants into simpler and less toxic substances. This reduction in toxicity directly contributes to reducing health hazards associated with hazardous waste (Gupta & Prakash, 2020; Yadav et al., 2022). Bioremediation offers a sustainable and environmentally friendly approach to hazardous waste management, effectively reducing health hazards associated with such waste (Yaashikaa & Kumar, 2022). These advanced treatment technologies contribute to ensuring environmental sustainability by reducing the number of hazardous substances in the environment, improving natural pollutant breakdown, allowing resource recovery, and reducing waste volume. Advanced oxidation processes (AOP) are effective in treating wastewater and reducing the toxicity of hazardous waste. These processes can be tailored to target specific pollutants, allowing for selective treatment and degradation of hazardous compounds while leaving less harmful substances untouched. AOPs are effective in treating a wide range of hazardous waste, including industrial wastewater, contaminated groundwater, and persistent organic pollutants (Bull & Zeff, 2021; Elkacmi & Bennajah, 2019). AOP is also efficient for organic contaminated soil remediation. Chemical oxidation changes the organic contamination into harmless chemical substances after direct introduction into the contaminated source.

6.14 CASE STUDIES OF SUCCESSFUL IMPLEMENTATION OF EFFECTIVE WASTE MANAGEMENT METHODS

The implementation of environmental policies in developed countries has a significant impact on the efficiency of solid waste management. According to Paes and de Oliveira (2021), Curitiba, a city in Brazil, is renowned for its innovative waste management system, which has led to high recycling rates and reduced health risks associated with waste. The city created a comprehensive waste collection and recycling program that emphasized community engagement and environmental education. By implementing a comprehensive waste separation system, Curitiba was able to achieve an impressive recycling rate of 70%, thereby substantially reducing the amount of waste sent to landfills. In addition, they also established an efficient composting program that transformed organic waste into nutrient-rich compost for agricultural use. They emphasize the need of implementing a comprehensive waste management strategy. This requires implementing a combination of waste reduction, recycling, composting, and energy recovery strategies. By contemplating the entire waste management cycle and implementing integrated methods, they can maximize waste diversion rates and reduce environmental impacts.

A study by Shenyoputro and Jones (2023) discussed the achievement of a small town in Japan in waste management by adopting a zero-waste policy. The town implements a strict waste separation system with 45 different categories, promoting recycling and composting. Through community engagement and education programs, they have minimized waste sent to landfills, reducing health hazards and environmental pollution. The case study highlights the significance of community participation and ongoing education in waste management programs. It emphasizes the importance of patience, long-term commitment, and an integrated strategy for behavioral change. The success of the town demonstrates the need of customizing waste management programs to local cultural contexts, encourages community ownership, and gives continuing assistance and resources to residents.

Another study by Pollans (2012) also reported the successful of Zero Waste Program in San Francisco. San Francisco has been a pioneer in waste management with its Zero Waste Program. The city implemented a comprehensive system that encourages recycling, composting, and reducing waste sent to landfills. As a result, San Francisco diverted over 80% of its waste from landfills by 2012, significantly reducing health hazards associated with waste disposal. The case study emphasizes the need for stakeholder engagement, collaboration, and effective leadership in waste management programs. It highlights the importance of clear communication, focused education efforts, and incentives to encourage behavior change. The success of San Francisco demonstrates the importance of governmental will, community participation, and ongoing innovation in trash management practices. While these case studies show successful implementation of new waste management systems, more research may be needed to acquire more thorough information and assess their long-term usefulness.

6.15 CONCLUSION

Waste management has emerged as a crucial issue, needing ongoing research and development to address the health risks connected with inappropriate garbage disposal. As we look ahead, several critical directions should be explored to improve waste management practices and protect public health. There are a variety of potential future studies and innovation directions in waste management. It is critical to investigate new technologies and approaches that can improve waste management practices even more. This includes improvements in waste sorting and separation, the creation of more efficient recycling methods, and the incorporation of digital technologies to improve waste tracking and management. Policymakers, industry leaders, and the general public must collaborate to address these challenges effectively. Waste management should be prioritized as a vital public health issue, with rules enacted to encourage sustainable waste practices. Organizations in the waste management industry should invest in research and development to drive innovation in waste management solutions. The general population should be taught the need for appropriate waste disposal and recycling to build a sustainable culture. To conclude the health risks linked with solid waste need immediate action. We can preserve human health and the environment by prioritizing new waste management solutions, doing additional research, and encouraging stakeholder collaboration. It is our joint responsibility to act to secure a sustainable and healthy future for future generations.

REFERENCES

Aamir, M., Yin, S., Guo, F., Liu, K., Xu, C., & Liu, W. (2019). Congener-specific mother–fetus distribution, placental retention, and transport of C 10–13 and C 14–17 chlorinated paraffins in pregnant women. *Environmental Science & Technology*, 53(19), 11458–11466. https://doi.org/10.1021/acs.est.9b02116

Adhikari, B. K., Barrington, S., Martinez, J., & King, S. (2009). Effectiveness of three bulking agents for food waste composting. *Waste Management*, 29(1), 197–203. https://doi.org/10.1016/j.wasman.2008.04.001

Aghdam, E. F., Fredenslund, A. M., Chanton, J., Kjeldsen, P., & Scheutz, C. (2018). Determination of gas recovery efficiency at two Danish landfills by performing downwind methane measurements and stable carbon isotopic analysis. *Waste Management*, 73, 220–229. https://doi.org/10.1016/j.wasman.2017.11.049

Alagöz, A. Z., & Kocasoy, G. (2008). Determination of the best appropriate management methods for the health-care wastes in İstanbul. *Waste Management*, 28(7), 1227–1235. https://doi.org/10.1016/j.wasman.2007.05.018

Alaswad, S. O. (2019). Evaluation of soil contamination with industrial waste effluent in Riyadh, Saudi Arabia. *Pakistan Journal of Biological Sciences*, 23(1), 9–16. https://doi.org/10.3923/pjbs.2020.9.16

Al-Masry, M. M. M. (2019). The concept of (cradle to cradle) theory in the design of recycled products. *International Journal of Design and Fashion Studies*, 2(2), 20–23. https://doi.org/10.21608/ijdfs.2019.180042

Amenaghawon, A. N., Anyalewechi, C. L., Okieimen, C. O., & Kusuma, H. S. (2021). Biomass pyrolysis technologies for value-added products: a state-of-the-art review. *Environment Development and Sustainability*, 23(10), 14324–14378. https://doi.org/10.1007/s10668-021-01276-5

Anju, A., Ravi S. P., & Bechan S. (2010). Water pollution with special reference to pesticide contamination in India. *Journal of Water Resource and Protection*, 02(05), 432–448. https://doi.org/10.4236/jwarp.2010.25050

Anyaoku, C. C., & Baroutian, S. (2018). Decentralized anaerobic digestion systems for increased utilization of biogas from municipal solid waste. *Renewable and Sustainable Energy Reviews*, 90, 982–991. https://doi.org/10.1016/j.rser.2018.03.009

Awasthi, A. K., Wang, M., Awasthi, M. K., Wang, Z., & Li, J. (2018). Environmental pollution and human body burden from improper recycling of e-waste in China: A short-review. *Environmental Pollution*, 243, 1310–1316. https://doi.org/10.1016/j.envpol.2018.08.037

Barayang, L. P. C., Cruz, I. C. A., De Vera, K. J. G., Inumerable, J. C. F., & Alam, Z. F. (2022). An assessment of biomedical waste (BMW) production and management and its impact on the environment and disease transmission amidst the COVID-19 pandemic in the Philippines. *Biomedical & Pharmacology Journal*, 15(3), 1573–1581. https://doi.org/10.13005/bpj/2495.

Bull, R. A., & Zeff, J. D. (2021). Hydrogen peroxide in advanced oxidation processes for treatment of industrial process and contaminated groundwater. In John A. R., *Chemical Oxidation* (pp 26–36). Taylor & Francis. https://doi.org/10.1201/9781003210597-3.

Cameselle, C., Gouveia, S., Eddine, D., & Belhadj, B. (2013). Advances in electrokinetic remediation for the removal of organic contaminants in soils. In M. N. Rashed (ed.), *Organic Pollutants – Monitoring, Risk and Treatment*, 10, 54334. InTech eBooks. https://doi.org/10.5772/54334

Cao, Y., Harada, K. H., Hitomi, T., Niisoe, T., Wang, P., Shi, Y., Yang, H.-R., Takasuga, T., & Koizumi, A. (2017). Lactational exposure to short-chain chlorinated paraffins in China, Korea, and Japan. *Chemosphere*, 173, 43–48. https://doi.org/10.1016/j.chemosphere.2016.12.078

Cesaro, A., & Belgiorno, V. (2015, September). Medical waste generation and management in different sized facilities. In Proceedings of the 14th International Conference on Environmental Science and Technology, Rhode, Greece (pp. 3–5).

Chandra, S., Chauhan, L. K. S., Dhawan, A., Murthy, R. C., & Gupta, S. K. (2006). In vivo genotoxic effects of industrial waste leachates in mice following oral exposure. *Environmental and Molecular Mutagenesis*, 47(5), 325–333. https://doi.org/10.1002/em.20210

Chapman, C., Taylor, R., & Deegan, D. (2011). Thermal plasma processing in the production of value added products from municipal solid waste (MSW) derived sources. In Proceedings 2nd International Slag Valorisation Symposium, Leuven, Belgium (pp. 147–161).

Chen, H., Lam, J. C. W., Zhu, M., Wang, F., Zhou, W., Du, B., Zeng, L., & Zeng, E. Y. (2018). Combined effects of dust and dietary exposure of occupational workers and local residents to short- and medium-chain chlorinated paraffins in a mega E-waste recycling industrial park in south China. *Environmental Science & Technology*. https://doi.org/10.1021/acs.est.8b02625

Cheremisinoff, N. P. (2003). *Source Reduction and Waste Minimization*. In Elsevier eBooks (pp. 1–22). Butterworth-Heinemann.

Conn, W. D. (1989). Managing household hazardous waste. *Journal of the American Planning Association*, 55, 192–203.

Cook, E., Velis, C. A., & Black, L. (2022). Construction and demolition waste management: A systematic scoping review of risks to occupational and public health. *Frontiers in Sustainability*, 3. https://doi.org/10.3389/frsus.2022.924926.

Critchley, K. (2013). A retrospective study examining climatic influence on the respiratory health of Qatar residents. *Qatar Foundation Annual Research Forum* 2013(1). https://doi.org/10.5339/qfarf.2013.biop-060.

Deegan, D. E., Chapman, C. D., Ismail, S. A., Wise, M. L. H., Ly, H., & Phillips, P. S. (2007). A radical new, environmentally acceptable approach to hazardous waste management in the UK – a case study of plasma arc technology. *Journal of Solid Waste Technology & Management*, 32(4), 246. http://nectar.northampton.ac.uk/1214/

Dong, J., Tang, Y., Nzihou, A., Chi, Y., Weiss-Hortala, E., Ni, M., & Zhou, Z. (2018). Comparison of waste-to-energy technologies of gasification and incineration using life cycle assessment: Case studies in Finland, France and China. *Journal of Cleaner Production*, 203, 287–300. https://doi.org/10.1016/j.jclepro.2018.08.139

Elkacmi, R., & Bennajah, M. (2019). Advanced oxidation technologies for the treatment and detoxification of olive mill wastewater: A general review. *Journal of Water Reuse and Desalination*, 9(4), 463–505. https://doi.org/10.2166/wrd.2019.033

Erickson, T. B., Brooks, J., Nilles, E. J., Pham, P. N., & Vinck, P. (2019). Environmental health effects attributed to toxic and infectious agents following hurricanes, cyclones, flash floods and major hydrometeorological events. *Journal of Toxicology and Environmental Health, Part B*, 22(5–6), 157–171. https://doi.org/10.1080/10937404.2019.1654422

Exposto, N. L. a. S. M., Bakta, N. I. M., Wirawan, N. I. M. A., & Sujaya, N. I. N. (2022). Benefits of medical waste management in the facility health services. *Journal of Medical and Health Studies*, 3(3), 75–82. https://doi.org/10.32996/jmhs.2022.3.3.11.

Fowle, S. E., Constantine, C. E., Fone, D., & McCloskey, B. (1996). An epidemiological study after a water contamination incident near Worcester, England in April 1994. *Journal of Epidemiology & Community Health*, 50(1), 18–23. https://doi.org/10.1136/jech.50.1.18

García-Pérez, J., Lope, V., López-Abente, G., González-Sánchez, M., & Fernández-Navarro, P. (2015). Ovarian cancer mortality and industrial pollution. *Environmental Pollution*, 205, 103–110. https://doi.org/10.1016/j.envpol.2015.05.024

Gentil, E., Christensen, T. H., & Aoustin, E. (2009). Greenhouse gas accounting and waste management. *Waste Management & Research the Journal for a Sustainable Circular Economy*, 27(8), 696–706. https://doi.org/10.1177/0734242x09346702

Gonawala, S. S., & Jardosh, H. (2018). Organic waste in composting: A brief review. *International Journal of Current Engineering and Technology*, 8(1). https://doi.org/10.14741/ijcet.v8i01.10884

Gupta, C., & Prakash, D. (2020). Novel bioremediation methods in waste management: Novel bioremediation methods. In *Waste Management: Concepts, Methodologies, Tools, and Applications* (pp. 1627–1643). IGI Global eBooks. https://doi.org/10.4018/978-1-7998-1210-4.ch075.

Gusat, D., Bud, I., & Pasca, I. (2019). Modelling the waste disposal deposit – (Part I). Scientific Bulletin Series D: Mining, Mineral Processing, *Non-Ferrous Metallurgy, Geology and Environmental Engineering*, 33(1), 51–61. https://doi.org/10.37193/SBSD.2019.1.06

Gutberlet, J., & Uddin, S. M. N. (2017). Household waste and health risks affecting waste pickers and the environment in low- and middle-income countries. *International Journal of Occupational and Environmental Health*, 23(4), 299–310. https://doi.org/10.1080/10773525.2018.1484996

Haji Nawawi, N. H. B., & Jaafar, M. N. (2021). Assessment of air pollution control technologies to reduce SOx emission from thermal oxidizer for oil and gas industry. *IOP Conference Series: Materials Science and Engineering*, 1195(1), 012046. https://doi.org/10.1088/1757-899X/1195/1/012046

Han, X., Chen, H., Shen, M., Deng, M., Du, B., & Zeng, L. (2021). Hair and nails as noninvasive bioindicators of human exposure to chlorinated paraffins: Contamination patterns and potential influencing factors. *Science of the Total Environment*, 798, 149257. https://doi.org/10.1016/j.scitotenv.2021.149257

Hasan, M. A., Ahmad, S., & Mohammed, T. (2021). Groundwater contamination by hazardous wastes. *Arabian Journal for Science and Engineering*, 46(5), 4191–4212. https://doi.org/10.1007/s13369-021-05452-7

Havinal, R. (2022). Electronic waste: Growing challenges and management. *IJCRT*, 10(4). https://ijcrt.org/papers/IJCRT2204018.pdf.

Hoang, N. H., & Fogarassy, C. (2020). Sustainability evaluation of municipal solid waste management system for Hanoi (Vietnam) – why to choose the 'waste-to-energy' Concept. *Sustainability*, 12(3), 1085. https://doi.org/10.3390/su12031085

Hussain, C. M., Paulraj, M. S., & Nuzhat, S. (2021). *Source Reduction and Waste Minimization*. Environmental sciences. Elsevier.

Jain, P., Wally, J., Townsend, T. G., Krause, M., & Tolaymat, T. (2021). Greenhouse gas reporting data improves understanding of regional climate impact on landfill methane production and collection. *PLOS ONE*, 16(2), e0246334. https://doi.org/10.1371/journal.pone.0246334

Jibril, J. D. azimi, Sipan, I. B., Sapri, M., Shika, S. A., Isa, M., & Abdullah, S. (2012). 3R s critical success factor in solid waste management system for higher educational institutions. *Procedia – Social and Behavioral Sciences*, 65, 626–631. https://doi.org/10.1016/j.sbspro.2012.11.175

Karim, S. M. R., Sharif, S. I., & Anik, M. a. R. (2018). Negative impact and probable management policy of E-Waste in Bangladesh. arXiv (Cornell University). https://doi.org/10.48550/arxiv.1809.10021.

Kristanti, R. A., Ningsih, F., Yati, I., Kasongo, J., Mtui, E., & Rachana, K. (2023). Role of fungi in biodegradation of bisphenol A: A review. *Tropical Aquatic and Soil Pollution*, 3(2), 131–143.

Labrada, G. V., Kumar, S., Predicala, B., & Nemati, M. (2019). Nanotechnology-based control of hazardous air pollutants emission: Pilot scale trials for simultaneous capture of H_2S, NH_3, and odours from livestock facilities. *WIT Transactions on Ecology and the Environment*, 236, 171–178. https://doi.org/10.2495/AIR190171

Lad, D., Chauhan, R., & Gole, P. (2020). A study on solid waste management awareness amongst youngsters of Mumbai. *EPRA International Journal of Multidisciplinary Research (IJMR)*, 6, 116–119. https://doi.org/10.36713/epra4115

Ławińska, O., Korombel, A., & Zajemska, M. (2022). Pyrolysis-based municipal solid waste management in poland—SWOT analysis. *Energies*, 15(2), 510. https://doi.org/10.3390/en15020510

Leem, J.-H., Hong, Y.-C., Lee, K.-H., Kwon, H.-J., Chang, Y.-S., & Jang, J.-Y. (2003). Health survey on workers and residents near the municipal waste and industrial waste incinerators in Korea. *Industrial Health*, 41(3), 181–188. https://doi.org/10.2486/indhea lth.41.181

Li, C., Yang, L., Shi, M., & Liu, G. (2019). Persistent organic pollutants in typical lake ecosystems. *Ecotoxicology and Environmental Safety*, 180, 668–678. https://doi.org/ 10.1016/j.ecoenv.2019.05.060

Li, H., Nitivattananon, V., & Li, P. (2015). Municipal solid waste management health risk assessment from air emissions for China by applying life cycle analysis. *Waste Management & Research: The Journal for a Sustainable Circular Economy*, 33(5), 401– 409. https://doi.org/10.1177/0734242X15580191

Liu, Y., Li, S., Wang, Q., Zheng, X., Zhao, Y., & Lu, W. (2023). Occupational health risks of VOCs emitted from the working face of municipal solid waste landfill: Temporal variation and influencing factors. *Waste Management*, 160, 173–181. https://doi.org/ 10.1016/j.wasman.2023.02.001

Luo, Q., Huang, L., Liu, Y., Xue, X., Zhou, F., & Hua, J. (2021). Monitoring study on dust dispersion properties during earthwork construction. *Sustainability*, 13(15), 8451. https:// doi.org/10.3390/su13158451.

Mahajan, R. (2023). Environment and health impact of solid waste management in developing countries: A review. *Current World Environment*, 18(1), 18–29. https://doi.org/10.12944/ CWE.18.1.3

Marchese, F., Fiore, S., Ruffino, B., Soldi, G., Demaio, M., & Luciani, P. (2013). An hazardous waste landfill in northern Italy: 25 years of activity. In *SARDINIA* (pp. 1–14). Eurowaste. https://hdl.handle.net/11583/2516285.

Mattiello, A., Chiodini, P., Bianco, E., Forgione, N., Flammia, I., Gallo, C., Pizzuti, R., & Panico, S. (2013). Health effects associated with the disposal of solid waste in landfills and incinerators in populations living in surrounding areas: A systematic review. *International Journal of Public Health*, 58(5), 725–735. https://doi.org/10.1007/s00 038-013-0496-8

Maulini-Duran, C., Puyuelo, B., Artola, A., Font, X., Sánchez, A., & Gea, T. (2014). VOC emissions from the composting of the organic fraction of municipal solid waste using standard and advanced aeration strategies. *Journal of Chemical Technology & Biotechnology*, 89(4), 579–586. https://doi.org/10.1002/jctb.4160

Mazalan, M., Cassendra, P. C. B., Wai, S. H., Jeng, S. L., Zarina, A. M., Haslenda, H., Sherien, E., Gabriel, L. H. T., & Chin, S. H. (2018). Review on the suitability of waste for appropriate waste-to-energy technology. *Chemical Engineering Transactions*, 63, 187–192. https://doi.org/10.3303/cet1863032

Mester, T., Szabó, G., Sajtos, Z., Baranyai, E., Szabó, G., & Balla, D. (2022). Environmental hazards of an unrecultivated liquid waste disposal site on soil and groundwater. *Water (Switzerland)*, 14(2), 226. https://doi.org/10.3390/w14020226

Moitra, S., Tabrizi, A. F., Khadour, F., Henderson, L., Melenka, L., & Lacy, P. (2022). Occupational exposure to insulating materials and risk of coronary artery diseases. medRxiv (Cold Spring Harbor Laboratory). https://doi.org/10.1101/2022.12.12.22283365.

Moya, D., Aldás, C., López, G., & Kaparaju, P. (2017). Municipal solid waste as a valuable renewable energy resource: A worldwide opportunity of energy recovery by using Waste-To-Energy Technologies. *Energy Procedia*, 134, 286–295. https://doi.org/10.1016/j.egy pro.2017.09.618

Nadal, M., Marquès, M., Mari, M., Rovira, J., & Domingo, J. L. (2020). Trends of polychlorinated compounds in the surroundings of a municipal solid waste incinerator in Mataró (Catalonia, Spain): Assessing health risks. *Toxics*, 8(4), 111. https://doi.org/10.3390/toxics8040111

Noh, N. M., & Mydin, M. A. O. (2018). Waste minimization strategy and technique: Towards sustainable waste management. *Analele Universității "Eftimie Murgu"*, 25(1), 91–98.

North, E. J., & Halden, R. U. (2013). Plastics and environmental health: The road ahead. *Reviews on Environmental Health*, 28(1), 1–8. https://doi.org/10.1515/reveh-2012-0030

Odukkathil, G., & Vasudevan, N. (2013). Toxicity and bioremediation of pesticides in agricultural soil. *Reviews in Environmental Science and Bio/Technology*, 12(4), 421–444. https://doi.org/10.1007/s11157-013-9320-4

Olagunju, E., Badmus, O., Ogunlana, F., & Babalola, M. (2018). Environmental Impact Assessment of Waste dumpsite using Integrated Geochemical and Physico-Chemical approach: A case study of Ilokun Waste Dumpsite, Ado - Ekiti, Southern Nigeria. *Civil Engineering Research Journal*, 4(2), 001-0013. https://doi.org/10.19080/CERJ.2018.04.555631. https://doi.org/10.19080/CERJ.2018.04.555631

Oli, A. N., Ekejindu, C. C., Adje, D. U., Ezeobi, I., Ejiofor, O. S., Ibeh, C. C., & Ubajaka, C. F. (2016). Healthcare waste management in selected government and private hospitals in Southeast Nigeria. *Asian Pacific Journal of Tropical Biomedicine*, 6(1), 84–89. https://doi.org/10.1016/j.apjtb.2015.09.019

Onyekwelu, I. L., & Aghamelu, O. P. (2019). Impact of organic contaminants from dumpsite leachates on natural water sources in the Enugu Metropolis, southeastern Nigeria. *Environmental Monitoring and Assessment*, 191(9), 543. https://doi.org/10.1007/s10661-019-7719-2

Paes, M. X., & de Oliveira, J. A. P. (2021). *Integrated Management of Municipal Solid Waste in Brazil: A Case Study in São Paulo City*. IOP Conference Series: Materials Science and Engineering, 1196.

Pollans, L. B. (2012). Greening Infrastructural Services: The Case of Waste Management in San Francisco. http://thriftstorejunkies.com/2011/01/from-trash-literally-to-treasures/

Posnack, N. G. (2014). The adverse cardiac effects of di(2-ethylhexyl)phthalate and bisphenol A. *Cardiovascular Toxicology*, 14(4), 339–357. https://doi.org/10.1007/s12012-014-9258-y

Pourali, M. (2010). Application of plasma gasification technology in waste to energy—Challenges and opportunities. *IEEE Transactions on Sustainable Energy*, 1(3), 125–130. https://doi.org/10.1109/tste.2010.2061242

Qasim, M., Xiao, H., He, K., Noman, A., Liu, F., Chen, M., Hussain, D., Jamal, Z. A., & Li, F. (2020). Impact of landfill garbage on insect ecology and human health. *Acta Tropica*, 211, 105630. https://doi.org/10.1016/j.actatropica.2020.105630

Qiao, L., Gao, L., Zheng, M., Xia, D., Li, J., Zhang, L., Wu, Y., Wang, R., Cui, L., & Xu, C. (2018). Mass fractions, congener group patterns, and placental transfer of short- and medium-chain chlorinated paraffins in paired maternal and cord serum. *Environmental Science & Technology*, 52(17), 10097–10103. https://doi.org/10.1021/acs.est.8b02839

Rajasekhar, M., Rao, N. V., Rao, G. C., Priyadarshini, G., & Kumar, N. J. (2015). Energy generation from municipal solid waste by innovative technologies – plasma gasification. *Procedia Materials Science*, 10, 513–518. https://doi.org/10.1016/j.mspro.2015.06.094

Rao, M. S., & Wankhede, S. (2011). Management of hospital wastes with potential pathogenic microbes. In *Microorganisms in Environmental Management: Microbes and Environment* (pp. 365–401). Springer. https://doi.org/10.1007/978-94-007-2229-3_17

Rodrigo, M., Oturan, N., & Oturan, M. A. (2014). Electrochemically assisted remediation of pesticides in soils and water: A review. *Chemical Reviews*, 114(17), 8720–8745. https://doi.org/10.1021/cr500077e

Rogers, D. E., Rohwer, M. B., & Brent, A. C. (2002). Life Cycle Check as a decision support tool for medical waste management in underdeveloped areas of Africa. University of Pretoria. https://repository.up.ac.za/handle/2263/4925.

Rushton, L. (2003). Health hazards and waste management. *British Medical Bulletin*, 68(1), 183–197. https://doi.org/10.1093/bmb/ldg034

Sang, Y., Yu, W., He, L., Wang, Z., Ma, F., Jiao, W., & Gu, Q. (2021). Sustainable remediation of lube oil-contaminated soil by low temperature indirect thermal desorption: Removal behaviors of contaminants, physicochemical properties change and microbial community recolonization in soils. *Environmental Pollution*, 287, 117599. https://doi.org/10.1016/j.envpol.2021.117599

Sarker, B., Keya, K. N., Mahir, F. I., Nahiun, K. M., Shahida, S., & Khan, R. A. (2021). Surface and ground water pollution: Causes and effects of urbanization and industrialization in South Asia. *Scientific Review*, 73, 32–41. https://doi.org/10.32861/sr.73.32.41.

Shah, S. a. A., Longsheng, C., Solangi, Y. A., Ahmad, M., & Ali, S. (2021). Energy trilemma based prioritization of waste-to-energy technologies: Implications for post-COVID-19 green economic recovery in Pakistan. *Journal of Cleaner Production*, 284, 124729. https://doi.org/10.1016/j.jclepro.2020.124729

Shenyoputro, K., & Jones, T. E. (2023). Reflections on a two-decade journey toward zero waste: A case study of Kamikatsu town, Japan. *Frontiers in Environmental Science*, 11. https://doi.org/10.3389/fenvs.2023.1171379

Singh, A., Pal, R., Gangwar, C., Gupta, A., & Tripathi, A. (2015). Release of heavy metals from industrial waste and e-waste burning and its effect on human health and environment. *International Journal of Emerging Research in Management*, 4(12), 51.

Streeter, J. L. (2016). Adoption of SO2 emission control technologies – An application of survival analysis. *Energy Policy*, 90, 16–23. https://doi.org/10.1016/j.enpol.2015.11.035.

Thakare, S., & Nandi, S. (2016). Study on potential of gasification technology for municipal solid waste (MSW) in Pune city. *Energy Procedia*, 90, 509–517. https://doi.org/10.1016/j.egypro.2016.11.218

Tiembre, I., Koné, B. A., Dongo, K., Tanner, M., Zinsstag, J., & Cissé, G. (2009). [Epidemiologic and clinical aspects of toxic waste poisoning in Abidjan]. *Sante (Montrouge, France)*, 19(4), 189–194. https://doi.org/10.1684/san.2009.0163

Tomasko, J., Stupak, M., Parizkova, D., Polachova, A., Sram, R. J., Topinka, J., & Pulkrabova, J. (2021). Short- and medium-chain chlorinated paraffins in human blood serum of Czech population. *Science of the Total Environment*, 797, 149126. https://doi.org/10.1016/j.scitotenv.2021.149126

Uzunlulu, G., Uzunlulu, M., Gencer, A., Özdoğru, F., & Seven, S. (2022). Knowledge on medical waste management among health care personnel: A report from Turkey. *Cyprus Journal of Medical Sciences*, 7(4), 552–558. https://doi.org/10.4274/cjms.2020.1107

Van Fan, Y., Klemeš, J. J., Lee, C. T., & Perry, S. (2018). Anaerobic digestion of municipal solid waste: Energy and carbon emission footprint. *Journal of Environmental Management*, 223, 888–897. https://doi.org/10.1016/j.jenvman.2018.07.005

van Mourik, L. M., Toms, L.-M. L., He, C., Banks, A., Hobson, P., Leonards, P. E. G., de Boer, J., & Mueller, J. F. (2020). Evaluating age and temporal trends of chlorinated paraffins in pooled serum collected from males in Australia between 2004 and 2015. *Chemosphere*, 244, 125574. https://doi.org/10.1016/j.chemosphere.2019.125574

Vehlow, J. (2015). Air pollution control systems in WtE units: An overview. *Waste Management*, 37, 58–74. https://doi.org/10.1016/j.wasman.2014.05.025

Verkhovets, I. A., Tuchkova, L. E., & Tikhoykina, I. M. (2020). Ecological and economic assessment of the impact of industrial and consumer waste on the quality of gray forest soil. *IOP Conference Series: Earth and Environmental Science*, 459(2), 022052. https://doi.org/10.1088/1755-1315/459/2/022052

Vinti, G., Bauza, V., Clasen, T., Medlicott, K., Tudor, T., Zurbrügg, C., & Vaccari, M. (2021). Municipal solid waste management and adverse health outcomes: A systematic review. *International Journal of Environmental Research and Public Health*, 18(8), 4331. https://doi.org/10.3390/ijerph18084331

Visvanathan, C., Tränkler, J., Kuruparan, P., & Xiaoning, Q. (2002). Influence of landfill operation and waste composition on leachate control: lysimeter experiments under tropical conditions. In 2nd Asia Pacific Landfill Symposium in Seoul, Korea (pp. 441–447). http://faculty.ait.ac.th/visu/public/uploads/participation/ILOWC.pdf.

Wang, N., Shi, M., Wu, S., Guo, X., Zhang, X., Ni, N., Sha, S., & Zhang, H. (2022). Study on volatile organic compound (VOC) emission control and reduction potential in the pesticide industry in China. *Atmosphere*, 13(8), 1241. https://doi.org/10.3390/atmos13081241

Wei, Y., Li, J., Shi, D., Liu, G., Zhao, Y., & Shimaoka, T. (2017). Environmental challenges impeding the composting of biodegradable municipal solid waste: A critical review. *Resources, Conservation and Recycling*, 122, 51–65. https://doi.org/10.1016/j.resconrec.2017.01.024

Witt, M., Goniewicz, M., Pawłowski, W., Goniewicz, K., & Biczysko, W. (2017). Analysis of the impact of harmful factors in the workplace on functioning of the respiratory system of firefighters. *Annals of Agricultural and Environmental Medicine*, 24(3), 406–410. https://doi.org/10.5604/12321966.1233561

World Health Organization. Regional Office for Europe. (2015, November 5–6). *Waste and Human Health: Evidence and Needs: WHO Meeting Report*. Bonn, Germany: World Health Organization. Regional Office for Europe.

Wu, Q., Leung, J. Y., Geng, X., Chen, S., Huang, X., Li, H., Huang, Z., Zhu, L., Chen, J., & Lu, Y. (2015). Heavy metal contamination of soil and water in the vicinity of an abandoned e-waste recycling site: Implications for dissemination of heavy metals. *The Science of the Total Environment*, 506–507, 217–225. https://doi.org/10.1016/j.scitotenv.2014.10.121

Xiaodong, L., Su, S., & Huang, T. (2015). Health damage assessment model for construction dust. *Journal of Tsinghua University*, 55, 50–55.

Xu, Y., Seshadri, B., Bolan, N., Sarkar, B., Ok, Y. S., Zhang, W., Rumpel, C., Sparks, D., Farrell, M., Hall, T., & Dong, Z. (2019). Microbial functional diversity and carbon use feedback in soils as affected by heavy metals. *Environment International*, 125, 478–488. https://doi.org/10.1016/j.envint.2019.01.071

Yaashikaa, P. R., & Kumar, P. S. (2022). Bioremediation of hazardous pollutants from agricultural soils: A sustainable approach for waste management towards urban sustainability. *Environmental Pollution*, 312, 120031. https://doi.org/10.1016/j.envpol.2022.120031

Yadav, A. N., Suyal, D. C., Kour, D., Rajput, V. D., Rastegari, A. A., & Singh, J. (2022). Bioremediation and waste management for environmental sustainability. *Journal of Applied Biology & Biotechnology*, 1–5. https://doi.org/10.7324/jabb.2022.10s201.

Yang, X., Yan, Y., Xue, C., Du, X., & Ye, Q. (2018). Association between increased small airway obstruction and asbestos exposure in patients with asbestosis. *The Clinical Respiratory Journal*, 12(4), 1676–1684. https://doi.org/10.1111/crj.12728

Yujun, T., Hongyan, W., Tao, L., & Chuansheng, P. (2020). The study on cost of application of International Emission Control Areas for China. *Journal of Resources and Ecology*, 11(4), 388. https://doi.org/10.5814/j.issn.1674-764x.2020.04.007

Zhao, C., Dong, Y., Feng, Y., Li, Y., & Dong, Y. (2019). Thermal desorption for remediation of contaminated soil: A review. *Chemosphere*, 221, 841–855. https://doi.org/10.1016/j.chemosphere.2019.01.079

7 Biological Hazardous Waste Management of Animal Husbandry and Food Processing Industry

Eulis Tanti Marlina, Yuli Astuti Hidayati, Ellin Harlia, and Norli Ismail

7.1 INTRODUCTION

Livestock waste, also known as livestock waste or manure, typically includes urine and feces, bedding material, waste from feeding, and water. The primary components of animal husbandry waste are urine and feces. Meanwhile, waste from the food processing industry mostly comprises wastewater from washing, blood, chemicals, and packaging. Animal and food processing waste have high amounts of organic material in the form of protein, fat, and carbohydrates so that microbes can utilize them as a source of nutrition for their growth. If not handled properly, the waste produced by livestock has the potential to harbor pathogenic bacteria. Several pathogenic bacteria are also normal microflora in the digestive tract of livestock, including *Escherichia coli* and *Salmonella* sp. During defecation, these pathogenic bacteria will come together with the feces and contaminate groundwater. Often, livestock waste contains several drugs harmful to the environment, including antibiotics and hormones. When livestock are ill and treated with antibiotics, residues of those antibiotics may be present in their feces. Antibiotic residues pollute the environment and can cause antibiotic resistance in humans. The characteristics of livestock waste are considerably influenced by the feed consumed.

The food and livestock processing industry produces waste that differs from that of farms. Processing milk, meat, and leather involves using chemical compounds that create hazardous waste, such as corrosives. Proper waste management is crucial to prevent harm to human health.

DOI: 10.1201/9781003543176-7

7.2 CHARACTERISTICS OF LIVESTOCK INDUSTRY WASTE

Livestock is a valuable human food source, and it is essential to use it efficiently. As the human population grows, the demand for animal food increases. However, we currently only use about 30–35% of the meat produced from one animal, with the remaining 65–70% being considered waste (Arvanitoyannis and Ladas, 2008). With the help of technology, these by-products can benefit both humans and the environment. It is important to note that there is a difference between livestock waste and by-products. Waste includes biomass like feces and urine, while by-products come from the remains of the cutting process, such as skin, bones, blood, offal, fur, horns, and organs.

Animal waste, or manure, refers to animal excreta or waste products. The characteristics of animal waste can vary depending on the type of animal, its diet, and its physiology (Said, 2019; Shober and Maguire, 2018). The characteristics of animal waste can have important implications for its management and use, as well as for the environmental impacts of animal agriculture. Proper handling and managing animal waste can help minimize its negative impacts and maximize its potential benefits as a source of nutrients and organic matter for crops and soil.

7.2.1 MOISTURE LEVELS

Animal waste contains varying moisture levels, depending on the type of animal and its diet. For example, chicken manure tends to have a higher moisture content than cow manure. Chickens have a high protein diet consisting primarily of plant-based feed, typically high in moisture content (Hicks and Verbeek, 2016; Singh and Kim, 2021). Cow diets, conversely, consist primarily of fibrous plant material that is drier in nature. Chickens have a faster digestive system than cows, that is, food passes through them more quickly, resulting in higher moisture content in their manure. Chickens require more water intake than cows to maintain their health and well-being. This increased water intake results in higher moisture content in their manure.

7.2.2 ORGANIC MATTER

Animal waste is rich in organic matter, made up of carbon-based compounds that provide nutrients to plants and microorganisms. It consists of animal excreta and bedding material, primarily composed of organic compounds such as carbohydrates, proteins, and lipids. These organic compounds are broken down by the digestive system of animals and converted into simpler compounds eliminated from the body as waste. Organic matter in animal waste is an essential source of nutrients for plants and soil microorganisms (Shober and Maguire, 2018; Goldan et al., 2023). The organic matter in animal waste is a valuable resource that can improve soil fertility, support sustainable agriculture, and promote renewable energy production. Animal waste is a valuable source of nutrients such as nitrogen, phosphorus, and potassium that are crucial for the growth of plants. When animal waste is applied to soil as a fertilizer, the organic matter it contains provides a food source for soil microorganisms, which break down into nutrients that plants can absorb. The organic matter in animal waste also improves soil structure, water-holding capacity, and nutrient retention, leading to healthier and more productive soils. The organic matter in animal waste can also be utilized as a renewable energy source through anaerobic digestion and composting processes. These processes break down the organic matter in animal waste and produce biogas or compost, which can be used as a renewable energy source or soil amendment.

7.2.3   pH

Animal waste can have varying pH levels, depending on the animal species and diet. It is an essential factor to consider in waste management as it influences the solubility and availability of nutrients and the potential for odor and pathogen emissions (Lorimor et al., 2004; Hill et al., 2005). Therefore, maintaining an appropriate pH level is crucial for efficient waste management. Animal waste is mainly influenced by the type of animal and its diet. Animals that consume more fibrous plant material, such as cows and horses, tend to have waste with a higher pH (more alkaline) due to the high concentration of bicarbonate and other buffering compounds in their saliva and rumen. Nevertheless, animals that consume more protein-rich diets, such as poultry and swine, tend to have waste with a lower pH (more acidic) due to the production of acidic metabolic waste products. For example, chicken manure typically has a lower pH (more acidic) than cow manure due to the higher protein content of the chicken diet. Similarly, pig manure has a lower pH than cow manure due to the high protein content of the pig diet. Conversely, cow manure tends to have a higher pH due to the high content of bicarbonate in its digestive system. The pH of animal waste can have significant implications for waste management (Moyo et al., 2022). A low pH can increase the solubility and availability of certain nutrients, such as phosphorus, which can lead to nutrient pollution in waterways. A high pH can contribute to the formation of ammonia gas, which is a potent greenhouse gas that can also contribute to odor emissions.

7.2.4   Odor

Animal waste can produce strong odors due to volatile organic compounds (VOCs) and other chemical compounds released during decomposition. These compounds can be produced by the breakdown of organic matter in the waste, as well as by microbial activity and chemical reactions. Some of animal waste's most common odor-causing compounds include ammonia, hydrogen sulfide, and volatile fatty acids (VFAs). These compounds are released due to microbial activity, and their concentrations can be influenced by temperature, moisture, and pH. The strong odors produced by animal waste can be a nuisance for nearby residents and have negative health impacts on both animals and humans (Schiffman et al., 2004). High ammonia and hydrogen sulfide exposure levels can cause respiratory irritation, headaches, and other health problems (Dufour et al., 2012; Penakalapati et al., 2017). Additionally, releasing VOCs from animal waste can contribute to air pollution and greenhouse gas emissions, likely leading to broader environmental impacts. Effective management practices can help reduce the production of odors from animal waste. This can include strategies such as proper storage and handling techniques, ventilation and odor control systems, and additives or treatments to reduce microbial activity and odor production. By implementing these practices, it is possible to minimize the impact of animal waste on nearby communities and promote sustainable and responsible waste management.

7.2.5 PATHOGENS

Animal waste can contain various pathogens, including bacteria, viruses, and parasites, which can pose a health risk to humans and other animals if not properly managed. Several reasons why these pathogens pose a risk to human health are described as follows:

1. **Zoonotic diseases:** Many pathogens found in livestock waste are zoonotic, meaning they can be transmitted between animals and humans. When humans come into contact with contaminated livestock waste directly or indirectly, they can contract these diseases. For example, *Salmonella* and *Campylobacter* can cause severe gastrointestinal infections in humans, leading to diarrhea, abdominal pain, and fever (Sobsey et al., 2006; Blaiotta et al., 2016; Delahoy et al., 2018).
2. **Foodborne illnesses:** Pathogens present in livestock waste can contaminate food and water sources. If these contaminated products are consumed without proper cooking or sanitation, they can cause foodborne illnesses. For instance, *E. coli* contamination in meat or dairy products can result in food poisoning, leading to symptoms such as diarrhea, vomiting, and in severe cases, kidney damage (Ucar et al., 2016; Heredia and Garcia, 2018).
3. **Environmental contamination:** Improper livestock waste management can lead to environmental contamination, such as leaching pathogens into water sources or spreading pathogens through air or soil. This contamination can increase the risk of exposure to humans living near livestock operations and contaminate crops or recreational areas, potentially causing disease outbreaks (Alegbeleye and Sant'Ana, 2020).
4. **Antibiotic resistance:** Livestock waste often contains high levels of antibiotics and antibiotic-resistant bacteria. Exposure to these antibiotic-resistant pathogens in the environment can contribute to the spread of antibiotic resistance, making it more difficult to treat bacterial infections in humans (Chen et al., 2019; He et al., 2020; Ibekwe et al., 2023).

7.2.5.1 Bacteria

Bacteria are microorganisms found in livestock waste. Several bacteria play a positive role as decomposers in decomposing waste organic matter into useful simple compounds, such as nutrients for soil fertility, and methane gas as an energy source. However, a few bacteria have a negative role, namely pathogens for humans and livestock.

Several types of pathogenic bacteria can infect humans in the livestock industry, such as *E. coli, Salmonella* spp., *Campylobacter* spp., *Listeria monocytogenes, Clostridium perfringens,* and *Leptospira* spp. Certain strains of *E. coli,* such as *E. coli* O157:H7, are pathogenic and can cause severe illness in humans (Dungan et al., 2012; Manyi-Loh et al., 2016). These bacteria are typically found in the intestines of healthy cattle and can contaminate the environment through cattle feces. *Salmonella* spp. can be found in many types of livestock, but poultry is a major source. *L. monocytogenes* can cause a severe infection known as listeriosis, especially in pregnant women, the elderly, and individuals with weakened immune systems (WHO, 2018). It can be found in soil and water, and animals can become carriers without appearing ill. *Campylobacter* spp. spores can cause food poisoning (Silva et al., 2011; WHO, 2020).

7.2.5.2 Viruses

Similar to bacteria, viruses from livestock waste can also pose significant health risks to humans. The modes of transmission can be quite similar. Several types of viruses found in livestock waste can potentially infect humans, such as influenza viruses, noroviruses and rotaviruses, hepatitis E virus (HEV), poliovirus, and coronaviruses (Sobsey et al., 2006; Delahoy, 2018).

Pigs and birds can excrete influenza viruses and HEV in their feces but the transmission is different (Kasorndorkbua et al., 2004). HEV can infect humans through the consumption of undercooked or raw pork, or direct contact with pig feces. Influenza viruses can infect humans through direct contact with contaminated feces (Louten, 2016; CDC, 2022). Poliovirus can be excreted in the feces of infected animals and humans (Farbu, 2017). It can cause polio in humans, which can lead to paralysis in severe cases. Coronaviruses are not typically found in livestock waste; they can theoretically be excreted in the feces of infected animals (Bonilauri and Rugna, 2021). Severe acute respiratory syndrome coronavirus 2 (SARS-CoV-2), the virus responsible for coronavirus disease 2019 (COVID-19), has been found in the feces of infected humans and some animals, raising the possibility of fecal–oral transmission.

7.2.5.3 Fungi

Fungi in livestock waste can pose significant health risks to humans. Several fungi found in livestock waste and associated environments can potentially infect humans such as *Histoplasma capsulatum, Aspergillus* spp., *Cryptococcus neoformans, Dermatophytes,* and *Blastomyces dermatitidis* (Wan et al., 2021).

Fungi *H. capsulatum* can cause histoplasmosis, a lung disease that can be severe in people with weakened immune systems (Mittal et al., 2018). People can become infected by inhaling fungal spores from the environment, such as from contaminated soil in chicken coops. *Aspergillus* spp. can cause various human diseases, ranging from allergic reactions to severe lung infections (Mousavi et al., 2016). Farmers and other people who work with livestock can be exposed to Aspergillus spores through contaminated feed or bedding. *C. neoformans* can cause cryptococcosis, a potentially severe lung or brain infection, especially in people with weakened immune systems (Denham and Brown, 2018). People can become infected by inhaling fungal spores from the environment. *Dermatophytes* cause skin, hair, and nail infections, collectively known as ringworm (Smith and McGinnis, 2006; Kruithoff et al., 2024). These fungi can be found in various animals and transmitted to humans through direct contact with infected animals or contaminated objects. *B. dermatitidis* is found in moist soil and decomposing organic matter. It can cause blastomycosis, a potentially severe lung disease, in humans. People can become infected by inhaling the fungal spores from the environment.

7.2.5.4 Parasites

Livestock waste containing parasites can be dangerous to human health. Parasites can cause disease in various ways, including damaging tissues, competing for nutrients, and triggering harmful immune responses (Dame-Korevaar et al., 2021). Several parasites in livestock waste can infect humans, such as *Cryptosporidium, Giardia, Toxoplasma gondii, Taenia* sp., and *Fasciola hepatica* (Youssef, 2014). *Cryptosporidium* causes the diarrheal disease, cryptosporidiosis. Humans can become infected with fecal-contaminated soil, water, or food. *Giardia* causes giardiasis, a diarrheal disease. Akin to *Cryptosporidium*, humans can become infected through contact with fecal-contaminated soil, water, or food. Toxoplasmosis, caused by *T. gondii*, is often severe for pregnant women and people with weak immune systems (Oyeyemi et al., 2020). This parasite is usually found in various animals' tissue, including livestock, but cats are the primary host. Humans can contract the infection by consuming undercooked, contaminated meat or by contacting cat feces. *Taenia saginata* is associated with cattle, while *Taenia solium* is linked to pigs, and both can infect humans who consume undercooked, contaminated meat. Fascioliasis in humans is caused by Fasciola hepatica, which can be contracted through ingestion of contaminated water or water plants, such as watercress, exposed to the eggs of liver flukes shed by livestock. This infection can lead to liver disease (Mas-Coma et al., 2013).

The following is the mechanism by which microorganisms and parasites from waste can infect humans:

1. **Water contamination:** If livestock waste is not effectively managed, it can contaminate surface and groundwater supplies. This is a common issue where manure is used as a fertilizer or large-scale livestock operations are located near water sources. People can become infected by drinking or bathing in contaminated water or consuming food irrigated with contaminated water.
2. **Food contamination:** Manure used as a fertilizer can contaminate crops, especially if not properly composted or treated. If these crops are consumed raw or not adequately cooked, some microorganisms can enter the human body and cause infections.
3. **Direct contact:** People who work with livestock, like farmers and veterinarians, can contact animal waste directly and get exposed to harmful microorganisms. If they have a cut or open wound, it can provide an entry point for microorganisms.
4. **Airborne transmission:** Some microorganisms can become aerosolized (i.e., suspended in the air) and inhaled, which is another potential route of

exposure. This is a particular concern in confined animal feeding operations (CAFOs) where large numbers of animals are kept in close proximity.

5. **Vectors:** Flies and other insects can transfer bacteria from animal waste to human food sources, causing contamination.
6. *Fomites* are inanimate objects like farm tools or clothing that can contaminate animal waste and indirectly transmit bacteria to humans.
7. **Inhalation:** Some microorganisms, like fungi, produce spores that can become airborne and be inhaled by humans, especially in confined or poorly ventilated spaces. This is a common route of exposure for farmers and others who work closely with livestock.

Once the microorganisms enter the human body, they can cause disease through various mechanisms. Some microorganisms produce toxins that damage cells and tissues, others can invade cells and disrupt their normal functioning, and others can trigger an overactive immune response that results in illness. The severity of the disease can vary greatly, depending on the type of microorganisms and the individual's health status. Preventing microorganism infections from livestock waste involves a combination of good hygiene practices, proper waste management, personal protective equipment (PPE), and, in some cases, vaccination. Proper waste management practices, good personal hygiene, use of PPE, and, in some cases, vaccination can help reduce the risk of these bacterial infections (Wyeth, 2013; CDC, 2021).

A summary of some of the critical biological hazards from livestock waste and waste from the food processing industry is presented in Table 7.1.

7.2.6 Texture

The texture of animal waste can range from solid to liquid, depending on the type of animal and the conditions under which the waste is stored. For instance, dairy cow manure tends to be more liquid than beef cow manure. The texture of livestock waste can also be affected by management practices, such as the type of housing, bedding material used, or waste handling and storage methods. Proper waste management practices can help maintain optimal waste texture for efficient handling, storage, and subsequent utilization or disposal of livestock waste. The texture of livestock waste can be influenced by several factors, including the following:

1. **Animal species and diet:** Different livestock species have varying digestive systems and diets, which can affect the texture of their waste. For example, ruminants like cattle and sheep have a complex digestive system that can break down fibrous plant material more thoroughly, resulting in waste with a drier and more fibrous texture compared to monogastric animals like pigs or poultry.
2. **Feed composition:** The composition of the animals' feed plays a signifi cant role in the texture of their waste. Diets rich in fiber, such as forage or roughage, tend to produce waste with a higher fiber content and a coarser texture. In contrast, diets high in concentrates or grains can produce waste with a softer and less fibrous texture.

TABLE 7.1

Livestock and food industry waste pose biological hazards

Hazard Type	Examples	Transmission Route	Associated Diseases
Bacteria (Livestock Waste)	*E. coli, Salmonella, Campylobacter, Listeria monocytogenes, Clostridium perfringens, Leptospira*	Direct contact with feces or contaminated environment, consumption of contaminated food or water	Gastroenteritis, listeriosis, leptospirosis
Viruses (Livestock Waste)	*Influenza, Norovirus, Rotavirus, Hepatitis E, Poliovirus, Coronaviruses*	Direct contact with feces or contaminated environment, consumption of contaminated food or water	Influenza, gastroenteritis, hepatitis E, polio, COVID-19
Fungi (Livestock Waste)	*Histoplasma capsulatum, Aspergillus* species, *Cryptococcus neoformans, Dermatophytes, Blastomyces dermatitidis*	Inhalation of spores from contaminated environment, direct contact with infected animals or contaminated objects	Histoplasmosis, aspergillosis, cryptococcosis, ringworm, blastomycosis
Parasites (Livestock Waste)	*Cryptosporidium, Giardia, Toxoplasma gondii, Taenia saginata, Taenia solium, Echinococcus granulosus, Echinococcus multilocularis, Fasciola hepatica*	Direct contact with feces or contaminated environment, consumption of contaminated food or water	Cryptosporidiosis, giardiasis, toxoplasmosis, taeniasis, echinococcosis, fascioliasis
Bacteria (Food Processing Waste)	*Salmonella, Listeria monocytogenes, E. coli*	Direct contact with waste or contaminated environment, consumption of contaminated food or water	*Salmonellosis*, listeriosis, *E. coli* infection
Viruses (Food Processing Waste)	Norovirus, Hepatitis A	Direct contact with waste or contaminated environment, consumption of contaminated food or water	Norovirus gastroenteritis, hepatitis A
Fungi (Food Processing Waste)	*Aspergillus* sp.	Inhalation of spores from contaminated environment	Aspergillosis

3. **Water intake:** Adequate water intake is important for maintaining regular bowel movements and hydration levels in livestock. Insufficient water intake can lead to dry and compacted waste, while increased water consumption can result in looser and more watery waste.
4. **Microbial activity:** The microbial population in the animal's gut and the fermentation processes that occur during digestion can influence the texture of the waste. Microbes play a crucial role in breaking down feed components and extracting nutrients. The activity of these microbes can affect the extent of digestion and the consistency of the waste produced.
5. **Health and digestive disorders:** Health conditions or digestive disorders in livestock can impact the texture of their waste. For example, gastrointestinal infections or inflammation can lead to diarrhea or loose stools, while conditions like constipation or impaction can result in dry and hardened waste.

7.3 LIVESTOCK WASTE MANAGEMENT

Waste management involves collecting, transporting, processing, disposing, and recycling waste materials.

7.3.1 COLLECTING

Collecting livestock waste typically involves a combination of manual labor and mechanical equipment, depending on the size of the operation and the type of animal. Some common methods for collecting livestock waste include manure scrapers, manure forks and shovels, flushing systems, and vacuum systems. A manure scraper is a mechanical device used to scrape manure and other waste materials from the floors of animal housing facilities. The manure is then collected in a pit or storage area and can be further processed or disposed of (US Department of Agriculture, 2009; Parihar et al., 2019).

Manure forks and shovels are typically used to manually remove waste materials from animal housing facilities and transport them to a collection area. This method is commonly used for smaller operations or in areas where mechanical equipment is impractical.

Flushing systems are typically used in dairy operations to flush manure and other waste materials from milking parlors and holding areas. The waste is then collected in a pit or storage area and can be further processed or disposed of.

Vacuum systems are commonly used in poultry operations to collect manure and other waste materials from chicken houses. The waste is transported through a network of pipes and collected in a storage area or transported directly to a processing facility.

7.3.2 TRANSPORTING

Transporting livestock waste can involve various methods depending on the type and volume of waste, the distance it needs to be transported, and the infrastructure available. Some standard methods of transporting livestock waste include spreader trucks, dump trucks or trailers, conveyor systems, and pipeline systems (US EPA, 2002; Watts et al., 2016).

Spreader trucks are typically used to transport liquid or slurry waste from animal housing facilities to storage or application sites. These trucks usually have a tank that can hold large volumes of waste and a pump or spray system to distribute the waste onto fields or other areas. Dump trucks or trailers can transport solid waste, such as manure or bedding materials, from animal housing facilities to storage or application sites. These trucks typically have a large, open bed that can be raised to dump the waste at the desired location. Conveyor systems can transport waste materials, such as manure or feed, within animal housing facilities or from one area of the operation to another. These systems can help reduce the need for manual labor and improve efficiency in waste management. Pipeline systems can transport liquid waste, such as manure or wastewater, from animal housing facilities to storage or treatment sites. These systems can be particularly useful in operations where waste needs to be transported long distances or across rugged terrain.

7.3.3 PROCESSING

Processing livestock waste typically involves one or more methods: composting, anaerobic digestion, drying, and chemical treatment. Composting is a natural process using bacteria and other microorganisms to break organic matter into a stable, nutrient-rich soil amendment (Sanchez et al., 2017; Palaniveloo et al., 2020). Livestock waste, such as manure and bedding materials, can be composted by creating piles or windrows that are turned regularly to promote decomposition. Anaerobic digestion is a process that uses bacteria to break down organic matter in the absence of oxygen, producing biogas that can be used for energy and a nutrient-rich digestate, which can be used as a soil amendment (Ileleji et al., 2015; Roder and Welfle, 2019). Livestock waste can be processed through anaerobic digestion in an enclosed tank or digester.

Drying livestock waste involves removing moisture to reduce weight and volume and make it easier to handle and transport. Various technologies can dry livestock waste, including natural, mechanical, and thermal drying (Tun and Juchelkova, 2019; Purnama et al., 2022). Chemical treatment can stabilize or sterilize livestock waste, making it safe for disposal or reuse. This method typically involves adding chemicals, such as lime or hydrogen peroxide, to the waste to neutralize pathogens and reduce odors.

The processing method choice will depend on the type and volume of waste and the desired end use. For example, composting and anaerobic digestion can produce a valuable agricultural soil amendment, while drying and chemical treatment may be more appropriate for waste disposal. Proper processing of livestock waste is important to prevent environmental contamination, reduce odor and nuisance issues, and promote sustainable and responsible agricultural practices.

7.4 BIOLOGICAL HAZARDS FROM WASTE OF PROCESSING LIVESTOCK PRODUCTS

7.4.1 MILK PROCESSING WASTE

The waste generated during milk processing refers to the by-products, residues, and materials that are not utilized in the final dairy products and are considered waste in the context of milk processing. Milk processing waste can occur at various stages of dairy production, including milk collection, pasteurization, separation, homogenization, and packaging. Some examples of the waste that is produced during the processing of milk are described as follows:

1. **Milk solids and whey:** Milk processing involves separating milk into its components, such as milk solids and liquid whey. The milk solids obtained during processes like cheese or yogurt production are used in the final product. However, excess or lower-grade milk solids and whey can be considered waste if not utilized further.
2. **Cream separation residue:** During the separation of milk to obtain cream, a residue known as skim milk or skimmed milk is produced. While skim milk is often used as a standalone product or as an ingredient in other dairy products, any excess or lower-grade skim milk can be considered waste.
3. **Packaging materials:** Milk processing facilities generate packaging waste such as plastic bottles, cartons, caps, and labels. If not effectively managed, these materials contribute to the overall waste stream.
4. **Cleaning and sanitation waste:** Maintaining hygiene and sanitation in milk processing facilities requires cleaning agents, sanitizers, and other chemicals. The wastewater generated during cleaning processes can hold residues of these substances and may require appropriate treatment to ensure environmental safety.
5. **Spillage and product defects:** During milk processing and packaging, there can be spillage, breakage, or product defects that render the milk unsuitable for sale. These discarded or damaged products can contribute to milk processing waste.

The waste from milk processing can pose biological hazards to human health: Pathogenic bacteria, such as *Salmonella, E. coli, Campylobacter*, and *Listeria*. These bacteria can contaminate the waste if not handled or stored correctly. If the waste comes into contact with other food products or the environment, it can spread bacterial contamination, potentially causing foodborne illnesses if consumed by humans (Marlina et al., 2016).

Spoilage bacteria and molds can also be present in milk processing waste. These microorganisms can grow and proliferate in the waste without proper sanitation and storage practices. Consumption of spoiled milk products or exposure to contaminated waste can lead to gastrointestinal issues or allergic reactions in susceptible individuals (Martin et al., 2021).

Mycotoxins are toxic substances that can contaminate milk and milk products produced by certain molds. If the milk processing waste contains mycotoxin-contaminated materials, there is a risk of mycotoxin presence in the waste. Exposure to mycotoxins through ingestion or inhalation can have adverse health effects, including liver damage, kidney damage, or even carcinogenic effects (Awuchi et al., 2021).

Enzymes and allergens: Milk processing waste may contain residual enzymes used in processing, such as rennet or lactase. Milk processing waste may also contain allergenic proteins, such as lactoglobulin or casein. Individuals with specific allergies or sensitivities may risk experiencing allergic reactions or adverse effects if exposed to these enzymes or allergenic proteins (Khan and Selamoglu, 2020; Geiselhart et al., 2021).

Antibiotic residues: If antibiotics are used in milk production to treat or prevent diseases in dairy animals, there is a potential risk of antibiotic residues in milk processing waste. These residues can occur if milk from treated animals is not adequately separated and discarded. Exposure to antibiotic residues in the waste can contribute to developing antibiotic resistance or trigger adverse reactions in susceptible individuals (Sachi et al., 2019; Rahman et al., 2021).

7.4.2 Meat Processing Waste

The waste generated during meat processing refers to by-products, residues, and materials not used for human consumption. Meat processing waste is generated at all stages of production, including slaughter, butchering, and further processing:

1. **Trimmings and offcuts:** Trimmings and offcuts are portions of meat removed during the butchering process to obtain specific cuts or remove excess fat and connective tissue. Although some trimmings and offcuts may be used for secondary meat products or processed into ingredients like ground meat, excess or lower-quality trimmings are often considered waste (Thies et al., 2021)

2. **Bones and carcass remnants:** After the butchering process, bones and carcass remnants, such as heads, feet, and internal organs, are left behind. These materials are typically not used directly for human consumption and are considered waste. However, they can have value in other applications, such as rendering into tallow or bone meal for various industries (Barbut, 2016; Nastasijevic et al., 2023)

3. **Blood and other bodily fluids:** During the slaughter and processing of animals, blood and other bodily fluids are generated. While some blood may be collected for further processing into blood meals or blood-based products, excess blood, and fluids can be considered waste if not utilized or treated correctly (Verma et al., 2022)

4. **Packaging materials:** Meat processing facilities generate packaging waste such as plastic trays, wrapping materials, labels, and containers. These materials contribute to the overall waste stream if not appropriately managed or recycled (Ncube et al., 2020)

5. **Cleaning and sanitation waste:** Maintaining hygiene and sanitation in meat processing facilities involves using cleaning agents, sanitizers, and other chemicals (FAO, 2005). The wastewater generated during cleaning processes can contain residues of these substances and may require proper treatment to ensure environmental safety.

   The waste generated from meat processing can potentially pose biological hazards to human health from pathogenic bacteria such as *Salmonella, E. coli, Campylobacter*, and *Listeria*, spoilage bacteria and molds, parasites such as *Taenia* spp., *Ascaris* spp., or *Trichinella*.

7.4.3 LEATHER PROCESSING WASTE

The waste generated during leather processing refers to the by-products, residues, and materials that are not utilized in the final leather products and are considered waste in the context of leather processing. Leather processing waste can occur at various stages of the production process, including rawhide preparation, tanning, dyeing, and finishing. The following are some instances of waste produced during the process of leather manufacturing:

1. **Trimmed edges and excess leather:** During the cutting and shaping of leather hides, trimmed edges and excess leather pieces are generated. These materials may not meet the specifications for the final product and are often considered waste.
2. **Fleshing and subcutaneous tissues:** This involves removing flesh and subcutaneous tissues from the hide generates waste. These tissues are typically discarded as waste, although efforts are made to utilize them for other applications, such as rendering for fats or using them as a source of protein for animal feed.
3. **Hair and wool waste:** In the case of hides and skins with hair or wool, the removal of hair or wool during processing produces waste materials. Hair and wool can be collected and recycled in other industries, such as textiles or insulation.
4. **Chemical and wastewater:** Leather processing involves various chemicals, including tanning agents, dyes, and finishing agents. The wastewater generated during the treatment and rinsing processes can contain residues of these chemicals and may require proper treatment before being discharged to ensure environmental safety.

The waste generated from leather processing can potentially pose biological hazards to human health. Some possible health risks are linked with the waste materials generated during the process of leather manufacturing: Pathogenic bacteria, including species like *Staphylococcus aureus* and *E. coli*, fungal growth, such as *Aspergillus*, and *Penicillium*, can lead to respiratory problems, skin irritation, or allergic reactions, and chemical residues, such as tanning agents, dyes, and finishing chemicals (Oruko et al., 2019). Prolonged or excessive exposure to these chemical residues can harm human health, including skin irritation, respiratory issues, and systemic toxicity.

Leather processing generates wastewater containing organic matter, chemicals, and other contaminants (Fei and Liu, 2016; Dowlath et al., 2021; Khan, 2022). If this wastewater is not properly treated before disposal, it can contaminate water sources, leading to environmental and health risks. Pathogens and chemical pollutants from the wastewater can enter the ecosystem and potentially impact human health through contaminated water supplies or agricultural produce. Leather processing waste can involve handling animal hides and skins, which may carry zoonotic diseases. Zoonotic diseases are infectious diseases that can be transmitted between animals and humans. Although the risk is generally low in properly managed leather processing facilities, potential pathogens such as anthrax or certain types of bacteria and viruses may be present in the waste. They could pose a risk if proper precautions are not taken.

Proper waste management practices are crucial for leather processing facilities to minimize the biological hazards they pose. Such practices include implementing good hygiene practices, using PPE (University of Washington, 2022), treating wastewater prior to disposal, and complying with regulations and guidelines set by environmental and health authorities. Proper handling, treatment, and disposal of leather processing waste are essential to reduce risks to human health and the environment.

The leather industry is making strides toward sustainability and waste reduction by promoting the reuse and recycling of waste materials. For instance, leather scraps can be repurposed for smaller leather goods or used as raw materials to produce composite materials. Proper waste management practices, including recycling, reuse, and responsible disposal, are implemented by leather processing facilities to minimize their waste footprint and promote sustainable practices within the industry.

7.5 REDUCING THE IMPACT OF BIOLOGICAL HAZARDS FROM LIVESTOCK WASTE

Efforts to reduce the impact of biological hazards associated with waste, regardless of the specific industry, typically focus on implementing effective waste management practices and adopting preventive measures.

7.5.1 COMPOSTING

Establishing protocols for the safe handling and disposal of waste is crucial, including proper segregation, containment, and waste storage to prevent cross-contamination. Adequate waste disposal methods, such as composting, should be employed to minimize the release of biological hazards into the environment. Composting is an effective method for reducing biological hazards associated with organic waste. Composting is a naturally occurring phenomenon that works under controlled conditions in which air, temperature, and moisture content are regulated for the growth of microorganisms and multiplication through which organic material is converted into a more usable form of organic matter (Chapman, 2005). Composting involves generating heat, especially during the thermophilic phase, where temperatures can reach levels that effectively kill pathogens. Adequate temperatures and retention times help in the inactivation and reduction of pathogenic bacteria, viruses, and parasites present in the organic waste, which reduces the risk of potential infections or diseases associated with these pathogens. The temperature in each composting phase is presented in Figure 7.1. Eliminating or reducing pathogenic livestock waste through composting has been widely carried out (Haug, 1993; Kalbasi et al., 2005; Paluszak et al., 2011; Cancelado et al., 2014; Marlina et al., 2016 Otaki and Kazama, 2018; Subirats et al., 2022).

Composting typically progresses through different phases, each contributing to the elimination of pathogens. The phases of composting are generally categorized as mesophilic, thermophilic, and maturation. Temperature, which determines the degree

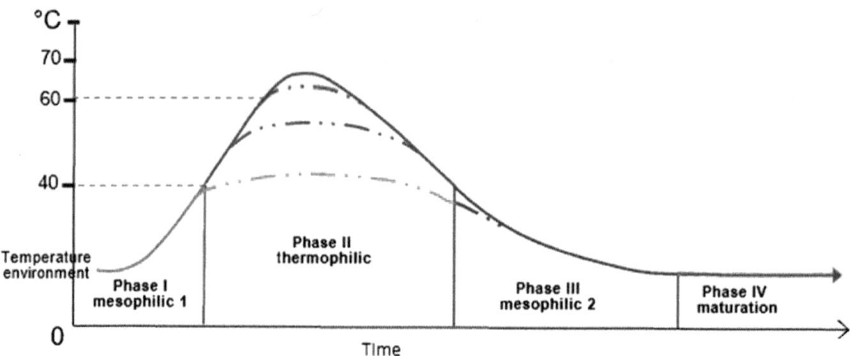

FIGURE 7.1   The temperature in the phases of the composting process.

of decomposition and pathogen inactivation, is the most crucial parameter of the composting process. Exposure to an average temperature of 55–60°C for 1 or 2 days is generally adequate for eliminating nearly all pathogenic viruses, bacteria, fungi, and protozoa (Haug, 1993; Kalbasi et al., 2005).

1. Mesophilic phase:
   - Temperature range: 68–104°F (20–40°C)
   - Duration: The mesophilic phase can last from a few days to a few weeks.
   - Initial breakdown: During this phase, mesophilic microorganisms, including bacteria and fungi, start breaking down readily available organic matter in the compost pile.
   - Pathogen reduction: The mesophilic phase may contribute to some initial reduction of pathogens. However, pathogen reduction during this phase is relatively limited compared to subsequent phases.
2. Thermophilic phase:
   - Temperature range: 122–160°F (50–71°C)
   - Duration: The thermophilic phase typically lasts several weeks.
   - Intense microbial activity: The increased temperatures promote the growth and activity of thermophilic microorganisms, such as heat-tolerant bacteria and actinomycetes.
   - Pathogen inactivation: The high temperatures achieved during the thermophilic phase are crucial in eliminating pathogens. The heat generated by microbial activity helps inactivate and destroy many pathogenic bacteria, viruses, and parasites present in the compost pile.
3. Maturation phase:
   - Temperature: Gradually cools down to ambient temperature
   - Duration: The maturation phase can last several weeks to months.
   - Stabilization and curing: During this phase, the compost pile cools down, and microbial activity slows down. The compost continues to undergo biological and chemical changes, further breaking down organic matter and allowing for the maturation and stabilization of the compost.

- Continued pathogen reduction: While temperatures have decreased, composting reduces pathogen during maturation. The activity of mesophilic microorganisms and natural decay processes further break down any remaining pathogens.

Pathogen elimination during composting occurs primarily because of the combination of high temperatures reached during the thermophilic phase, microbial competition, antagonism, and the passage of time. The elevated temperatures and prolonged exposure to microbial activity help inactivate and destroy many pathogens present in the initial organic waste materials.

It is noteworthy that the duration and effectiveness of pathogen elimination can vary depending on factors such as composting method, waste composition, pile size, management practices, and initial pathogen load. Proper monitoring, management, and adherence to recommended guidelines for composting can help ensure effective pathogen elimination and the production of safe, high-quality compost.

7.5.2 Anaerobic Digestion

Anaerobic digestion has proven to be an effective method for managing animal waste due to its several benefits and positive outcomes. Anaerobic digestion can be integrated into livestock operations as part of a comprehensive waste management system. It offers a sustainable solution for managing large volumes of animal waste, reducing the environmental impact of traditional waste disposal methods. By converting waste into valuable resources such as biogas and nutrient-rich digestate, anaerobic digestion facilitates a circular economy approach to waste management.

Anaerobic digestion can help livestock operations comply with environmental regulations and standards related to waste management. Farmers and producers can show their commitment to responsible waste handling, pathogen reduction, odor control, and environmental sustainability by implementing anaerobic digestion systems.

One of the significant benefits of anaerobic digestion is the production of biogas, which mainly consists of methane and carbon dioxide. This biogas can be captured, treated, and utilized as a renewable energy source. Methane, a potent greenhouse gas, is a valuable resource that can be used for electricity generation, heating, or as a vehicle fuel. Utilizing biogas helps reduce dependence on fossil fuels, mitigates greenhouse gas emissions, and contributes to climate change mitigation efforts.

Anaerobic digestion significantly reduces pathogens present in animal waste. The combination of elevated temperatures, microbial activity, and the anaerobic environment helps inactivate or destroy bacteria, viruses, parasites, and other harmful microorganisms. Studies have shown that anaerobic digestion can substantially reduce pathogen levels, making the end product safer for handling and use.

Animal waste can produce strong odors, which can be a nuisance and cause air pollution in surrounding areas. Anaerobic digestion helps control these odors by

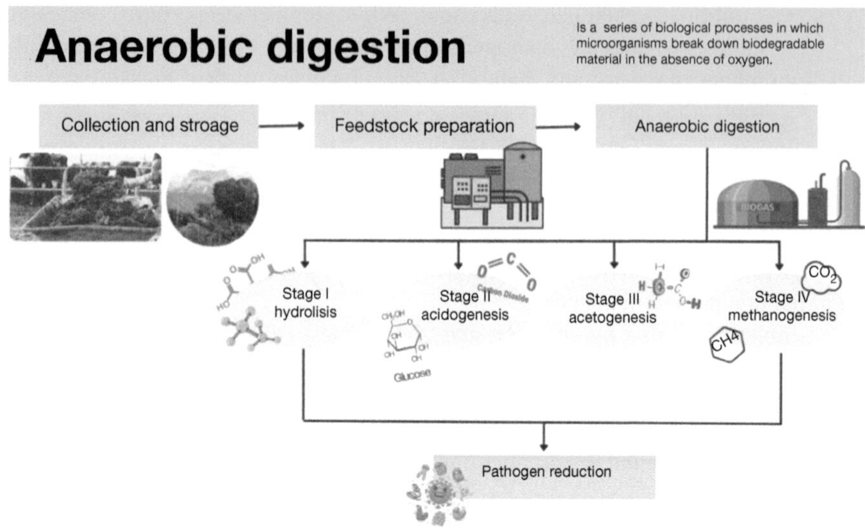

FIGURE 7.2   The stages of anaerobic digestion process.

breaking down organic matter and reducing the release of volatile compounds responsible for foul smells. The process also captures and utilizes biogas, which contains methane, a potent greenhouse gas, and odor-causing component. By capturing methane, anaerobic digestion minimizes the release of odorous compounds, contributing to improved air quality.

Livestock waste is rich in nutrients such as nitrogen and phosphorus. Anaerobic digestion eases the separation and concentration of these nutrients in the digestate. The stabilized digestate can then be utilized as a valuable fertilizer or soil amendment, providing a sustainable method of nutrient recycling. By using the nutrients in the digestate, anaerobic digestion helps reduce the environmental impact of excess nutrient runoff into water bodies, minimizing water pollution and eutrophication.

Each stage in the anaerobic digestion process contributes to reducing or eliminating pathogens in livestock waste (Figure 7.2).

1. **Hydrolysis:** In this initial stage, complex organic compounds in the waste, including proteins, lipids, and carbohydrates, are broken down into simpler molecules by hydrolytic bacteria. Pathogens are often associated with the organic matter in the waste, and as the compounds break down, the pathogens become more exposed and vulnerable to subsequent stages.

2. **Acidogenesis:** Acidogenic bacteria further metabolize the hydrolyzed products, producing VFAs, alcohols, and other compounds. The VFAs, especially short-chain fatty acids, have antimicrobial properties and can inhibit the growth of pathogens. This helps to reduce the population of pathogens in the waste.

3. **Acetogenesis:** Acetogenic bacteria convert the VFAs and alcohols produced during acidogenesis into acetic acid, hydrogen, and carbon dioxide. With its low pH, acetic acid creates an acidic environment that is hostile to many

pathogens. The presence of hydrogen also plays a role in pathogen reduction, as some hydrogen-utilizing microorganisms can outcompete and suppress pathogenic populations.

4. **Methanogenesis:** Methanogenic archaea utilize acetic acid, hydrogen, and other compounds to produce methane and carbon dioxide. The anaerobic conditions within the digester further inhibit the growth of certain pathogenic microorganisms that require oxygen. Additionally, the high temperatures often maintained in the anaerobic digester (around 35–55°C or 95–131°F) help inactivate or kill pathogens. The combined effects of elevated temperature, anaerobic conditions, and the by-products of methanogenesis contribute to significant pathogen reduction.

The retention time of the waste in the anaerobic digester is a crucial factor in pathogen reduction. A longer retention time allows for extended waste exposure to the conditions within the digester, including elevated temperature and microbial activity. This increased exposure enhances the inactivation or destruction of pathogens, further reducing their viability.

After exiting the anaerobic digester, the digested material, called digestate, may undergo additional stabilization processes such as dewatering, composting, or storage. These processes further contribute to pathogen reduction by providing extra time and conditions for the remaining organic matter to decompose and for pathogens to become inactive or die off.

Some pathogens, such as bacteria, viruses, parasites, and helminths, can be reduced through anaerobic digestion (Subirats et al., 2022; Benito et al., 2011; Marlina et al., 2011; Harlia et al., 2009; Benito et al., 2016). Anaerobic digestion can help reduce bacterial pathogens such as *E. coli*, *Salmonella* spp., *Campylobacter* spp., *Clostridium* spp., and *Listeria* spp. (Zhao and Liu, 2019; Avery et al., 2014; Seruga et al., 2020). These bacteria are typically found in livestock waste and can cause diseases in humans and animals. Various viruses can be reduced during anaerobic digestion. This includes enteric viruses such as norovirus, rotavirus, adenovirus, and hepatitis A, which are commonly associated with fecal contamination and can cause gastrointestinal illnesses (Sassi et al., 2018; Roos. 2020). Parasitic pathogens like *Cryptosporidium* spp. and *Giardia* spp. are often found in livestock waste. These parasites can cause gastrointestinal infections in humans. Anaerobic digestion can help inactivate or destroy the infectious stages of these parasites (Zhao and Liu, 2019; Jiang et al., 2020). Certain parasitic worms or helminths, including roundworms (*Ascaris* spp.) and hookworms, can also be reduced through anaerobic digestion. These worms are commonly found in livestock waste, and their eggs can pose health risks if not properly managed (Else et al., 2020).

The effectiveness of pathogen reduction in anaerobic digestion can vary depending on various factors, including the operating conditions of the digester (temperature, retention time, pH, etc.), the initial concentration of pathogens, and the specific characteristics of the waste being treated. Proper system design, operational control, and monitoring are essential to ensure effective pathogen reduction and minimize potential risks associated with the remaining microorganisms.

It is important to remember that while composting and anaerobic digestion can significantly reduce pathogens, it might not remove all microorganisms. Therefore, it is still crucial to handle, store, and use the digestate properly to minimize any potential risks associated with the remaining microorganisms. Consistent monitoring and adherence to the best management practices are imperative to ensure the final product's safety and prevent potential recontamination.

7.6 CONCLUSION

Livestock waste can be hazardous to human health if not managed correctly or if excessive exposure occurs. Farm animals can be carriers of various zoonotic diseases, namely diseases that can be transmitted from animals to humans. Humans can become infected through direct contact with animals or contaminated animal waste. It is important to manage livestock waste carefully to prevent the spread of zoonotic diseases and protect human health.

Livestock waste, such as feces and urine, contains chemicals and nutrients that can pollute water and groundwater if not managed properly. The high content of nutrients such as nitrogen and phosphorus in animal waste can cause the flow of excess nutrients into waters, which can trigger excessive algae growth (eutrophication) and disrupt aquatic ecosystems.

Animal waste can produce harmful gases, such as ammonia (NH_3) and methane (CH_4), that threaten human health and the environment. Ammonia can cause respiratory problems and health issues in humans, while methane is a greenhouse gas contributing to global climate change. In large farms with high livestock populations, the environment can become dangerous for livestock and workers. Exposure to toxic gases, such as hydrogen sulfide (H_2S), and contaminated air particles can cause respiratory and health problems among the workers. The use of antibiotics and hormones in animal husbandry can lead to residues in animal products consumed by humans, which can pose a potential health risk. Overuse of antibiotics in animal husbandry can also contribute to the development of antibiotic resistance, which is a global health problem.

To minimize the health risks caused by livestock waste, adopting effective waste management practices on farms is crucial. This includes proper waste processing, utilization of organic fertilizers, regular monitoring of water quality, and careful utilization of antibiotics and hormones. Furthermore, livestock workers and farmers need to follow strict hygiene and safety guidelines to prevent harmful exposure.

REFERENCES

Alegbeleya, O.O., and A. S. Sant'Ana, 2020. Manure-borne pathogens as an important source of water contamination: An update on the dynamics of pathogen survival/transport as well as practical risk mitigation strategies. *International Journal of Hygiene and Environmental Health*, 227, 113524. doi: 10.1016/j.ijheh.2020.113524

Arvanitoyannis, I. S., and D. Ladas, 2008. Meat waste treatment methods and potential uses. *International Journal of Food Science & Technology*, 43(3), 543–559. doi: 10.1111/j.1365-2621.2006.01492.x.

Avery L. M., K. Y. Anchang, V. Tumsweige, N. Strachan, and P. J. Goude. 2014. Potential for pathogen reduction in anaerobic digestion and biogas generation in Sub-Saharan Africa. *Biomass and Bioenergy, 70*, 112–124. doi: 10.1016/j.biombioe.2014.01.053

Awuchi, C. G., E. N. Ondari, C. U. Ogbonna, A. K. Upadhyay, K. Baran, C. O. R. Okpala, M. Korzeniowska, and R. P. F. Guiné. 2021. Mycotoxins affecting animals, foods, humans, and plants: Types, occurrence, toxicities, action mechanisms, prevention, and detoxification strategies—a revisit. *Foods, 10*, 1279. doi: 10.3390/foods10061279

Barbut, S. 2016. Poultry: Processing. *Encyclopedia of Food and Health* pp. 458–463. doi: 10.1016/B978-0-12-384947-2.00557-2

Benito, A. K., Y. A. Hidayati, E. T. Marlina, and E. Harlia. 2016. The effect of anaerobic fermentation processing of cattle waste for biogas as a renewable energy resources on the number of contaminant microorganism. AIP Conference Proceedings, Vol. 1712, 050017. AIP Publishing. doi: 10.1063/1.4941900

Benito, A. K., Y. A. Hidayati, U. D. Rusdi, and E. T. Marlina. 2011. Detection of total bacteria and total coliform in sludge of biogass mixture formation from cattle and horse faeces. 2010. *Jurnal Ilmiah Ilmu-Ilmu Peternakan*, Februari, 2010, XIII(5), 269–272.

Blaiotta, G., A. D. Cerbo, N. Murru, R. Coppola, and M. Aponte. 2016. Persistence of bacterial indicators and zoonotic pathogens in contaminated cattle wastes. *BMC Microbiology*, 16–87. doi: 10.1186/s12866-016-0705-8

Bonilauri, P. and G. Rugna. 2021. Animal coronaviruses and SARS-COV-2 in animals, what do we actually know? *Lofe (Basel), 11*(2), 123.

Cancelado, S. V., J. C. C. Cepeda, C. C. Fernandez, A. M. Varani, and L. M. Carareto Alves. 2014. Microbiological quality assessment of a compost produced from animal waste and vegetables. Vol. 191, 1469–1479. Department of Technology, FCAV, Universidade Estadual Paulista, Brazil, WIT Press.

Carroll, H. and Underwood, K. 2023. Cattle bedding and food safety. *South Dakota State Extension*. https://extension.sdstate.edu/cattle-bedding-and-food-safety

CDC. 2021. *Guidance for Workers Handling Human Waste or Sewage*. Centers for Disease Control and Prevention, National Center for Emerging and Zoonotic Infectious Diseases (NCEZID), Division of Foodborne, Waterborne, and Environmental Diseases at CDC.

CDC. 2022. *Bird Flu Virus Infections in Humans*. www.cdc.gov/flu/avianflu/avian-in-hum ans.htm

Chapman, D., 2005. *Dead Bird Composting: Final Report for Contract USDA-43-2D81-1-561*. Submitted to USDA/SCS through Soil Conservation Service, Department Animal and Dairy Sciences, Auburn.

Chen, Z., W. Zhang, L. Yang, R. D. Stedtfeld, A. Peng, C. Gu, S. A. Boyd, and H. Li. 2019. Antibiotic resistance genes and bacterial communities in cornfield and pasture soils receiving swine and dairy manures. *Environmental Pollution, 248*, 947–957.

Daiane H., M. R. Garcia, C. D. O. Stopiglia, C. M. Magagnin, T. C. Daboit, G. Vetoratto, J. Schwartz, T. G. Amaro, M. L. Scroferneker. 2015. Dermatophytosis: A 16-year retrospective study in a metropolitan area in southern Brazil. *Journal of Infection in Developing Countries, 9*(8), 865–871. doi: 10.3855/jidc.5479

Dame-Korevaar, A., I. J. M. M. Boumans, A. F. G. Antonis, E. van Klink, E. M. de Olde. 2021. Microbial health hazards of recycling food waste as animal feed. *Future Foods, 4*, 100062. https://doi.org/10.1016/j.fufo.2021.100062

Delahoy, M. J., B. Wodnik, L. McAliley, G. Penakalapati, J. Swarthout, M. C. Freeman, and K. Levy. 2018. Pathogens transmitted in animal feces in low-and middle-income countries. *International Journal of Hygiene and Environmental Health, 221*, 661–676. doi: 10.1016/j.ijheh.2018.03.005

Denham, S. T. and J. C. S. Brown. 2018. Mechanisms of pulmonary escape and dissemination by *Cryptococcus neoformans*. *Journal of Fungi*, 4(1), 25. doi: 10.3390/jof4010025

Dolliver, H., S. Gupta, and S. Noll. *Antibiotic Degradation During Manure Composting*. Technical Report: Organic Compounds in the Environment.

Dowlath M. J. H., S. K. Karuppannan, P. Rajan, S. B. M. Khalith, S. Rajadesingu, and K. D. Arunachalam. 2021. Application of advanced technologies in managing wastes produced by leather industries—an approach toward zero waste technology. *Concepts of Advanced Zero Waste Tools*. Elsevier, pp. 143–179.

Dufour, A., J. Bartram, R. Bos, and V. Gannon. 2012. *Animal Waste, Water Quality, and Human Health*. London-New York: WHO, EPA, IWA Publishing.

Dungan, R. S., M. Klein, and A. B. Leytem. 2012. Quantification of bacterial indicators and zoonotic pathogens in dairy wastewater ponds. *Applied and Environmental Microbiology*, 78(22), 8089–8095.

Else, K.J., J. Keiser, C. V. Holland, R. K. Grencis, D. B. Sattelle, R. T. Fujiwara, L. L. Bueno, S. O. Asaolu, O. A. Sowemim, and P. J. Cooper. 2020. Whipworm and roundworm infections. *Nature Reviews, Disease Primers*, 6, 44. doi: 10.1038/s41572-020-0171-3

FAO. 2005. Code of hygienic practice for meat. CAC/RCP 58-2005. www.fao.org/fao-who-codexalimentarius/sh-proxy/en/?lnk=1&url=https%253A%252F%252Fworkspace.fao.org%252Fsites%252Fcodex%252FStandards%252FCXC%2B58-2005%252FCXP_058e.pdf

Farbu. 2017. Post-polio syndrome. *Reference Module in Neuroscience and Biobehavioral Psychology*. doi: 10.1016/B978-0-12-809324-5.01939-8

Fei, Y. and C. Liu. 2016. Detoxification and resource recovery of chromium-containing wastes. *Environmental Materials and Waste, Resource Recovery and Pollution Prevention*. Academic Press, pp. 265–284. doi: 10.1016/B978-0-12-803837-6.00012-3

Ferraz, P. F. P., G. A. S. Ferraz, L. Leso, M. Klopcic, M. Barbari, and G. Rossi. 2020. Properties of conventional and alternative bedding materials for dairy cattle. *Journal of Dairy Science*, 103(9), 8661–8674. doi: 10.3168/jds.2020-18318

Geiselhart, S., A. Podzhilkova, and K. Hoffmann-Sommergruber. 2021. Cow's milk processing—friend or foe in food allergy? *Foods*, 10, 572. https://doi.org/10.3390/foods10030572

Goldan, E., V. Nedeff, N. Barsan, M. Culea, M. Panainte-Lehadus, E. Mosnegutu, C. Tomozei, D. Chitimus, and O. Irimia. 2023. Assessment of manure compost used as soil amendment—a review. *Processes*, 11, 1167. doi: 10.3390/pr11041167

Harlia, E., Y. Astuti, and D. Suryanto. 2009. The effect of anaerobic fermentation to various animal feces on the egg and infective worm larva in biogas sludge. *Hemera Zoa, the Indonesian Journal of Veterinary Science & Medicine*, 1(1), 17–20.

Haug, R. T. 1993. *The Practical Handbook of Compost Engineering*. Lewis Publishers.

He, Y., Q. Yuan, J. Mathieu, L. Stadler, N. Senehi, R. Sun, and J. J. Alvarez. 2020. Antibiotic resistance genes from livestock waste: Occurrence, dissemination, and treatment. *Nature Partner Journals Clean Water*, 4, 1–11.

Heredia, N. and S. Garcia. 2018. Animal as sources of food-borne pathogens: A review. *Animal Nutrition*, 4, 250–255.

Hicks, T. M. and C. J. R. Verbeek. 2016. *Meat Industry Protein By-products: Sources and Characteristics. Protein Byproducts*. Academic Press, pp. 37–61.

Hill, D. D., W. E. Owens, and P. B. Tchounwou. 2005. Impact of animal waste application on runoff water quality in field experimental plots. *International Journal of Environmental Research and Public Health*, 2(2), 314–321.

Ibekwe, A. M., A. S. Bhattacharjee, D. Phan, D. Ashworth, M. P. Schmidt, S. E. Murinda, A. Obayiuwana, M. A. Murry, G. Schwartz, T. Lundquist, J. Ma, H. Karathia, B. Fanelli, N.

A. Hasan, and C. H. Yang. 2023. Potential reservoirs of antimicrobial resistance in livestock waste and treated wastewater that can be disseminated to agricultural land. *Science of the Total Environment*, 872, 162194. doi: 10.1016/j.scitotenv.2023.162194

Ileleji, K. E., C. Martin, and D. Jones. 2015. Chapter 17—Basics of energy production through anaerobic digestion of livestock manure. *Bioenergy, Biomass to Biofuels*. Academic Press, pp. 287–295. doi: 10.1016/B978-0-12-407909-0.00017-1

Jiang, Y., S. H. Xie, C. Dennehy, P. G. Lawlor, Z. H. Hu, G. X. Wu, X. M. Zhan, and G. E. Gardiner. 2020. Inactivation of pathogens in anaerobic digestion systems for converting biowastes to bioenergy: A review. *Renewable and Sustainable Energy Reviews*, *120*, 109654. doi: 10.1016/j.rser.2019.109654

Kalbasi, A., S. Mukhtar, S. E. Hawkins, and B. W. Auvermann. 2005. Carcass composting for management of farm mortalities: A review. *Compost Science and Utilization*, *13*, 180–193.

Kasorndorkbua, C., D. K. Guenette, F. F. Huang, P. J. Thomas, X.-J. Meng, and P. G. Halbur. 2004. Routes of transmission of swine hepatitis E virus in pigs. *Journal of Clinical Microbiology*, *42*(11), 5047–5052.

Khan, J. 2022. Effect of wastewater from industries on freshwater ecosystem: threats and remedies. *Microbial Consortium and Biotransformation for Pollution Decontamination. Advances in Environmental Pollution Research*. Elsevier, pp. 41–57. https://doi.org/10.1016/B978-0-323-91893-0.00010-9

Khan, M. U. and Z. Selamoglu. 2020. Use of enzymes in dairy industry: A review of current progress. *Razi Vaccine & Serum Research Institute*, 75(1), 131–136. doi: 10.22092/ARI.2019.126286.1341.

Kruithoff, C., A. Gamal, T. S. McCormick, M. A. Ghannoum. 2024. Dermatophyte infections worlwide: Increase in incidence and associated antifungal resistance. *Life*, 14(1), 1. https://doi.org/10.3390/life14010001

Lorimor, J., W. Powers, and A. Sutton. 2004. *Manure Characteristics. Manure Management Systems Series.* MWPS. www.canr.msu.edu/uploads/files/ManureCharacteristicsMWPS-18_1.pdf

Louten, J. 2016. Virus transmission and epidemiology. *Essential Human Virology*, 2016, 71–92.

Mahfooz. S. A., A. Saghir, and A. Ashar. 2006. Composting: A unique solution to animal waste management. *Journal of Agriculture & Social Science* 2(1), 2006. www.fspublisher.org

Manyi-Loh, C. E., S. N. Mamphweli, E. L. Meyer, G. Makaka, M. Simon, and A. I. Okoh. 2016. An overview of control of bacterial pathogen in cattle manure. *International Journal of Environmental Research and Public Health*, *13*(9), 843. doi: 10.3390/ijerph13090843

Marlina, E. T., E. Harlia, and Y. A. Hidayati. 2011. Reduction of coliform through small-scale biogas processing of beef cattle faeces mixed with wood shavings. https://pustaka.unpad.ac.id/archives/81243

Marlina, E. T., T. B. A. Kurnani, D. Z. Badruzzaman, and A. Firman. 2016. Detection of pathogenic bacteria and heavy metal on liquid organic fertilizer from dairy cattle waste. *Proceedings of International Seminar on Livestock Production and Veterinary Technology*. doi: 10.14334/Proc.Intsem.LPVT-2016-p.520-525

Martin, N. H., P. Torres-Frenzel, and M. Wiedmann. 2021. Invited review: Controlling dairy product spoilage to reduce food loss and waste. *Journal of Dairy Science*, *104*, 1251–1261. doi: 10.3168/jds.2020-19130

Mas-Coma, S., V. H. Agramunt, and M. A. Valero. 2013. Direct and indirect affection of the central nervous system by Fasciola infection. *Handbook of Clinical Neurology*. Elsevier, pp. 297–310. doi: 10.1016/B978-0-444-53490-3.00024-8

Misselbrook, T. H. and J. M. Powell. 2005. Influence of bedding material on ammonia emissions from cattle excreta. *Journal of Dairy Science*, 88(12), 4304–4312.

Mittal, J., M. G. Ponce, I. Gendlina, and J. D. Nosanchuk. 2018. *Histoplasma capsulatum*: Mechanisms for pathogenesis. *Fungal Physiology and Immunopathogenesis*, pp 157–191.

Mousavi, B., M. T. Hedayati, N. Hedayati, M. Ilkit, and S. Syedmousavi. 2016. *Aspergillus* species in indoor environments and their possible occupational and public health hazards. *Current Medical Mycology*, 2(1), 36–42.

Moyo, L. B., G. S. Simate, and T. Mutsatsa. 2022. Biological acidification of pig manure using banana peel waste to improve the dissolution of particulate phosphorus: A critical step for maximum phosphorus recovery as struvite. *Heliyon*, 8(8), e10091. https://doi.org/10.1016/j.heliyon.2022.e10091

Nastasijevic, I., M. Boskovic, and M. Glisic. 2023. Chapter 29—Abattoir hygiene. *Present Knowledge in Food Safety*. Academic Press, pp. 412–438. doi: 10.1016/B978-0-12-819 470-6.00002-0

Ncube, L. K., A. U. Ude, E. N. Ogunmuyiwa, R. Zulkifli, and I. N. Beas. 2020. Environmental impact of food packaging materials: A review of contemporary development from conventional plastics to polylactic acid based materials. *Materials*, 13, 4994. doi: 10.3390/ma13214994

Oruko, R. O., J. O. Odiyo and J. N. Edokpayi. 2019. The role of leather microbes in human health. *Role of Microbes in Human Health and Diseases*. Intechopen. doi: 10.5772/intechopen.81125

Oswaldo, P. 2021. *Compost and Composting: The Basics*. https://opanatura.com/compost-and-composting/ August, 2021.

Otaki, M., and Kazama, S. 2018. Fate of pathogens in composting process. *Resource-Oriented Agro-Sanitation Systems*, 61–77. doi: 10.1007/978-4-431-56835-3_5

Oyeyemi, O. T., Oyeyemi, I.T., Adesina, I. A., Tiamiyu, A. M., Oluwafemi, Y. D., Nwuba, R. I. and R. F. Grenfell. 2020. Toxoplasmosis in pregnancy: A neglected bane but a serious threat in Nigeria. *Parasitology*, 147, 127–134. doi: 10.1017/S0031182019001525

Palaniveloo, K., M. A. Amran, N. A. Norhashim, N. M. Fauzi, F. Peng-Hui, L. Hui-Wen, Y. Kai-Lin, L. Jiale, M. G. Chian-Yee, L. Jing-Yi, B. Gunasekaran, and S. A. Razak. 2020. Food waste composting and microbial community structure profiling. *Processes*, 8, 723. www.mdpi.com/journal/processes

Paluszak, Z., J. Bauza-Kaszewska, and K. Skowron. 2011. Effect of animal by-products composting process on the inactivation of indicator bacteria. *Bulletin-Veterinary in Pulawy*, 55(1), 45–49.

Parihar, S. S., K. P. S. Saini, G. P. Lakhani, A. Jain, B. Roy, S. Ghosh, and B. Aharwal. 2019. Livestock waste management: A review. *Journal of Entomology and Zoology Studies*, 7(3), 384–393.

Penakalapati, G., J. Swarthout, M. J. Delahoy, L. McAliley, B. Wodnik, K. Levy, and M. C. Freeman. 2017. Exposure to animal feces and human: A systematic review and proposed research priorities. *Environmental Science Technology*, 51, 11537–11552. http://dx.doi.org/10.1021/acs.est.7b02811

Prasad, R. and K. Stanford, 2019. *Nutrient Content and Composition of Poultry Litter*. www.aces.edu/blog/topics/farming/nutrient-content-and-composition-of-poultry-litter/

Purnama, I., S. P. Sari, and H. Kuncoro. 2022. Comparative study of waste moisture content drying method between hot air blowing method and boiling method in a heated room. *International Journal Science and Technology*, 1(3), 1–8.

Rahman, Md. S., M. M. Hassan, and S. Chowdhury. 2021. Determination of antibiotic residues in milk and assessment of human health risk in Bangladesh. *Heliyon*, 7, e07739 doi: 10.1016/j.heliyon.2021.e07739

Roder, M. and A. Welfle. 2019. Bioenergy. *Managing Global Warming, An Interface of Technology and Human Issues*. Academic Press, pp. 379–398. doi: 10.1016/B978-0-12-814104-5.00012-0

Roos, Y. H. 2020. Water and pathogenic viruses inactivation—food engineering perspectives. *Food Engineering Reviews*, *12*, 251–267. https://doi.org/10.1007/s12393-020-09234-z

Sachi, S., J. Ferdous, M. H. Sikder, S. M., and A. K. Hussani. 2019. Antibiotic residues in milk: Past, present, and future. *Journal of Advanced Veterinary and Animal Research*, *6*(3), 315–332.

Said, M. I. 2020. Livestock waste and its role in the composting process: A review. *The 2nd International Conference of Animal Science and Technology. IOP Conference Series: Earth and Environmental Science*. IOP Publishing, Vol. 492, 012087. doi: 10.1088/1755-1315/492/1/012087

Said, M. I. 2019. Characteristics of by-product and animal waste: A review. *Large Animal Review*, *25*(6), 243–250.

Sanchez, O., D. A. Ospina, and S. Montoya. 2017. Compost supplementation with nutrients and microorganism in composting process. *Waste Management*, 69, pp. 136–153. doi: 10.1016/j.wasman.2017.08.012

Sassi, H. P., L. A. Ikner, S. Abd-Elmaksoud, C. P. Gerba, I. L. Pepper. 2018. Comparative survival of viruses during thermophilic and mesophilic anaerobic digestion. *Science of the Total Environment*, *615*, 15–19. https://doi.org/10.1016/j.scitotenv.2017.09.205

Schiffman, S. S., J. M. Walker, P. Dalton, T. S. Lorig, J. H. Raymer, D. Shusterman, and C. M. Williams. 2004. Potential health effects of odor from animal operations, wastewater treatment, and recycling of byproducts. *Journal of Agromedicine*, *9*(2), 397–403.

Seruga, P., M. Krzywonos, Z. Paluszak, A. Urbanowska, H. Pawlak-Kruczek, Ł. Niedzwiecki, and H. Pinkowska. 2020. Pathogen reduction potential in anaerobic digestion of organic fraction of municipal solid waste and food waste. *Molecules*, *25*, 275. doi: 10.3390/molecules25020275, www.mdpi.com/journal/molecules

Shober, A. L. and R. O. Maguire. 2018. *Manure Management. Reference Module in Earth Systems and Environmental Sciences*. Elsevier Inc.

Silva, J., D. Leite, M. Fernandes, C. Mena, P. A. Gibbs, and P. Teixeira. 2011. *Campylobacter* spp. as a foodborne pathogen. A review. *Frontiers in Microbiology*, *2*, 200. doi: 10.3389%2Ffmicb.2011.00200

Singh, A. K. and W. K. Kim. 2021. Effect of dietary fiber on nutrients utilization and gut health of poultry: A review of challenges and opportunities. *Animals*, *11*, 181. https://doi.org/10.3390/ani11010181

Smith, M. B. and M. R. McGinnis. 2006. Chapter 77—Dermatophytosis. *Tropical Infectious Diseases. Principles, Pathogens, and Practice* (2nd Ed.). Churchill Livingstone, vol. 2, pp. 884–891. doi: 10.1016/B978-0-443-06668-9.50082-X

Sobsey, M. D., L. A. Khatib, V. R. Hill, E. Alocilja, and S. Pillai. 2006. Pathogens in animal wastes and impacts of waste management practices on their survival, transport and fate. In *Animal Agriculture and Environment: National Center for Manure and Animal Waste Management White Papers*. J. M. Rice, D. F. Caldwell, and F. J. Humenik, eds. ASABE, Publication Number 913C0306, pp. 609–666.

Subirats, J., H. Sharpe, and E. Topp. 2022. Fate of *Clostridia* and other spore-forming Firmicute bacteria feedstock anaerobic digestion and aerobic composting. *Journal of Environmental Management*, *309*, 114643. doi: 10.1016/j.envman.2022.114643

Thies, A. J., F. Schneider, and J. Efken. 2021. The meat we do not eat. A survey of meat waste in German hospitality and food service businesses. *Sustainability*, *13*(9), 5059. https://doi.org/10.3390/su13095059

Tun, M. M. and D. Juchelkova. 2019. Drying methods for municipal solid waste quality improvement in the developed and developing countries: A review. *Environmental Engineering Research*, 24(4), 529–542. https://doi.org/10.4491/eer.2018.327

Ucar, A., M. V. Yilmaz, and F. P. Cakiroglu. 2016. Food safety – problems and solutions. *Significance, Prevention and Control of Food Related Diseases*. doi: 10.5772/63176

University of Washington. 2022. Guidelines for personal protective equipment (PPE). *Environmental Health and Safety*. www.ehs.washington.edu

US Department of Agriculture. 2009. *Agricultural Waste Management Field Handbook*, Part 651. Issued August 2009. Natural Resources Conservation Service. https://directives.sc.egov.usda.gov/OpenNonWebContent.aspx?content=31529.wba

US EPA. 2002. Waste transfer stations: A manual for decision-making. United States Environmental Protection Agency, Solid Waste and Emergency Response (5306W) EPA530-R-02-002, June 2002.

Verma, A. K., P. Umaraw, P. Kumar, N. Mehta, and A. Q. Sazili. 2022. Processing of red meat carcasses. *Postharvest and Postmortem Processing of Raw Food Materials*. Woodhead Publishing, pp. 243–280. doi: 10.1016/B978-0-12-818572-8.00002-4

Wan, J., X. Wang, T. Yang, Z. Wei, S. Banerjee, V.-P. Friman, X. Mei, Y. Xu, and Q. Shen. 2021. Livestock manure type affects microbial community composition and assembly during composting. *Frontiers in Microbiology*, 12. doi: 10.3389/fmicb.2021.621126

Watts, P. J., R. J. Davis, O. B. Keane, M. M. Luttrell, R. W. Tucker, R. Stafford, and S. Janke. 2016. *Beef Cattle Feedlots: Design and Construction*. Meat & Livestock Australia. www.mlacom.au

WHO. 2018. Listeriosis. www.who.int/news-room/fact-sheets/detail/listeriosis

WHO. 2020. Campylobacter. www.who.int/news-room/fact-sheets/detail/campylobacter

Wyeth, J. 2013. Hand hygiene and the use of personal protective equipment. *British Journal of Nursing*, 22(16). doi: 10.12968/bjon.2013.22.16.920

Youssef, A. I., and S. Uga. 2014. Review of parasitic zoonoses in Egypt. *Tropical Medicine and Health*, 42(1), pp. 3–14. doi: 10.2149/tmh.2013-23.

Zhao, Q., and Y. Liu. 2019. Is anaerobic digestion a reliable for deactivation of pathogens in biosludge? *Science of the Total Environment*, 668, 893–902. doi: 10.1016/j.scitotenv.2019.03.063

8 Ecological and Human Health Risk Assessments in Water and Soil Samples from Beris Lalang Waste Dumpsite, Kelantan, Malaysia

Florence Rutselin Kelanit, Norli Ismail,
Mohd Hafiidz Jaafar, Nurul Ilyana Sansudin,
Hasmah Abdullah, and Widad Fadhullah

8.1 INTRODUCTION

Most countries have undoubtedly increased goods and services due to population growth, rapid urbanization, and increasing demands (Abdolkhaninezhad et al., 2022). These factors in turn caused an increase in solid waste disposal, which subsequently affected ecosystem services such as food security, water and climate regulation, and soil quality (Vaverková et al., 2019). This mounting solid waste requires further management, which is primarily through landfill as it is low cost and economical. However, this method of waste disposal creates a medium for environmental pollution due to the absence of pipeline installation for leachate management (Ahmad et al., 2021). In the case of open dumping, many environmental issues may occur through the biological and chemical reactions of the waste and during the time of heavy rain, leachate production will be higher. Consequently, this leachate as a carrier of many organic and inorganic pollutions is drained into the groundwater and subsequently into the nearest river, impacting the water quality (Lombardi et al., 2023).

In an open dumpsite, there is the potential to spread pollutants such as heavy metals to humans, animals, and plants as well as natural resources in the surrounding environment through the wind, water, and soil (Figure 8.1). Heavy metals in the air can be inhaled, and the ones in water and soil can be accidentally swallowed or eaten from plants and animals exposed to them (Sulaiman et al., 2020). Certain heavy metals can be toxic even at a minimal concentration such as arsenic, cadmium, nickel, and chromium labeled as class 1 carcinogens by International Agency for Research

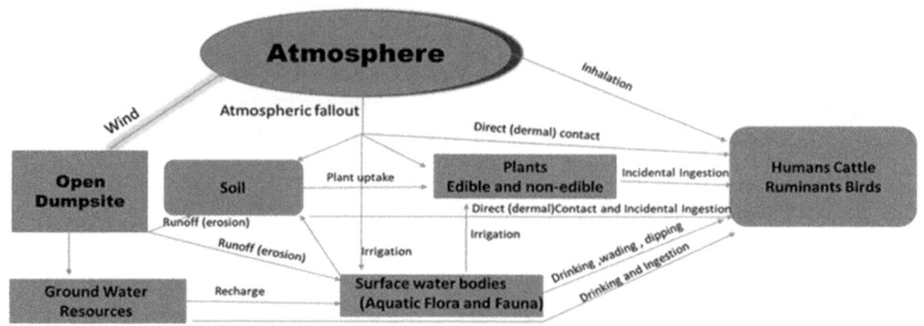

FIGURE 8.1 Illustration of how heavy metals are exposed to humans.

Source: Adapted from Ngole-Jeme and Fantke (2017).

on Cancer (IARC). Heavy metals contained in open waste disposal may be harmful to those exposed through skin with contact to contaminated water or soil. Another exposure pathway could be via accidental dust inhalation or by swallowing the water (Mohammadi et al., 2019).

Previous studies showed evidence of heavy metal pollution in surface water, surface soil, and groundwater near landfills (Boateng et al., 2019; Olagunju et al., 2020; Obiri-Nyarko et al., 2021; Moldovan et al., 2022). In addition, these heavy metals have been associated with potential health risks such as respiratory irritation, cancer, and central nervous system damage due to long-term exposure (Wu et al., 2018).

Owing to the concern of heavy metals to the environmental and human health, many previous studies have focused on risk assessments to address the potential ecological and human health risk of metal pollution in water, soil, and sediment systems (Ekere et al., 2020; Ahmad et al., 2021; Afolabi et al., 2023). Human health risk assessment was conducted to determine the nature and magnitude of adverse health effects in humans who may be exposed to toxic substances in a contaminated environment. Human exposure to heavy metals principally occurs via pathways of drinking water, food, and inhaled aerosol particles and dust (Mohammadi et al., 2019).

Pollution indices are also commonly applied in heavy metal risk assessment studies to determine whether the metal contamination is derived from natural or anthropogenic sources (Moldovan et al., 2022). Heavy metal pollution index (HPI) is a rating method that is used to assess the total effect of different metals on surface-water quality while heavy-metal evaluation index (HEI) shows the overall surface-water quality in terms of metal concentration (Ahirvar et al., 2023). Both single and combined indices are used in assessing the risk of heavy metals such as Index of Geoaccumulation (Igeo) and Potential Ecological Risk Index (PERI), which are found to be useful for soil quality assessment (Ekere et al., 2020; Moldovan et al., 2022).

Many studies were focused on heavy metals in the landfill or open dumpsites integrated with risk assessments and the use of pollution indices, but many did not use multivariate analysis. Additionally, albeit existing studies on heavy metal pollution in the environmental constituents such as groundwater, surface water, soil, and sediments worldwide, systematically gathered quantitative data on heavy metal concentrations, their pollution sources, and risk assessments are reported to be inadequate, particularly in Malaysia. The present study focused on the integrated assessments of heavy metal pollution, risk assessments, pollution indices, and the use of multivariate analysis using Beris Lalang landfill, the largest waste dumpsite in Kelantan, Malaysia as a case study. Although there are plans to provide a leachate treatment plant, it is imperative to monitor the condition of the site as there are informal scavenging activities by local waste workers and rearing of livestock in the vicinity of the landfill. It is, therefore, crucial to assess the potential threat of the site to human health and the environment.

Our previous study by Fadhullah et al. (2019) has reported exceeding acceptable levels of As and Pb in groundwater samples as per Ministry of Health standard and exceeding acceptable levels of Cr, Cu, and Pb in leachates according to Environmental Quality Act (1974) standard. There were also higher concentrations of Ni, Pb, and Zn in the surface water within the vicinity of Beris Lalang waste dumpsite. Thus, further studies that include ecological and human health risk assessments, as well as pollution indices, are needed. Specifically, this present study expanded the study by Fadhullah et al. (2019) by finding the association between heavy metals in water and soil, assessing the potential ecological risk in water and soil, and carcinogenic and non-carcinogenic risk assessments of water and soil from Beris Lalang waste dumpsite. Accordingly, the objectives of the present study were (1) to evaluate the relationship between heavy metals concentration in the water and soil samples, (2) to analyze the heavy metal pollution indices in water and the ecological risk assessment of heavy metals in soil, and (3) to estimate the human health risk assessments of heavy metals via inhalation, ingestion, and dermal exposure route.

8.2 METHODOLOGY

8.2.1 STUDY AREA

The waste dumpsite is in Bachok, Kelantan, a remote area about 500 m close to the residential settlement. The landfilling of waste started in 2010, and the site is still in operation, receiving approximately 350 tons of waste per day. The dumpsite covers an area of 30.5 hectares (305,000 m^2).

Beris Lalang is surrounded by oil palm plantations and there is a water channel that flows to the Gali River that ends at Kandis Beach, Bachok. There are two schools located about 1 km from the dumpsite, namely Beris Lalang Primary School and Sekolah Menengah Ugama Darul Iman. In addition, 7 km from the dumpsite, there are tourist attractions in Tok Bali. There are four areas for water sampling (Table 8.1) consisting of two surface water points (SW1 and SW2) and two groundwater points (GW1 and GW2). Meanwhile, the soil is taken from seven points, including Soil 1, Soil 2, Soil 3, Soil 4, Soil 4, Soil 5, and two samples from soil near the surface of water sampling points (SW1, SW2) (Fadhullah et al., 2019).

TABLE 8.1

Description of the selected sampling points during field samples collection from Beris Lalang waste dumpsite, Bachok, Kelantan

Station	Latitude	Longitude	Description
GW1	N 05°55.780′	E102°24.729′	Groundwater used for ablution by landfill workers and waste collectors
GW2	N 05°55.796′	E102°24.741′	Groundwater occasionally used for a dip by locals
L1	N 05°55.576′	E102°24.915′	Fresh leachates near lorry unloading the waste
L2	N 05°55.615′	E102°24.853′	Passageway of lorry unloading waste
L3	N 05°55.740′	E102°24.673′	Near Kota Bharu Municipal Council (MPKB)'s landfill supervisor office
L4	N 05°55.566′	E102°24.620′	Between the dumpsite's entrance and leachate collection pond
L5	N 05°55.654′	E102°24.640′	Exit entry before flowing to Gali River, a small stream
SW1	N 05°55.571′	E102°24.598′	After leachate discharge to Gali River
SW2	N 05°55.649′	E102°24.635′	Gali River opposite L5.

8.2.2 DATA COLLECTION

This study was a part of another study that previously focused on heavy metals in surface water, groundwater, and leachates which was reported by Fadhullah et al. (2019). However, pollution indices, risk assessments, and multivariate analysis were not estimated. Therefore, this study is an extension of the previous study by integrating risk assessments, pollution indices, and statistical analysis as well as covering the heavy metal concentrations within the soil samples. The data of metals in groundwater and surface water such as arsenic, cadmium, chromium, copper, nickel, lead, and zinc from the study by Fadhullah et al. (2019) were used to estimate the risk. Metal concentration in soil samples is added in this study from Beris Lalang, Bachok, Kelantan dumpsite.

Based on the given data, water, groundwater, and soil samples were analyzed using inductively coupled plasma mass spectrometry (ICPMS) model Agilent 7700. The water samples were filtered with 0.45 µm membrane filter for analysis of metals (Fadhullah et al., 2019). For heavy metal analysis in soil, soil samples were dried and ground into powdered samples. Samples were then digested using microwave digestion (Anton Paar Multiwave 3000) by adding HNO_3, HCl, and H_2O_2. Digested samples were then filtered and analyzed using ICPMS. The percent recovery of all metals was within the range of 92–104%. The calibration curve for the Agilent multi-element standards was accepted with an $R^2 > 0.99$.

8.2.3 ESTIMATION OF POLLUTION INDICES IN WATER SAMPLES

Using the heavy metal data in GW, and SW samples from Fadhullah et al. (2019), HPI and heavy metal assessment index (HEI) were calculated.

8.2.4 Heavy Metal Pollution Index

HPI is frequently used to assess water quality (Othman et al., 2018). It is calculated considering the maximum acceptable concentration of heavy metals. World Health Organization (WHO) standards are inversely proportional to relative weight (Wi) (WHO, 2011). HPI was calculated using Eq. (8.1) (Mohan et al., 1996)

$$HPI = \frac{\sum_{i=1}^{n} W_i Q_i}{\sum_{i=1}^{n} W_i} \qquad (8.1)$$

W_i: unit weight of the ith parameter
Q_i: sub-index of the ith parameter
n: number of parameters considered Q_i determined Eq. (8.2) (Mohan et al., 1996):

$$Q_i = \sum_{i=1}^{n} \frac{M_i - I_i}{S_i - I_i} \times 100 \qquad (8.2)$$

I_i: ideal values
M_i: heavy metal concentration of the ith parameter.

8.2.5 Heavy Metal Assessment Index

HEI was calculated to assess the overall water quality relative to heavy metal (Boateng et al., 2019) using the Eq. (8.3):

$$HEI = \sum_{i}^{n} \frac{H_c}{H_{MAC}} \qquad (8.3)$$

H_{mac}: Monitored value
H_c: Maximum admissible concentration of the ith parameter

8.2.6 Soil Samples' Metal Risk Assessments

Metal data from the soil samples were calculated for Igeo and PERI.

8.2.7 Index of Geo-accumulation

The Igeo of metal ions in soil was calculated using the following formula:

$$\log_2\left(\frac{Cn}{1.5 * Bn}\right)$$

where Cn is the concentration of metal n, and Bn is the background value for the metal n either directly measured in pre-industrialized sediment of the area or taken from literature (Razan et al., 2012). Factor 1.5 was used to minimize the effect of the possible variations in the background values attributed to lithogenic variations in the soil (Muller, 1969). The negative Igeo values found in the table are the results of relatively low levels of contamination for some metal ions, and the background variability factor (1.5) in the Igeo equation. The Igeo for each metal can be classified as uncontaminated (Igeo $\leq$ 0); uncontaminated to moderately contaminated (0 < Igeo $\leq$ 1); moderately contaminated (1 < Igeo $\leq$ 2); moderately to heavily contaminated (2 < Igeo $\leq$ 3); heavily contaminated (3 < Igeo $\leq$ 4); heavily to extremely contaminated (4 < Igeo $\leq$ 5); and extremely contaminated (Igeo $\geq$ 5) (Nikolaidis et al., 2010; Essien et al., 2022).

8.2.8 POTENTIAL ECOLOGICAL RISK INDEX

The evaluation of ecological risk for a given metal was made by adopting the potential risk of individual metal (Er) and PERI (Hakanson, 1980).

The formulas for the two calculations are as follows (Hakanson, 1980):

$$C_f^i = \frac{\bar{C}_D^i}{C_r^i}$$

$$E_r^i = T_r^i \times C_f^i$$

where $\bar{C}_D^i$ represents the mean content of the metal; C_r^i denotes the preindustrial reference value for the metal in the sediment; and T_r^i is the toxic response factor for a given metal. According to Hakanson (1980), (C_f^i) can be classified into four stages of contamination factors which are low contamination factor (C_f^i < 1); moderate contamination factor (1 $\leq C_f^i$ < 3); considerable contamination factor (3 $\leq C_f^i$ < 6); and very high contamination factor ($C_f^i \geq$ 6), while for E_r^i are low potential ecological risk (E_r^i < 40); moderate potential ecological risk (40 $\leq E_r^i$ < 80); considerable potential ecological risk (80 $\leq E_r^i$ < 160); high potential ecological risk(160 $\leq E_r^i$ < 320); and very high ecological risk ($E_r^i \geq$ 320).

8.2.9 HEALTH RISK ASSESSMENT

In the present work, exposure and risk assessments were conducted based on the USEPA methodology. In this study, among seven heavy metals, five heavy metals (Cr, Ni, As, Cd, and Pb) were estimated for their carcinogenic risk. Cu and Zn were not calculated due to lack of data on the slope factor. The values for each parameter in the equation are given in Tables 8.2 and 8.3.

$$ADD_i = \frac{C_i \times IR \times EF \times ED}{BW \times AT}$$

Carcinogenic risks were calculated for As, Cd, Cr, Ni, and Pb using the following equation:

$$Risk = ADD \times SF$$

TABLE 8.2
Reference doses for heavy metal

Heavy metal	Reference doses (mg/kg/day)			Slope factors (mg/kg/day)		
	Inhalation	Ingestion	Dermal	Inhalation	Ingestion	Dermal
Cr	2.86E-05	0.003	0.00006	42.0		
Ni	0.00009	0.02	0.00009	0.84	0.84	
Cu	0.00069	0.04	0.012			
Zn	0.3	0.3	0.06			
As	0.000015	0.0003	0.0009	15.1	1.5	3.66
Cd	0.000015	0.001	0.00001	6.3		
Pb	0.00352	0.000035	0.00052		8.50E-03	

TABLE 8.3
Exposure parameter used for assessment of carcinogenic and non-carcinogenic health risk in water

Parameter CDI	Unit mg/kg/d	Definition	Values of parameter		References
		Chemical daily intake of heavy metal	Children	Adults	
IR	mg/d	Ingestion rate	200	100	USEPA (2002); USDOE
EF	d/year	Exposure frequency	350	350	USEPA (2002)
ED	Year	Exposure duration	6	30	USEPA (2005)
PEF		Particle emission factor			USEPA (2002)
InhR	mg/d	Inhalation rate	7.6	20	USEPA (2011)
BW	kg	Body weight	15	62.65	Malaysia Adult Nutrition Survey (2009); USEPA (2011)
AT	days	Average time (carcinogenic) (ED × 365)	2550	2550	USEPA (2002, 2011)
AT	days	Average time (non-carcinogenic) (ED × 365)	2190	10950	USEPA (2002, 2011)
CF	kg/mg	Units concentration factor			USEPA (2002)
SA	cm^2	Skin surface area available for exposure	3300	2800	USEPA (2011)
AF	mg / cm^2	Adherence factor	0.04	0.02	USEPA (2002, 2011)
ABS	cm^2	Dermal absorption factor	0.13	0.13	USEPA (2011)

Carcinogenic risk to health represents risk value $< 10^{-6}$, while a risk value $> 1 \times 10^{-4}$ suggests a high potential promoting cancer. A range of the risk from 1×10^{-6} to 1×10^{-4} signifies an acceptable risk to human health (Mohammadi et al., 2019). Non-carcinogenic risk will be calculated using the hazard quotient index:

$$Hazard\,Quotient\,(HQ)\,for\,non-carcinogenic\,effect = \frac{ADD}{RfD}$$

$$Hazard\,index\,(HI_i) = \sum HQ_i$$

8.2.10 STATISTICAL ANALYSES

Data were analyzed using the software SPSS version 27.0 (SPSS Inc., Chicago, IL, USA). Pearson's correlation analysis was conducted to determine the relationship among the metals. Multivariate statistical analysis like principal component analysis (PCA) was performed to determine grouping of variables (Field, 2009) and to identify possible sources of the metals (Sulaiman et al., 2020). PCA was an effective technique for presenting beneficial information regarding heavy metal sources on water and soil (Boateng et al., 2019). The values of an eigenvalue >1 were selected. According to Sulaiman et al. (2020), there are three categories utilized for factor loading after rotation: weak (<0.50), moderate (0.50–0.75), or strong (>0.75). The main reason to use the Kaiser–Meyer–Olkim (KMO) measure of sampling adequacy and Bartlett's test of sphericity was to evaluate the data suitability for factor analysis (Tekler et al., 2019).

8.3 RESULTS AND DISCUSSION

8.3.1 RELATIONSHIP BETWEEN METALS IN SOIL AND WATER OF BERIS LALANG WASTE DUMPSITE

The results of the PCA in Beris Lalang water and soil samples area are presented in Figure 8.2. In this study, the usage of KMO is to test the feasibility of PCA. The result of PCA demonstrated an acceptable correlation between variables (KMO = 0.757) and a high dependence ($p < 0.01$). Generally, a p-value lower than 0.05 and a KMO value higher than 0.5 could be accepted as suitable for PCA (Field, 2009). Based on Figure 8.2, Cr, Pb, As, and Cd are clustered together, suggesting these heavy metals are linked to each other and come from uniform and same source. Pb, As, and Cd are commonly found in anthropogenic waste such as batteries, paint, electroplating industries, and manufacturing steel (Gunawardhana et al., 2016; Krasicka-Korczyńska et al., 2017). The presence of Cu far from other clusters may indicate that there is appreciable contamination of copper in the soil and water through leachate migration from the dumping site to other areas (Madzin et al., 2017).

According to Ngole and Ekosse (2012), the widespread usage of Cu in production such as coloring, plating, rinsing, and other industrial uses creates waste containing Cu that ends up in dumpsites. Furthermore, Cu is used in commercial companies to

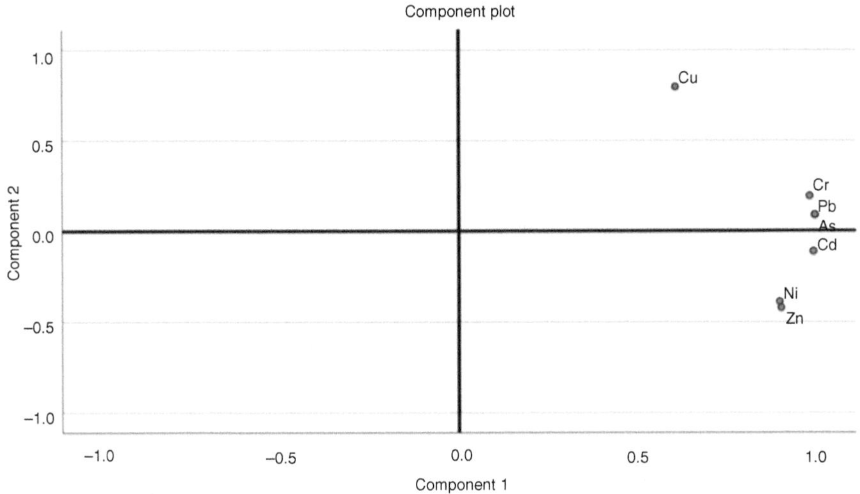

FIGURE 8.2 Component plot between heavy metals in the soil and water of Beris Lalang dumpsite.

produce electronic gadgets. Therefore, Cu tends to accumulate in municipal waste and leachate via disposal of materials from these sources. The presence of Zn and Ni together implies that both could be from the same anthropogenic origin, in particular domestic waste, municipal sewage, atmospheric accumulation, traffic sources, and industrial activities. Both are essential micronutrients for plants and microorganisms (Achary et al., 2016) but may cause non-carcinogenic health hazards such as headache and liver disease, when they exceed the acceptable level (USEPA, 2000).

Table 8.4 shows Pearson's correlation coefficients of studied heavy metals in water and soil in Beris Lalang dumpsite. Cd showed a very strong correlation with several heavy metals at $p < 0.05$, such as Pb–Cd ($r = 0.979$), Cd–As ($r = 0.979$), Cd–Cr (0.952), and Cd–Zn ($r = 0.947$). These correlations suggest that these metals are impacted by similar potential sources. Cd–Pb–As–Cr–Zn originate mostly from electronic waste such as old computers, TVs, and other scrapped electronics from local or international industries that are dumped in public waste landfills. The sources of cadmium that contaminate the aquatic environment include domestic wastewater, atmospheric deposition, and industrial discharges (Ali et al., 2014; Vaccari et al., 2018).

8.3.2 Groundwater and Surface Water Assessment Using Pollution Evaluation Indices, Heavy Metal Pollution Index, and Heavy Metal Evaluation Index

In this study, we computed HPI values for the total heavy metals sampled (Table 8.5) and the HPI value based on sampling points (Figure 8.3). The overall HPI value of sample calculation is approximately 257.257. An HPI value of more than 100 indicates that the groundwater and surface water are categorized as contaminated (Boateng et al., 2019). Most of the studies use comparison results of HPI critical values of

TABLE 8.4

Pearson correlation analysis between metals in water and soil

	Cr	Ni	Cu	Zn	As	Cd	Pb
Cr	1						
Ni	0.731*	1					
Cu	0.530*	0.195	1				
Zn	0.638*	0.714*	0.306	1			
As	0.577*	0.647*	0.259	0.399**	1		
Cd	0.952*	0.922*	0.506	0.947*	0.979*	1	
Pb	0.714*	0.532*	0.547*	0.761*	0.377**	0.979*	1

* Shows significant values of $p < 0.05$ (2-tailed).
** Shows significant values of $p < 0.01$ (2-tailed).

TABLE 8.5

Heavy metal pollution index (HPI) from the groundwater and surface water of Beris Lalang waste dumpsite

Metal samples	Mean conc. ppm (Mi)Vi	Standard-NWQS-Msia (Si)	Unit weightage (Wi)	Sub-index (Qi)	Wi*Qi
Cr	0.02877	0.05	20	57.54	1150.8
Ni	1.2193556	0.05	20	2438.711111	48774.2222
Cu	0.0144667	0.02	50	72.33333333	3616.66667
Zn	1.2598167	5	0.2	25.19633333	5.03926667
As	0.0100667	0.05	20	20.13333333	402.666667
Cd	0.0031	0.01	100	31	3100
Pb	0.0542917	0.05	20	108.5833333	2171.66667
			230.2		59221.0615
			257.2591724		

100 with their HPI value. The water with HPI >100 is harmful (Singh et al., 2017). However, Bodrud-Doza et al. (2016) used the water HPI classification as: excellent (0–25), good (26–50), poor (51–75), very poor (76–100), and unsuitable (100).

HPI was also calculated separately for each sampling location to compare the pollution load and assess the water quality of the selected locations (Figure 8.4). The highest value of HPI was found in sampling point SW1 (HPI of 653.89). The possible cause of water pollution in SW1 is that this location is the immediate receiving area for leachate discharge from the dumpsite. The heavy metals Cr, Ni, Cu, Zn, As, Cd, and Pd were found to have anthropogenic origin and mainly came from leachate discharge through municipal sewage, domestic waste, and chemical weathering of minerals (Singh et al., 2017). While the lowest HPI value was found in GW 1. Lower

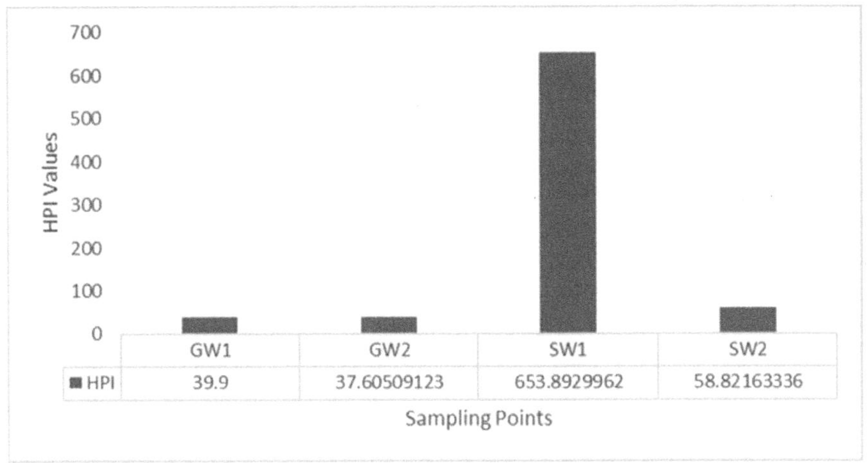

FIGURE 8.3 HPI values of groundwater (GW1 and GW2) and surface water (SW1 and SW2) at different sampling points.

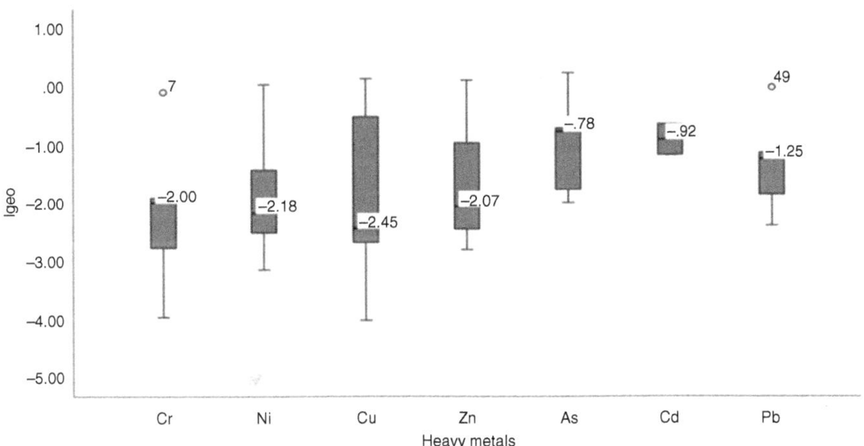

FIGURE 8.4 Geo-accumulation index (Igeo) of soil samples from Beris Lalang waste dumpsite.

HPI in GW compared to SW is related to its distance which was far from the leachate collection and discharge area. Additionally, this GW was dug for the purpose of collecting runoff and overflow from the dumpsite particularly during heavy rain and the monsoon season, Thus, the lowest HPI value may indicate the dilution effect due to seepage or percolation of rainwater (Reza et al., 2011).

In water sample, the HEI value was 27.95381 (Table 8.6). For the classification of overall HEI, there are three categories of low (HEI < 10), medium (HEI = 10–20), and high (HEI > 20) (Bodrud-Doza et al., 2016). According to the classification,

TABLE 8.6
Heavy metal evaluation index (HEI) in water samples from Beris Lalang waste dumpsite

Metal samples	Mean conc. ppm(Mi)Vi	Standard-NWQS-Msia (Si) Hmax	HEI
Cr	0.02877	0.05	0.5754
Ni	1.219355556	0.05	24.38711
Cu	0.014466667	0.02	0.723333
Zn	1.259825	5	0.251965
As	0.010108333	0.05	0.202167
Cd	0.0031	0.01	0.31
Pb	0.075191667	0.05	1.503833
		Total	27.95381

HEI water sample in Beris Lalang waste dumpsite for overall sampling points falls within high metal contamination. By observing the mean concentration of HEI, Ni has the highest HEI over four sampling points, 24.38711, which suggests high contamination in the site by Ni. High HEI value was driven by the exceeding Ni concentration compared to NWQS standard, which in turn caused nickel release into the environment from both natural sources and anthropogenic activity. The high value of Ni present in HEI in water samples at Beris Lalang may be the consequence of accumulation of Ni from weathering of rocks and soils (Cempel and Nikel, 2006). In addition, Ni can be found in food cans, metallic waste, and contaminated organic waste from urban areas because of the vast abundance of Ni in nature (Achary et al., 2016).

8.3.3 Soil Metal Assessments Using Index of Geo-accumulation and Potential Ecological Risk Index

The values of Igeo from all sampling sites are presented in Table 8.7. Igeo values indicated that the soil was not contaminated by Cr, Cd, and Pb (Igeo ≤ 0) and within the range of "no to moderate pollution" by Ni, Cu, As, and Zn (Igeo = 0–1). Similarly, Figure 8.4 illustrates the boxplot of Igeo from all locations combined in soil samples which are uncontaminated (class 0) because Igeo values were less than zero.

The PERI elucidated that Cd has the highest Er and Zn has the lowest. The overall ecological risk value was below 40 (Table 8.8), suggesting low potential ecological risk. The major contributor to the risk index was Cd compared to the other heavy metals such as Cr, Ni, Cu, Zn, As, and Pb that showed low potential ecological risk indices. Cd is released into the ecosystem through natural and human activities. The sources of Cd in landfill may come from industrial waste. In the last two decades, global usage of cadmium has increased due to the use of batteries (Chiamsathit et al., 2020). The total PERI value of <1 suggest low potential ecological risk of the corresponding heavy metals from soil.

TABLE 8.7

Geo-accumulation index (Igeo) of soil samples from seven locations collected

Sample location	Cr	Ni	Cu	Zn	As	Cd	Pb
S1	−1.88203308	−2.27448349	0.12235264	−1.687280787	−0.7691323	−1.183696497	−1.25213
S2	−3.246507211	−3.15119214	−2.82523827	−2.071379052	−2.0044789		−1.49142
S3	−3.963188583	−2.75362657	−2.54761389	−2.313319467	−1.8992412		−1.13384
S4	−1.945999262	−1.57610804	−4.02182257	−2.808891659	−1.6521896		−2.3985
S5	−2.001827068	−1.29222042	−2.44950371	−0.265066576	−0.6721647	−0.655929231	−1.15745
SW1	−2.306982736	−2.17605323	−0.97533593	−2.600786585	−0.7777262		−2.24542
SW2	−0.103853446	0.028656888	−0.07896488	0.100963912	0.22859874		−0.02705

TABLE 8.8

Potential ecological risk index of soil metals from Beris Lalang waste dumpsite

	Cr	Ni	Cu	Zn	As	Cd	Pb
Soil L1	36.6257	6.2006	81.6381	81.5093	13.2023	0.6603	44.0818
Soil L2	14.2245	3.3769	10.5823	62.4572	5.6076	<MDL	37.3445
Soil L3	8.6556	4.4483	12.8278	52.8142	6.0319	<MDL	47.8487
Soil L4	35.0372	10.0616	4.6171	37.4601	7.1585	<MDL	19.9145
Soil L5	33.7073	12.2497	13.7305	218.4421	14.1202	0.9520	47.0720
Soil SW1	27.2422	6.6295	38.1236	43.2031	13.1176	<MDL	22.1129
Soil SW2	31.9956	9.6469	18.8237	99.7018	11.4654	<MDL	29.6998
Mean	26.7840	7.5162	25.7633	85.0840	10.1005	0.8062	35.4392
Preindustrial values C_r^i	90.0	20.00	50.0	175.0	15.0	1.00	70.0
Toxic response factor T_r^i	2.00	5.00	5.00	1.00	10.0	30.0	5.00
C_i^f	0.2976	0.37581	0.515266	0.486194	0.673367	0.806167	0.506274
E_r^i	0.5952	1.8791	2.5763	0.4862	6.7337	24.1850	2.5314
PERI	38.9868						
$C_i^f < 40$	Low						

8.3.4 Health Risk Assessment of Heavy Metals in Soil

8.3.4.1 Non-carcinogenic Risk of Metals for Adults and Children

The highest hazard indices (HI) value across different sampling location was in soil 1 (approximately 0.03), while the lowest value was in soil 3 (0.01) as shown in Figure 8.5a. Despite showing higher HI in children than adults, nevertheless, the values were <1 indicating potential non-carcinogenic risk to both children and adults. In this study, we assessed the human exposure route to heavy metals from the soil around Beris Lalang dumpsite among children and adults from three different pathways. Among the three-exposure route, ingestion (Figure 8.5b) had the highest HI value compared to inhalation and dermal exposure. Children had an HI value of 90 compared to adults (10) via ingestion, indicating a possibility of non-cancer risks as the value of HI > 1 (Praveena and Lin, 2015). There might be a probability of accidental ingestion by both adults and children who spend more than 6 hours every day in waste dumpsites. This non-cancer risk may be associated with health implications including mental retardation, neurocognitive disorders, behavior disorders, respiratory problems, cancer, and cardiovascular diseases (Al Osman et al., 2019).

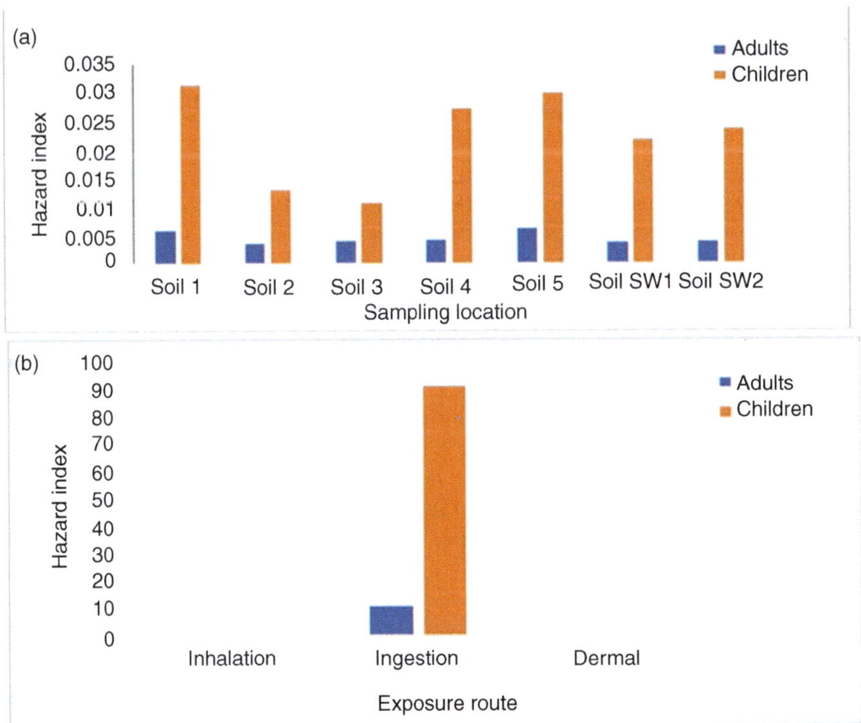

FIGURE 8.5 Hazard index (HI) (a) value for sampling locations and (b) exposure route.

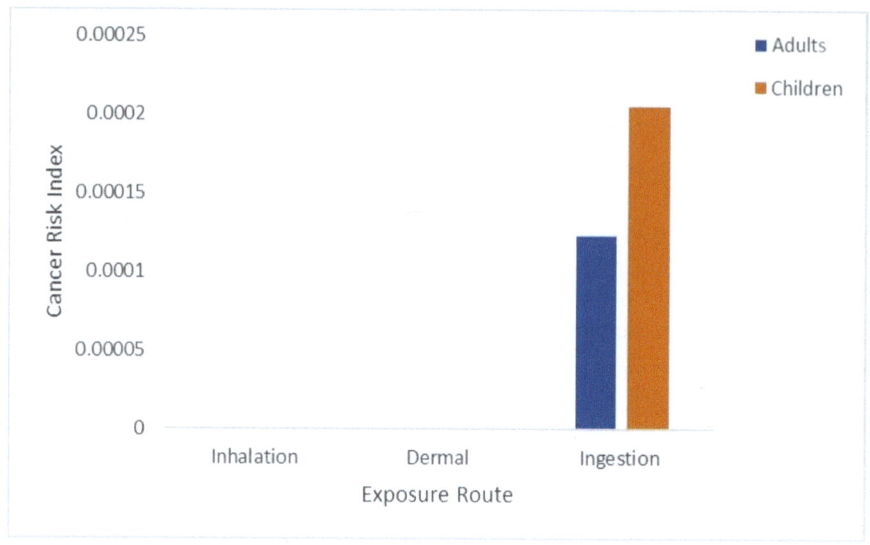

FIGURE 8.6 Three exposure routes to cancer risk among adults and children at Beris Lalang waste dumpsite, Kelantan, Malaysia.

8.3.4.2 Cancer Risk Index from Soil

From the calculated data, cancer risk among adults did not exceed 1×10^{-4}, suggesting that cancer risk is insignificant. Risks surpassing 1×10^{-4} are unacceptable, while risks below 1×10^{-6} are not considered to pose significant health effects (Fryer et al., 2007; Thongyuan et al., 2020)

Total cancer risk among adults from inhalation, ingestion, and dermal exposures was between 10^{-5} and 10^{-6}, indicating an acceptable or tolerable risk (Figure 8.6). However, overall total cancer risk for children exceeds 10^{-4}, showing an unacceptable risk. Ingestion was observed to be the highest exposure route for cancer risk compared to the other two routes: inhalation and dermal. This study points out that if incidental ingestion of soil occurs in children, the children of waste workers have a potential carcinogenic risk. This finding is consistent with a study by Ahmad et al. (2021) stating that soil ingestion is the main route of exposure to children.

8.3.5 Health Risk Assessments of Groundwater and Surface Water in Beris Lalang Waste Dumpsite

8.3.5.1 Hazard Index

According to the result of health risk assessment for groundwater and surface water, the highest values were in GW 2, followed by SW 2 and GW1 in both adults and children (Figure 8.7a). HI for children was higher compared to adults across sampling locations (Figure 8.7b). For the exposure through different pathways, HI value was approximately 0.17, higher for children than for adults, <0.1. However, the total

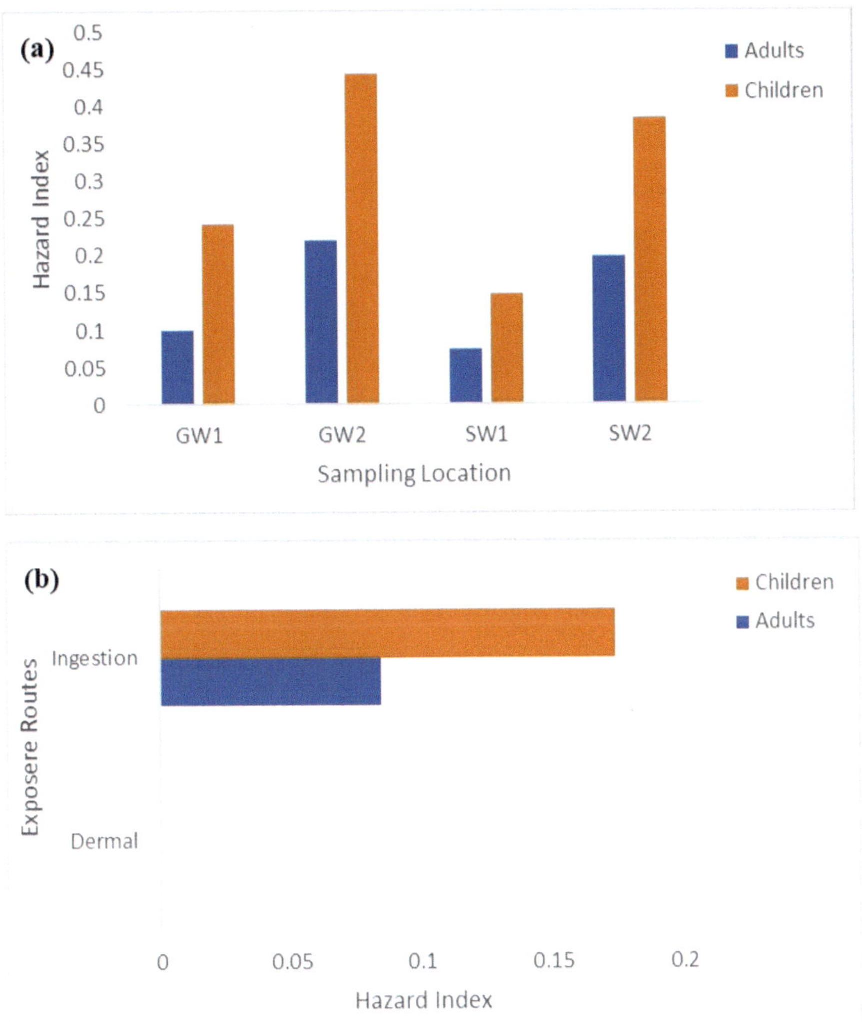

FIGURE 8.7 Hazard index (HI) (a) values for groundwater and surface water from different locations; (b) pathways contribution to the overall HI values.

value does not exceed the value of (1); therefore, there are no serious health effects (Boateng et al., 2019).

8.3.5.2 Cancer Risk Index in Water

In this study, the cancer risk index in water was highest at GW2 and was highest in children across all four sampling locations (Figure 8.8a). In terms of route of exposure, ingestion was higher than dermal exposure. The cancer risk was low in adults, which could not be seen in Figure 8.8. Higher cancer risk exposure in children is exacerbated when informal waste workers brought them together to their workplace

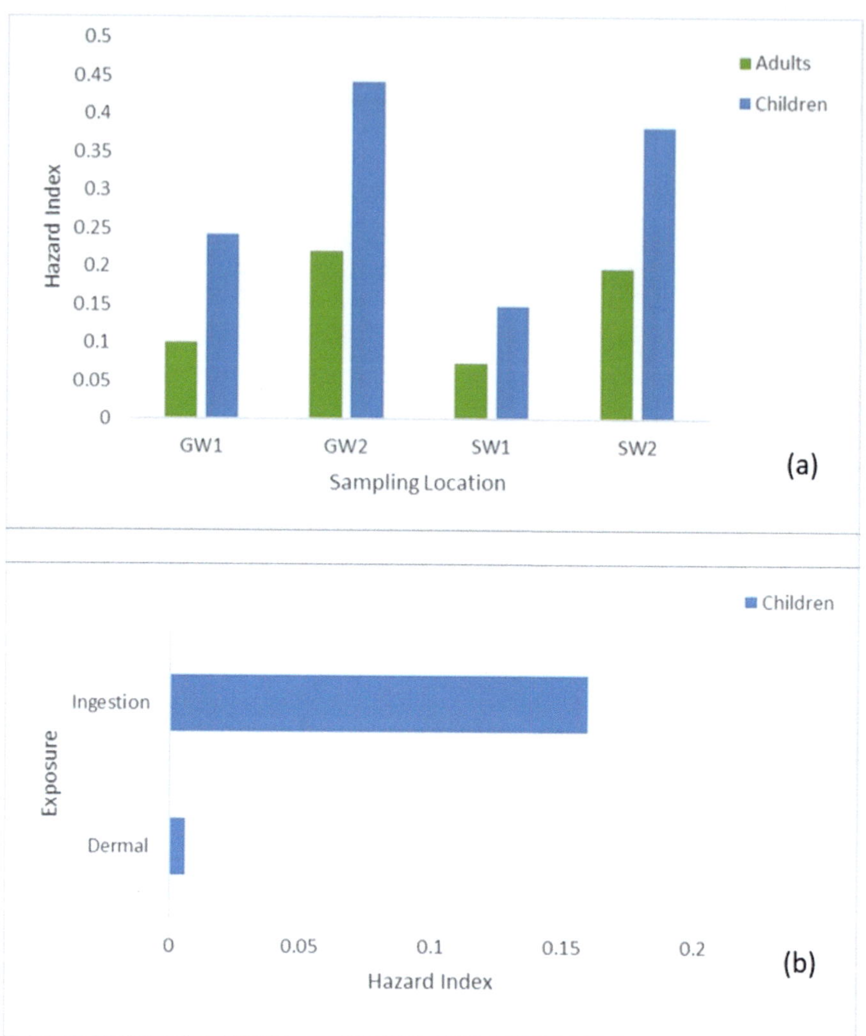

FIGURE 8.8 Cancer risk index value among adults and children from water samples for all metals based on (a) sampling location within the dumpsite and (b) overall HI pathway exposure.

(Thongyuan et al., 2020). Based on our observation, the informal waste workers were also eating and drinking at their worksite, which is unhygienic despite wearing some form of personal protection equipment, such as gloves, boots, or long-sleeve shirts during their working routine.

Ingestion was found as the primary pathway to heavy metal contamination for children (7–12 years old) living near a typical contamination area or who are working in waste dumpsite (Du et al., 2019). Even though based on the result of our study, the cancer risk is still less than the potential cancer risk index threshold of $<10^{-6}$, the need for immediate action before cancer risk increase is of utmost importance.

8.4 RECOMMENDATIONS AND LIMITATIONS OF THE STUDY

This paper highlights integrated assessments of metal pollution in water and soil samples, risk assessments, pollution indices, and the use of statistical analysis. This approach allows a more thorough assessment of the degree of plausible risk to the environment and the humans, in particular the waste workers and its nearby inhabitants. However, this assessment provides the baseline conditions of the existing situation, but it must be taken into account that this study is beyond conclusive and is subject to change temporally due to existing plans to include a proper weigh-bridge and a leachate treatment plant. Nevertheless, the study could be strengthened by including assessments in the air as landfills are also medium ground exposure through landfill gases. The fact that the risk of inhalation is present curtails further assessments in this field.

8.5 CONCLUSION

Open dumpsites present ecological and human risk in relation to exposure to heavy metals. Pearson's correlation and PCA showed that all other heavy metals except copper have a strong positive relationship with one another suggesting various sources of the heavy metals and that of copper. The condition of water in Beris Lalang was found to be unsuitable due to the HP1 > 100 while the soil is classified as not contaminated because PERI is <40. The population around the vicinity of Beris Lalang dumpsite are at low risk of developing cancer and non-cancer health problem as the overall health and cancer-related indices are at a lower level. However, the related risk exposure of heavy metals in children is higher than adults primarily via ingestion. Therefore, workers who work there must take necessary precautions to avoid using water from the area for any purposes. Furthermore, more emphasis should be laid on protecting children from exposure to the dumpsite due to their susceptibility to heavy metal pollution. Local authorities or other competent regulatory agencies should conduct regular monitoring studies to assess the levels of heavy metals and other compounds in the dumpsite to better understand the environmental and health-related risks the waste workers are exposed to.

REFERENCES

Abdolkhaninezhad, T., Monavari, M., Khorasani, N., Robati, M., Farsad, F. (2022). Analysis indicators of health-safety in the risk assessment of landfill with the combined method of fuzzy multi-criteria decision making and bow tie model. *Sustainability*, *14*, 15465. https://doi.org/10.3390/su142215465

Achary, M. S., Panigrahi, S., Satpathy, K. K., Prabhu, R. K., Panigrahy, R. C. (2016). Healthrisk assessment and seasonal distribution of dissolved trace metals in surface waters of Kalpakkam, southwest coast of Bay of Bengal. *Regional Studies in Marine Science*, *6*, 96–108. https://doi.org/10.1016/j.rsma.2016.03.017

Afolabi, O. O., Wali, E., Asomaku, S. O., Yemi-Jonathan, O. I. T., Ogbuehi, N. C., Bosco-Abiahu, L. C., Orji, M. C., Emelu, V. O. (2023). Ecotoxicological and health risk assessment of toxic metals and metalloids burdened soil due to anthropogenic influence. *Environmental Chemistry and Ecotoxicology*, 5, 29–38. ISSN 2590-1826, https://doi.org/10.1016/j.enceco.2022.12.002

Ahirvar, B. P., Das, P., Srivastava, V., Kumar, M. (2023). Perspectives of heavy metal pollution indices for soil, sediment, and water pollution evaluation: An insight. *Total Environment Research Theme*, 6(3), 1000039. https://doi.org/10.1016/j.tot ert.2023.100039

Ahmad, W., Alharthy, R.D., Zubair, M., Ahmed, M., Hameed, A., Rafique, S. (2021). Toxic and heavy metals contamination assessment in soil and water to evaluate human health risk. *Science Reports*, *11*, 17006. https://doi.org/10.1038/s41598-021-94616-4

Ali, S. M., Pervaiz, A., Afzal, B., Hamid, N., Yasmin, A. (2014). Open dumping of municipal solid waste and its hazardous impacts on soil and vegetation diversity at waste dumping sites of Islamabad city. *Journal of King Saud University-Science*, 26(1). https://doi.org/10.1016/j.jksus.2013.08.003

Al Osman, M., Yang, F., Massey, I. Y. (2019). Exposure routes and health effects of heavy metals on children. *BioMetals*, *32*(4), 563–573. https://doi.org/10.1007/s10534-019-00193-5

Boateng, T. K., Opoku, F., Akoto, O. (2019). Heavy metal contamination assessment of groundwater quality: A case study of Oti landfill site, Kumasi. *Applied Water Science, 9*(2), 33. https://doi.org/10.1007/s13201-019-0915-y

Bodrud-Doza, M., Islam, A. R. M. T., Ahmed, F., Das, S., Saha, N., Rahman, M. S. (2016). Characterization of groundwater quality using water evaluation indices, multivariate statistics and geostatistics in central Bangladesh. *Water Science*, *30*(1), 19–40. https://doi.org/10.1016/j.wsj.2016.05.001

Cempel, M., Nikel, G. (2006). Nickel: A review of its sources and environmental toxicology. *Polish Journal of Environmental Studies*, *15*(3), 375–382.

Chiamsathit, C., Charin, P., Thammarakcharoen, S. (2020). Heavy metal pollution index for assessment of seasonal groundwater supply quality in rural area, Kalasin, Thailand. *International Journal of Science 17*(1), 45–60.

Du, S. N. N., Choi, J. A., McCallum, E. S., McLean, A. R., Borowiec, B. G., Balshine, S., Scott, G. R. (2019). Metabolic implications of exposure to wastewater effluent in bluegill sunfish. *Comparative Biochemistry and Physiology Part – C: Toxicology and Pharmacology*, *224*(March), 108562. https://doi.org/10.1016/j.cbpc.2019.108562

Ekere, N. R., Ugbor, M. C. J., Ihedioha, J. N., Ukwueze, N. N., Abugu, H. O. (2020). Ecological and potential health risk assessment of heavy metals in soils and food crops grown in abandoned urban open waste dumpsite. *Journal of Environmental Health Science and Engineering*, *18*(2), 711–721. https://doi.org/10.1007/s40201-020-00497-6

Fadhullah, W., Kamaruddin, M. A., Ismail, N., Sansuddin, N., Abdullah, H. (2019). Characterization of landfill leachates and its impact to groundwater and river water quality: A case study in Beris Lalang waste dumpsite, Kelantan. *Pertanika Journal of Science and Technology*, *27*(2), 633–646.

Field, A. (2009). *Discovering Statistics Using SPSS*. 3rd ed. London: Sage Publications Ltd.

Fryer, S. L., McGee, C. L., Matt, G. E., Riley, E. P., Mattson, S. N. (2007). Evaluation of psychopathological conditions in children with heavy prenatal alcohol exposure. *Pediatrics*, *119*(3). https://doi.org/10.1542/peds.2006-1606

Gunawardhana, W. D. T. M., Jayawardhana, J. M. C. K., Udayakumara, E. P. N. (2016). Impacts of agricultural practices on water quality in Uma Oya Catchment Area in Sri Lanka. *Procedia Food Science*, 6, 339–343. https://doi.org/10.1016/j.profoo.2016.02.068

Hakanson, L. (1980). An ecological risk index for aquatic pollution control. A sedimentological approach. *Water Research, 14*(8), 975–1001. https://doi.org/10.1016/0043-1354(80)90143-8

Krasicka-Korczyńska, E., Dembek, R., Korczyński, M., Stosik, T. A. R. K. (2017). Natural and fodder values of community with Carex Nigra in The Bydgoszcz Canal Valley. *Acta Scientiarum Polonorum. Agricultura, 13*(4), 77–91. https://doi.org/10.15666/aeer/1503

Lombardi, M., Mauro, F., Fargnoli, M., Napoleoni, Q., Berardi, D., Berardi, S. (2023). Occupational risk assessment in landfills: research outcomes from Italy. *Safety, 9*(1), 3. https://doi.org/100.3390/safety90010003

Madzin, Z., Kusin, F. M., Yusof, F. M., Muhammad, S. N. (2017). Assessment of water quality index and heavy metal contamination in active and abandoned iron ore mining sites in Pahang, Malaysia. *MATEC Web of Conferences, 103*, 9. https://doi.org/10.1051/matecconf/201710305010

Malaysian Adult Nutrition Survey. (2009). Current status of food composition data in Malaysia. In *First Australian Food Metrology Symposium*, eds. I. Amin, S. Rusidah, M. K. Norhayati, M. N. Fairulnizal, and A. H. Norliza, Melbourne, Australia, 23 October 2012.

Mohammadi, A. A., Zarei, A., Majidi, S., Ghaderpoury, A., Hashempour, Y., Saghi, M. H., Alinejad, A., Yousefi, M., Hosseingholizadeh, N., Ghaderpoori, M. (2019). Carcinogenic and non-carcinogenic health risk assessment of heavy metals in drinking water of Khorramabad, Iran. *MethodsX, 6*, 1642–1651. https://doi.org/10.1016/j.mex.2019.07.017

Mohan, S. V., Nithila, P., Reddy S. J. (1996). Estimation of heavy metal in drinking water and development of heavy metal pollution index. *Journal of Environmental Science and Health A, 31*(2), 283–289.

Moldovan, A., Török, A. I., Kovacs, E., Cadar, O., Mirea, I. C., Micle, V. (2022). Metal contents and pollution indices assessment of surface water, soil, and sediment from the Aries, River Basin Mining Area, Romania. *Sustainability, 14*, 8024. https://doi.org/10.3390/su14138024

Muller, G. (1969). Index of geoaccumulation in sediments of the Rhine River. *Geology Journal, 2*, 109–118.

Ngole, V. M., Ekosse, G. I. E. (2012). Copper, nickel and zinc contamination in soils within the precincts of mining and landfilling environments. *International Journal of Environmental Science and Technology, 9*(3), 485–494. https://doi.org/10.1007/s13762-012-0055-5

Ngole-Jeme, V. M., Fantke, P. (2017). Ecological and human health risks associated with abandoned gold mine tailings contaminated soil. *PLoS ONE, 12*(2), e0172517. https://doi.org/10.1371/journal.pone.0172517

Nikolaidis, C., Zafiriadis, I., Mathioudakis, V., Constantinidis, T. (2010). Heavy metal pollution associated with an abandoned lead-zinc mine in the Kirki Region, NE Greece. *Bulletin of Environmental Contamination and Toxicology, 85*(3), 307–312. https://doi.org/10.1007/s00128-010-0079-9

Obiri-Nyarko, F., Duah, A. A., Karikari, A. Y., Agyekum, W. A., Manu, E., Tago, R. (2021). Assessment of heavy metal contamination in soils at the Kpone landfill site, Ghana: Implication for ecological and health risk assessment. *Chemosphere, 282*, 131007.

Olagunju, T. E, Olagunju, A. O., Akawu, I. H., Ugokwe C. U. (2020) Quantification and risk assessment of heavy metals in groundwater and soil of residential areas around Awotan Landfill, Ibadan, Southwest-Nigeria. *Journal of Toxicology and Risk Assessment, 6*, 033. https://doi.org/10.23937/2572-4061.1510033

Othman, F., Chowdhury, M.S.U., Wan Jaafar, W.Z., Mohammad Faresh, E.M. Shirazi, S.M. (2018). Assessing risk and sources of heavy metals in a tropical river basin: A case study of the Selangor River, Malaysia. *Polish Journal of Environmental Studies, 27*(4), 1659–1671. https://doi.org/10.15244/pjoes/76309.

Praveena, S. M., Lin, C. L. S. (2015). Assessment of heavy metal in self-caught saltwater fish from Port Dickson coastal water, Malaysia. *Sains Malaysiana, 44*(1), 91–99. https://doi.org/10.17576/jsm-2015-4401-13

Razan, J. I., Islam, R. S., Hasan, R., Hasan, S., Islam, F. (2012). A Comprehensive study of micro-hydropower plant and its potential in Bangladesh. *ISRN Renewable Energy, 2012*, 1–10. https://doi.org/10.5402/2012/635396

Reza, R., Singh, G., Jain, M. (2011). Application of heavy metal pollution index for ground water quality assessment in Angul District of Orissa, India. *International Journal of Research in Chemistry and Environment,* 1(2), 118–122.

Singh, R., Venkatesh, A. S., Syed, T. H., Reddy, A. G. S., Kumar, M., Kurakalva, R. M. (2017). Assessment of potentially toxic trace elements contamination in groundwater resources of the coal mining area of the Korba Coalfield, Central India. *Environmental Earth Sciences, 76*(16). https://doi.org/10.1007/s12665-017-6899-8

Sulaiman, F. R., Mohd Ghazali, F., Ismail, I. (2020). Health risks assessment of metal exposure from a tropical semi-urban soil. *Soil and Sediment Contamination: An International Journal, 29*(1), 14–25. https://doi.org/10.1080/15320383.2019.1669530

Tekler, Z. D., Low, R., Chung, S. Y., Low, J. S. C., Blessing, L. (2019). A waste management behavioural framework of 987 Singapore's Food Manufacturing Industry using Factor Analysis. *Procedia CIRP, 80*, 578–583.

USEPA (2000). *Guidance for Assessing Chemical Contaminant Data for Use in Fish Advisories.* Washington, DC: Office of Water, Office of Science and Technology.

USEPA (2002). *Supplemental Guidance for Developing Soil Screening Levels for Superfund Sites.* Washington, DC: Office of Emergency and Remedial Response. https://rais.ornl.gov/documents/SSG_nonrad_supplemental.pdf

USEPA – U.S. (2005). *Environmental Protection Agency. Guidelines for Carcinogen Risk Assessment.* March. www.epa.gov/sites/production/files/2013-09/documents/cancer_gui delines_final_3-25-05.pdf

USEPA – U.S. (2011). *Environmental Protection Agency. Guidelines for Carcinogen Risk Assessment.* March. www.epa.gov/sites/production/files/2013-09/documents/cancer_gui delines_final_3-25-05.pdf

Vaccari, M., Vinti, G., Tudor, T. (2018). An analysis of the risk posed by leachate from dumpsites in developing countries. *Environments, 5*(9), 99. https://doi.org/10.3390/environments5090099

Vaverková, M. D. (2019). Landfill impacts on the environment – review. *Geosciences, 9*(10), 431. https://doi.org/10.3390/geosciences9100431

Wu, C., Liu, J., Liu, S., Li, W., Yan, L., Shu, M., Zhao, P., Zhou, P., Cao, W. (2018). Assessment of the health risks and odor concentration of volatile compounds from a municipal solid waste landfill in China. *Chemosphere, 202*, 1–8. https://doi.org/10.1016/j.chemosphere.2018.03.068

WHO. (2011). *Guidelines for Drinking-Water Quality.* Geneva, Switzerland.

9 Wealth from Municipal Solid Waste

Ramnarayan Singh and Neha Verma Gour

9.1 INTRODUCTION

Municipal solid waste (MSW) refers to the waste generated by households, commercial establishments, institutions, and other non-industrial sources within a municipality or urban area. It consists of a variety of materials discarded by individuals and businesses, including household waste, construction and demolition (C&D) debris, commercial waste, packaging materials, food waste, street sweepings, and other items. According to the World Bank, the world generates approximately 2.01 billion metric tons of MSW annually, and this is expected to increase by approximately 70% to 3.4 billion metric tons by 2050 [1]. The average global waste generation per capita is around 0.74 kg per day. However, MSW generation per capita and recycling rates can vary greatly between countries [2]. Some countries have established efficient recycling systems, while others struggle to manage waste effectively. Globally, the recycling rate for MSW is estimated to be around only 13.5%. Landfilling is the most common method of MSW disposal worldwide. It is estimated that around 25% 33% of global waste is disposed of in landfills, contributing to environmental and health concerns [1].

In India, the generation of MSW has been increasing rapidly due to population growth, urbanization, and changes in lifestyle. The total generation of MSW in India was estimated to be around 160,038.9 tons per day (TPD) of which 152,749.5 TPD of waste is collected at a collection efficiency of 95.4%. In total, 79,956.3 TPD (50%) of waste is treated and 29,427.2 (18.4%) TPD is landfilled. 50,655.4 TPD which is 31.7% of the total waste generated remains un-accounted [3]. This figure is projected to reach 165 million tonnes per year by 2030. India is the third-largest waste generator in the world, after China and the United States. Only about 80%–85% of the total MSW generated in urban areas is collected, and the collection efficiency in rural areas is relatively low. A significant portion of the collected is either dumped in open landfill sites or disposed of in uncontrolled dumpsites, leading to environmental and health hazards. In earlier years, the percentage of waste treated through proper mechanisms like composting, recycling, and waste-to-energy (WTE) facilities was relatively low. However, with technological advancements and increasing awareness about sustainable waste management practices, this share has also increased significantly

DOI: 10.1201/9781003543176-9

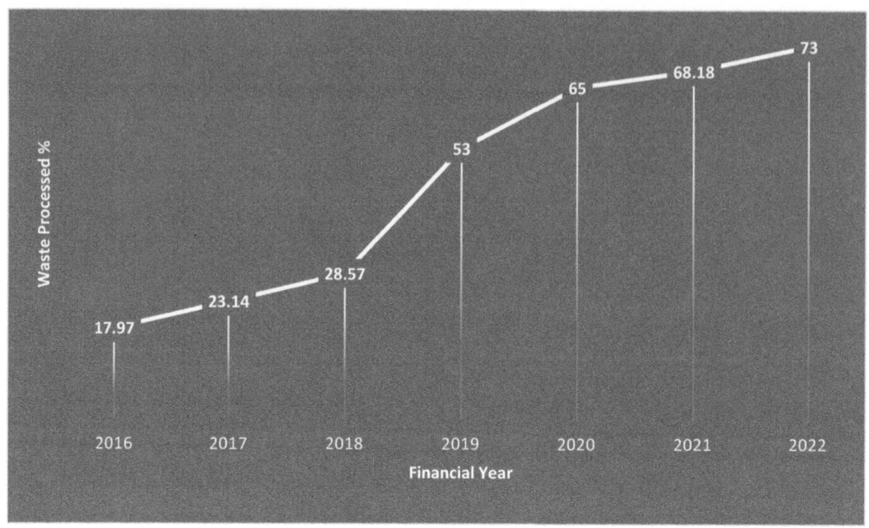

FIGURE 9.1 Year-wise waste processed in India.

Source: Kaza et al. [6].

(Figure 9.1). India generates a substantial amount of plastic waste, with estimates suggesting that around 26,000 tonnes of plastic waste is generated daily [4]. Effective management of plastic waste has become a major challenge for the country. The rapid growth of the electronic industry in India has resulted in a significant increase in electronic waste generation. It is estimated that India generated approximately 3 million tonnes of e-waste in 2018 [5].

Efficient management of MSW is crucial to address the environmental and health impacts associated with waste generation. The government and various stakeholders are working toward improving waste management practices, promoting recycling and WTE projects, and creating awareness among the public to reduce waste generation and promote sustainable waste management practices.

9.2 GENERATION OF MUNICIPAL SOLID WASTE

MSW is generated from a variety of sources, including urban, rural, and commercial domains. Urban areas are the most significant contributors to MSW generation. The urban sources include the waste generated in household, commercial, restaurants, shops, shopping malls, institutes, schools, hospitals, government offices, universities, construction activities, renovation projects, and demolition of buildings. Rural areas also contribute to MSW generation in a lesser extent compared to urban areas. The rural sources of waste include agricultural waste generated from agricultural activities such as crop residues, animal manure, and farm waste. It also includes household waste generated by households in rural areas, kitchen waste, packaging materials, and other similar waste streams. Composition and quantity of the waste generated from these sources may vary depending on the factors such as geographical location,

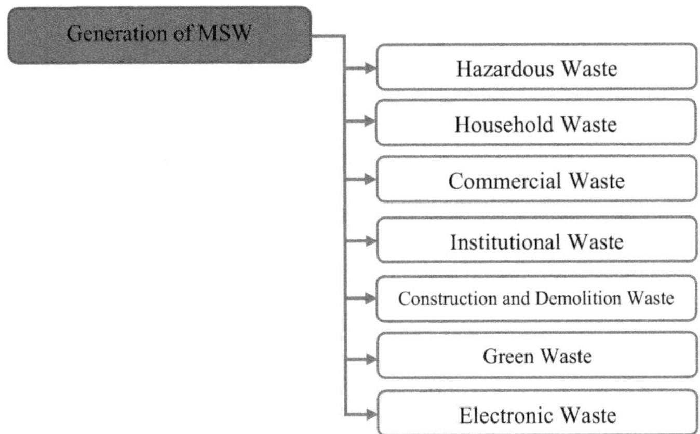

FIGURE 9.2   Sources of MSW generation.

cultural practices, economic activities, and waste management system available in the region. Classification and details of MSW from each domain are presented in Figure 9.2 and explained below.

Hazardous Waste: Hazardous waste refers to any waste material that poses a substantial threat to human health, the environment, or both. It contains substances that are toxic, flammable, corrosive, reactive, or otherwise hazardous in nature. Hazardous waste can originate from various sources, including industrial processes, manufacturing operations, healthcare facilities, laboratories, and households. It includes batteries, chemicals, paint, pesticides, and medical waste.

Household Waste: Household waste refers to the waste generated by individuals and families in their day-to-day activities within their homes. It consists of various materials and items that are discarded after use or no longer needed. This includes food waste like leftover food, spoiled food, peels, and scraps, packaging materials as cardboard boxes, plastic containers, metal cans, and glass bottles, paper, plastics, glass, textiles, and e-waste like discharged batteries, scrapped electronics toys, and gadgets.

Commercial Waste: Commercial waste refers to the waste generated by businesses, industries, offices, and other commercial establishments during their operations. It includes diverse types of waste materials that result from commercial activities. This waste includes office paper, cardboard, packaging materials, restaurant waste, and other materials.

Institutional Waste: Institutional waste refers to the waste generated by various institutions such as schools, hospitals, government offices, universities, and other similar establishments. These institutions produce a significant amount of waste due to their daily operations and activities. It includes paper, food waste, packaging, and other materials. The quantity of waste generated by institutions depends on factors such as the size of the institution, the number of people it serves, and the nature of its operations.

Construction and Demolition Waste: C&D waste refers to the waste generated during construction, renovation, and demolition activities. It includes a wide range of materials such as concrete, wood, metal, bricks, glass, plaster, plastics, and other construction materials.

Green Waste: Green waste refers to organic waste materials that are generated from various sources, predominantly from plants and vegetation. It is commonly known as garden waste or biodegradable waste. Green waste consists of materials such as leaves, grass clippings, branches, tree trimmings, flowers, weeds, and other organic matter derived from gardens, parks, landscapes, and agricultural activities. Agricultural activities generate a significant amount of green waste, such as crop residues, agricultural trimmings, and byproducts like straw or hay. These materials can be recycled or used for composting or bioenergy production.

Electronic Waste (E-waste): This refers to disposal of electronic devices such as computers, printers, copiers, and other electronic equipment, contributing to e-waste. Responsible recycling and proper disposal of e-waste are necessary to prevent environmental pollution and recover valuable resources. E-waste consists of discarded electronic devices such as computers, laptops, mobile phones, televisions, and other electronic equipment.

It is important to manage MSW effectively to minimize its impact on the environment and public health. Proper waste management practices include waste reduction, recycling, composting, and proper disposal methods such as landfilling and WTE processes.

9.3   CLASSIFICATION OF MUNICIPAL SOLID WASTE

In a broad sense, MSW can be classified into two major groups, one is biodegradable or organic and the other is non-biodegradable or inorganic. MSW may vary based on regional regulations and waste management practices. General overview of MSW classification is shown in Figure 9.3; however, specific waste streams and classification may differ depending on local circumstances.

Biodegradable Waste: Biodegradable waste, also known as organic waste, refers to any waste material that can be broken down naturally by biological processes, such as the action of bacteria, fungi, or other microorganisms. These waste materials primarily consist of organic compounds derived from plants, animals, and other living organisms. Biodegradable waste can be found in both residential and commercial settings.

Management of Biodegradable Waste: Proper management of biodegradable waste is imperative for minimizing its impact on the environment and promoting sustainability. Common methods of managing biodegradable waste include:

- Composting: Organic waste can be composted, creating nutrient-rich compost that can be used as a natural fertilizer.

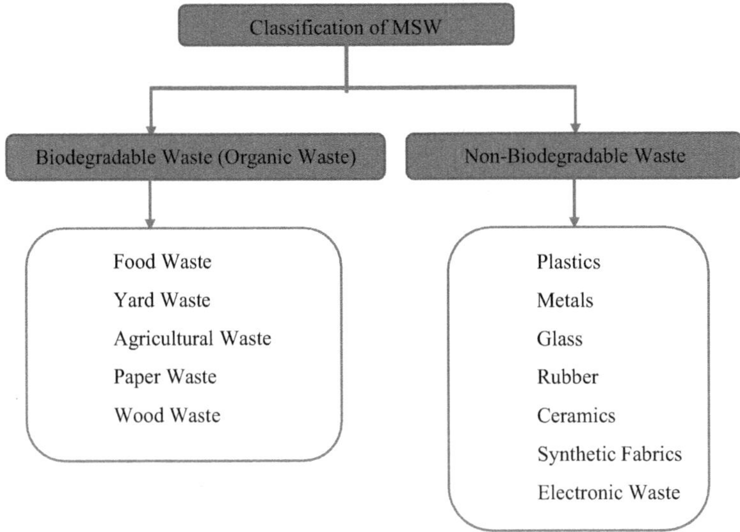

FIGURE 9.3 Classification of MSW.

- Anaerobic Digestion: This process involves the breakdown of organic waste in the absence of oxygen, producing biogas and nutrient-rich digestate.
- Vermicomposting: Using worms to decompose organic waste and produce nutrient-rich vermicompost.
- Environmental Importance: Effective management of biodegradable waste is crucial to minimize the generation of greenhouse gases, reduce landfill space utilization, and promote sustainable waste management practices.

It is important to note that while biodegradable waste can naturally decompose, it is still essential to practice proper waste management to ensure its efficient breakdown and minimize any potential negative environmental impacts.

Non-biodegradable Waste: Non-biodegradable waste refers to any waste material that cannot be broken down or decomposed naturally by biological processes over a relatively short period of time. Unlike biodegradable waste, which can be broken down by microorganisms, non-biodegradable waste persists in the environment for a long time and may cause environmental pollution.

Management of Non-Biodegradable Waste:

Proper waste management is essential to minimize the environmental impact of non-biodegradable waste. Strategies include:

- Recycling: Non-biodegradable materials such as plastics, glass, metals, and electronics can often be recycled and transformed into new products.
- Reusing: Encouraging the reuse of items, such as glass containers or durable products, reduces the need for single-use items and helps minimize non-biodegradable waste.

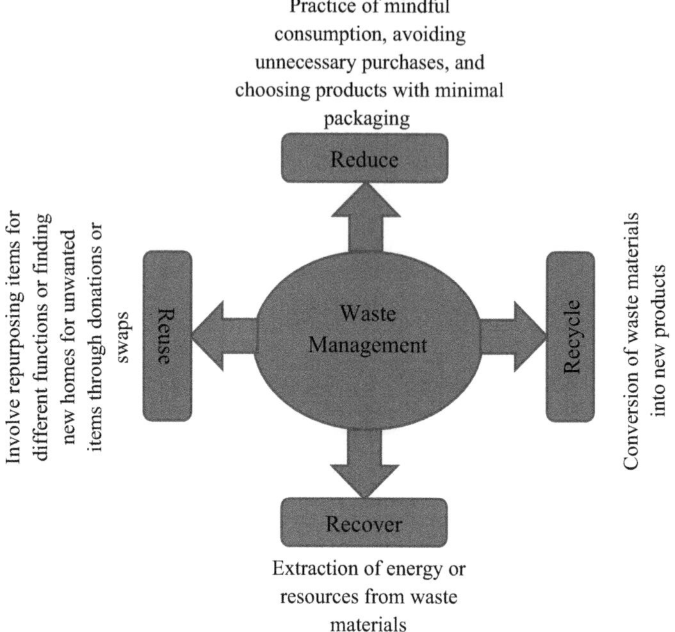

FIGURE 9.4   Waste management techniques.

- Proper Disposal: When recycling or reuse is not feasible, non-biodegradable waste should be appropriately disposed of in designated landfill facilities or through WTE processes.
- Awareness Programs: It is important to raise awareness about the proper management of non-biodegradable waste and promote sustainable practices to minimize its environmental impact and encourage the transition to more eco-friendly alternatives.

Figure 9.4 representing the 4 R's, individuals, businesses, and communities can minimize the amount of waste sent to landfills and contribute to a more sustainable approach to waste management. It is important to prioritize the reduction of waste generation and maximize the reuse, recycling, and recovery of materials to conserve resources, reduce pollution, and mitigate environmental impacts.

9.4   CHARACTERIZATION OF MUNICIPAL SOLID WASTE

Characterization of MSW involves assessing the composition, physical properties, and quantities of waste generated in a specific area. This information helps in developing effective waste management strategies, including recycling, treatment, and disposal. The characterization process involves the following aspects:

- **Waste Composition**: MSW characterization involves determining the types and proportions of different waste materials present in the waste stream. This includes organic waste (food scraps, yard waste), paper, plastics, metals, glass, textiles, wood, and other materials. It helps identify the potential for recycling, resource recovery, or energy generation from specific waste streams.

- **Physical Properties:** Physical properties such as density, moisture content, particle size, and bulk density are assessed. These properties affect waste handling, storage, and transportation requirements, as well as the efficiency of waste treatment and disposal processes.
- **Quantities:** Waste quantities are determined by measuring the weight or volume of waste generated over a specific period. This helps estimate the total waste generation rate and forecast future waste generation trends. It also helps in designing waste management infrastructure and evaluating the effectiveness of waste reduction initiatives.
- **Seasonal Variation:** Waste characterization may consider seasonal variations in waste composition and quantities. For example, during holiday seasons or festivals, there may be an increase in packaging waste or specific types of waste materials. Understanding these variations helps in planning waste management activities and resource allocation accordingly.
- **Hazardous and Special Waste:** Characterization includes identifying hazardous waste or special waste streams within the MSW. These may include electronic waste, batteries, fluorescent bulbs, medical waste, or other materials that require separate handling and disposal due to their potential environmental or health risks.
- **Source Segregation:** Waste characterization also examines the potential for source segregation, where waste is separated at the point of generation. This allows for more efficient recycling and diversion of specific waste streams, reducing the overall volume of waste that requires disposal.

The characterization of MSW is usually conducted through waste audits, sampling, and analysis. Data collected from these assessments is essential for waste management planning, infrastructure development, policy formulation, and implementation of waste reduction and recycling initiatives. It helps in optimizing resource allocation, promoting sustainable practices, and minimizing the environmental impacts of waste disposal.

9.5 CURRENT STATUS OF WASTE MANAGEMENT WORLDWIDE

Waste management practices worldwide demonstrate significant variations across regions and countries. According to the report "WHAT A WASTE 2.0" by The World Bank, the global waste volume is projected to reach a staggering 3.40 billion tonnes by 2050 [1]. This growth is not evenly distributed, with different patterns observed in high-income countries compared to low- and middle-income countries. In high-income countries, daily per capita waste generation is estimated to increase by 19% by 2050. Conversely, low- and middle-income countries are projected to experience a much higher growth rate, with waste generation expected to increase by approximately 40% or more. Particularly in low-income countries, the total quantity of waste generated is predicted to increase by more than three times by 2050.

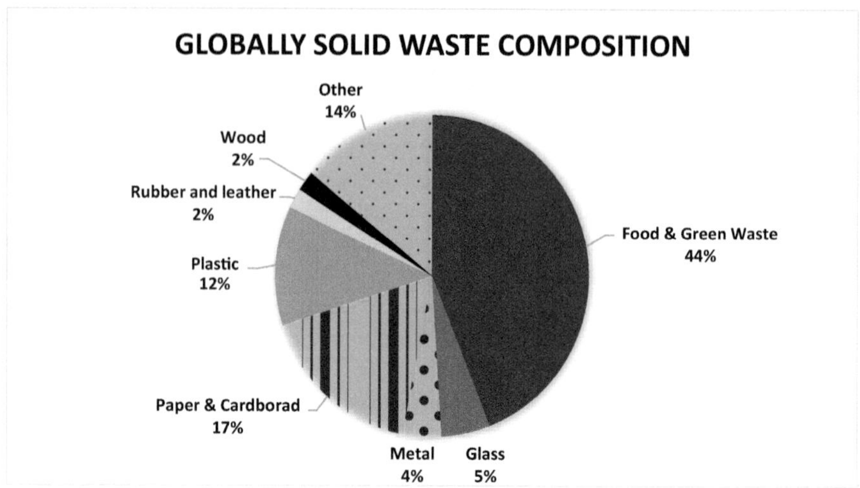

FIGURE 9.5   Global waste composition.

Source: The World Bank [1].

The East Asia and Pacific region currently hold the distinction of being the largest waste generator, accounting for approximately 23% of the global waste. In contrast, the Middle East and North Africa region produce the least amount of waste, contributing only 6% in absolute terms. The regions experiencing the fastest growth in waste generation are sub-Saharan Africa, South Asia, and the Middle East and North Africa. By 2050, these regions are expected to witness waste generation more than tripling, doubling, and doubling, respectively. Some nations have implemented advanced waste management systems, others are still struggling to address the growing waste generation and its environmental impacts. Figures 9.5 and 9.6 represent the global waste composition and waste treatment and disposal scenario, respectively.

The parameters affecting current global waste management scenario are specified as follows:

- **Waste Generation:** The global generation of waste has been steadily increasing due to population growth, urbanization, and changes in consumption patterns. Rapid industrialization and economic development have also contributed to higher waste volumes. MSW, industrial waste, electronic waste, and plastic waste are among the major waste streams of concern.
- **Collection and Disposal:** Many developed countries have well-established waste collection systems, including curbside collection, recycling programs, and landfill facilities. However, in developing regions, particularly in low-income countries, waste collection coverage and infrastructure are often inadequate, leading to uncontrolled dumping, open burning, and environmental pollution.

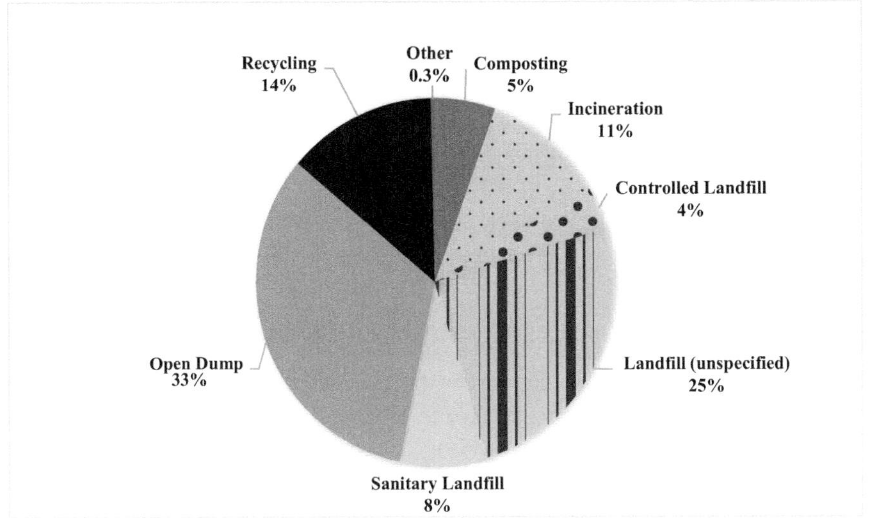

FIGURE 9.6   Waste treatment and disposal global scenario.

Source: The World Bank [1].

- **Recycling and Resource Recovery:** Recycling and resource recovery efforts vary globally. Some countries have implemented effective recycling programs, promoting waste segregation, material recovery, and reuse. Others still face challenges in establishing recycling infrastructure and creating awareness among the public. The recycling rates for different materials, such as paper, plastics, and metals, also vary widely.
- **WTE and Treatment:** WTE technologies, such as incineration and anaerobic digestion, are increasingly being used to recover energy from waste and reduce the volume of landfill-bound waste. Advanced waste treatment methods, including composting and bioremediation, are being adopted to manage organic waste and promote circular economy principles.
- **Plastic Waste Crisis:** The global plastic waste crisis has gained significant attention in recent years. Plastic pollution poses a major environmental challenge, with vast amounts of plastic waste entering oceans and ecosystems. Several countries have implemented measures to reduce single-use plastics, promote plastic recycling, and encourage alternatives to plastic packaging.
- **Circular Economy Approaches:** The concept of a circular economy, which aims to minimize waste generation, promote recycling and reuse, and prioritize sustainable resource management, is gaining momentum worldwide. Circular economy principles aim to maximize resource efficiency, reduce environmental impacts, and create economic opportunities through innovative waste management practices.
- **Policy and Legislation:** Waste management policies and regulations vary across countries. Some nations have comprehensive waste management frameworks

with strict regulations, waste reduction targets, and extended producer responsibility schemes. Others are in the process of developing or strengthening their waste management policies to address environmental concerns and promote sustainable practices.

- **Global Initiatives:** Numerous international organizations and initiatives are working toward improving waste management worldwide. The United Nations Environment Programme (UNEP), World Bank, and other organizations provide support and guidance to countries in implementing sustainable waste management practices, capacity building, and knowledge sharing.

9.6   CHALLENGES IN WASTE MANAGEMENT IN INDIA

Waste management in India is a complex and challenging issue due to the country's large population, rapid urbanization, and increasing consumption patterns. The management of waste involves collection, transportation, treatment, and disposal of various types of waste to minimize its environmental and health impacts. MSW management in India is primarily the responsibility of urban local bodies (ULBs) or municipal corporations. However, the existing waste management infrastructure and systems are often inadequate, leading to issues such as improper waste disposal, open dumping, and inadequate waste processing. The Government of India launched the "Swachh Bharat Mission" in 2014, aiming to achieve universal sanitation coverage and proper waste management practices across the country. The mission emphasizes the construction of toilets, solid waste management, and behavior change campaigns to promote cleanliness and hygiene. The segregation of waste into biodegradable, non-biodegradable, recyclable, and hazardous waste streams allows for appropriate treatment and disposal. However, waste segregation practices are still limited in many areas, and the collection of segregated waste is not consistently implemented. India primarily relies on the use of landfill sites for waste disposal, which can lead to environmental pollution and health hazards if not properly managed. WTE plants and composting facilities are also being implemented to convert waste into energy or organic fertilizer, but their adoption is still limited. The informal sector, including waste pickers and recyclers, plays a significant role in waste management in India. These individuals often collect recyclable materials from waste streams, contributing to resource recovery and recycling. However, they encounter challenges such as lack of recognition, low income, and unsafe working conditions. However, efforts are being made to address these challenges through policy reforms, technology adoption, public–private partnerships, and community participation.

To improve waste management in India, there is a need for comprehensive waste management policies, increased investment in infrastructure and technology, public awareness campaigns, and strengthened collaboration between government, private sector, and communities. Sustainable waste management practices (Figure 9.7), including waste reduction, recycling, and decentralized waste treatment options, can help minimize the environmental impact of waste and promote a circular economy approach.

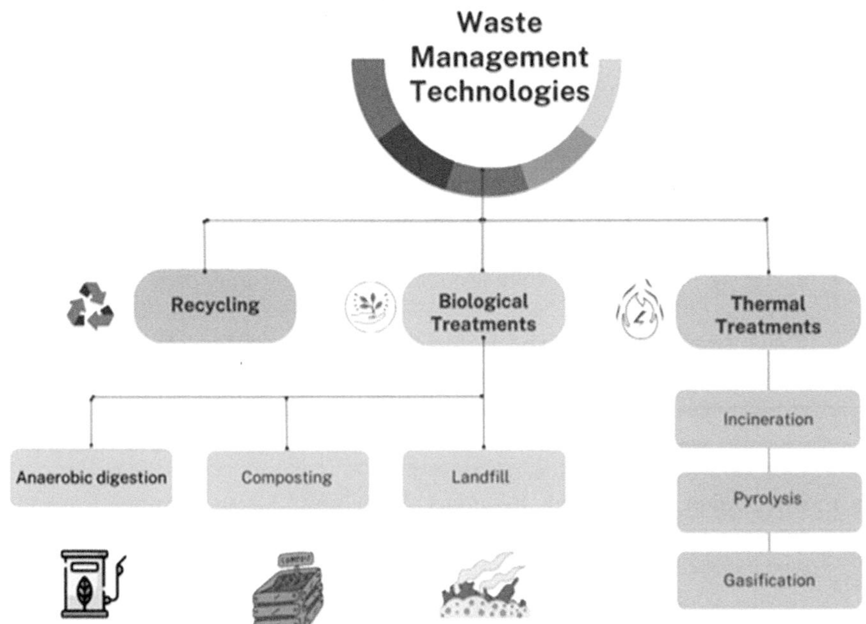

FIGURE 9.7 Waste management technologies.

Waste management in India faces several challenges, including:

- **Rapid Urbanization and Population Growth:** The rapid urbanization and population growth in India have led to increased waste generation. MSW management systems often struggle to keep up with the rising volumes of waste, resulting in inadequate collection and disposal infrastructure.
- **Lack of Infrastructure and Resources:** Several cities and towns in India lack proper waste management infrastructure, including waste collection systems, processing facilities, and sanitary landfills. Limited financial resources and inadequate investments in waste management further worsen the problem.
- **Inefficient Collection and Segregation:** Waste collection systems in India often suffer from inefficiencies, with irregular collection schedules and inadequate coverage. Proper segregation of waste at source is also a significant challenge, leading to mixed waste streams that are difficult to manage and recycle.
- **Open Dumping and Improper Disposal:** Open dumping of waste is still prevalent in many parts of India, leading to environmental pollution, health hazards, and land degradation. Additionally, improper disposal of hazardous waste poses risks to human health and the environment.

- **Limited Awareness and Behavioral Change:** Awareness about waste management practices, including waste segregation and recycling, is relatively low among the general population. Lack of public participation and resistance to behavioral change hinder the effectiveness of waste management initiatives.
- **Informal Sector Involvement:** The informal sector, including waste pickers and recyclers, plays a significant role in waste management in India. However, their activities are often unregulated, and they face challenges such as low income, lack of protective gear, and limited access to formal waste management systems.
- **E-Waste Management:** The rapid growth of electronic waste (e-waste) presents a specific challenge. E-waste contains hazardous substances and requires specialized recycling facilities and processes. India faces challenges in managing the increasing volumes of e-waste and ensuring proper disposal and recycling.
- **Policy Implementation and Enforcement:** Although India has implemented waste management policies and regulations, effective implementation and enforcement remain challenging. Coordination among various government agencies and stakeholders is crucial to ensure compliance and the smooth functioning of waste management systems.
- **Integration of Circular Economy Principles:** Shifting toward a circular economy approach, which emphasizes waste reduction, recycling, and resource recovery, requires significant infrastructure and policy changes. Incorporating circular economy principles into waste management practices is a complex challenge that requires systemic transformation.

Addressing these challenges requires a multi-faceted approach, including improved infrastructure, public awareness campaigns, capacity building, policy reforms, and stakeholder collaboration. It is essential to promote sustainable waste management practices, encourage public participation, and invest in innovative technologies and infrastructure to overcome the waste management challenges in India.

9.7   OPPORTUNITIES IN WASTE MANAGEMENT IN INDIA

Waste management in India presents several opportunities for innovation, investment, and sustainable development.

- **Recycling and Resource Recovery:** There is significant potential for expanding recycling and resource recovery initiatives in India. Establishing efficient recycling facilities for various waste streams such as plastics, paper, glass, and metals can help reduce the environmental impact of waste and create a circular economy.
- **Organic Waste Management:** Organic waste, including food waste and agricultural residues, provides opportunities for composting, biogas production, and organic fertilizer production. Proper management of organic waste can contribute to soil health, reduce greenhouse gas emissions, and generate renewable energy.

- **Waste-to-Energy:** The conversion of waste into energy, such as biogas or electricity, presents an opportunity to harness the potential of waste as a renewable energy source. Technologies such as anaerobic digestion and waste incineration can be explored for energy recovery from different waste streams.
- **E-Waste Management:** The growing e-waste sector offers opportunities for setting up recycling and refurbishment facilities. Proper handling, dismantling, and recycling of electronic waste can help recover valuable materials while minimizing the environmental and health risks associated with improper disposal.
- **Waste Management Technologies:** Innovative waste management technologies, such as waste segregation systems, sensor-based sorting, and waste-to-product conversion technologies, can improve the efficiency and effectiveness of waste management operations.
- **Public–Private Partnerships:** Collaborations between the public and private sectors can drive investments, expertise, and technology transfer in waste management. Public–private partnerships can enable the development of integrated waste management systems and encourage sustainable practices.
- **Waste Awareness and Education:** There is a need for increased awareness and education among the general public, businesses, and institutions regarding proper waste management practices. Opportunities exist for organizations to offer waste management training, awareness campaigns, and consultancy services.
- **Job Creation and Social Entrepreneurship:** The waste management sector has the potential to generate employment opportunities, particularly in the informal sector. Supporting social entrepreneurship and small-scale waste management enterprises can contribute to job creation and inclusive economic growth.
- **Policy and Regulatory Support:** The Indian government has implemented policies and regulations to promote sustainable waste management. Opportunities exist for policymakers to further strengthen and enforce regulations, provide incentives for investments in waste management, and support research and development initiatives.

By capitalizing on these opportunities, India can transform its waste management sector, promote environmental sustainability, generate renewable energy, conserve resources, and create a circular economy that contributes to both economic growth and social welfare.

There are various waste management technologies used in different industries to efficiently manage and treat diverse types of waste [7]. Some of the commonly employed waste management technologies include the following:

- **Controlled Landfilling:** Landfilling is a widely used waste management technology where solid waste is disposed of in designated landfill sites. The waste is compacted and covered with soil to minimize environmental impacts. Landfill gas can also be collected and utilized for energy generation.

- **Incineration:** Incineration involves the combustion of solid waste at high temperatures. This technology is used for the treatment of MSW and certain types of hazardous waste. Energy recovery from incineration can be achieved through the generation of steam or electricity [8].
- **Composting:** Composting is a biological process that converts organic waste into compost, a nutrient-rich soil amendment. It is commonly used for the treatment of organic waste, such as food waste, yard waste, and agricultural residues. Composting helps in reducing waste volume and producing a valuable end product.
- **Anaerobic Digestion:** Anaerobic digestion is a process that converts organic waste into biogas and digestate through the action of microorganisms in the absence of oxygen. Biogas can be used for energy generation, and the digestate can be used as a fertilizer.
- **Mechanical Biological Treatment (MBT):** MBT is a combination of mechanical and biological processes used for the treatment of mixed waste. It involves sorting, shredding, composting, and other treatment steps to recover recyclable materials and produce a stabilized waste residue.
- **Pyrolysis and Gasification:** Pyrolysis and gasification are thermal treatment technologies that convert solid waste into syngas, bio-oil, and char through the application of heat in the absence of oxygen. These technologies are used for the treatment of various waste streams, including plastics, biomass, and certain types of hazardous waste.
- **Waste-to-Energy:** WTE technologies encompass various processes, such as incineration, gasification, and anaerobic digestion, which convert waste into energy in the form of electricity, heat, or fuel. WTE technologies can help reduce the reliance on fossil fuels and generate renewable energy.
- **Recycling and Material Recovery:** Recycling technologies involve the sorting, separation, and processing of recyclable materials, such as paper, plastics, metals, and glass, to produce secondary raw materials. These materials can be used in the manufacturing of new products, reducing the need for virgin resources.
- **Chemical Treatment:** Chemical treatment technologies, such as chemical precipitation, coagulation, and neutralization, are used for the treatment of specific types of hazardous waste and industrial wastewater.
- **Advanced Technologies:** Advancements in waste management technologies include innovative processes such as plasma gasification, microwave treatment, and electrochemical methods. These technologies are still in the experimental or early stages of implementation but show promise for the future.

The choice of waste management technology depends on factors such as the type of waste, its characteristics, regulatory requirements, environmental considerations, and economic feasibility. A combination of these technologies is often used in integrated waste management systems to maximize resource recovery and minimize environmental impacts. Energy recovery from MSW offers several benefits, including:

- **Reduction in landfill space requirements:** By diverting waste from landfills and extracting energy from it, the volume of waste that needs to be disposed of in landfills is significantly reduced.
- **Renewable energy generation:** The energy recovered from MSW is considered renewable as it is derived from organic and combustible waste materials.
- **Reduction in greenhouse gas emissions:** Energy recovery from MSW helps reduce methane emissions from landfills, which is a potent greenhouse gas contributing to climate change.
- **Resource conservation:** By using the energy potential of waste, energy recovery reduces the reliance on fossil fuels for electricity generation and conserves natural resources.
- **Waste management integration:** Energy recovery can be integrated into comprehensive waste management systems, where the focus is on reducing, reusing, recycling, and recovering energy from waste, promoting a more sustainable approach to waste management.
- It is important to note that proper management of emissions and environmental controls are necessary in energy recovery facilities to minimize any potential environmental impacts [9]. Additionally, waste reduction, recycling, and other sustainable waste management practices should be prioritized.

9.8 PLANTS WITH WASTE-TO-ENERGY IN INDIA

India has been actively working toward WTE initiatives to address the growing waste management challenges and harness the potential of waste as a renewable energy source. These WTE plants in India are part of the country's efforts to address waste management challenges, promote sustainable practices, and utilize waste as a valuable resource for energy generation.

Following are the notable WTE plants in India:

Okhla WTE Plant, Delhi: The Okhla WTE plant is located in Delhi and is one of the largest WTE facilities in India. It processes MSW and converts it into electricity, reducing the burden on landfill sites. Handling capacity of this plant is 1950 metric tonnes of MSW per day and it is able to generate 23 MW of green energy. Timarpur-Okhla Waste Management Company Limited (TOWMCL) scientifically handles MSW of Delhi. Tonnes of mixed waste with no value were studied and understood, and mechanisms were developed to segregate it to get some value out of it. The plant handles MSW, by eliminating the environmental threats that the waste would have caused, if disposed of at the landfills. The Okhla plant, operated by Jindal Urban Infrastructure Limited (JUIL), employs gasification technology to convert MSW into energy. The plant has a capacity to process approximately 2000 metric tons of MSW per day. It is one of the largest WTE plants in India. The plant employs advanced gasification technology, where the municipal waste is subjected to high temperatures in a controlled environment to produce syngas. Syngas is a mixture of hydrogen, carbon monoxide, and other gases. The syngas produced through the gasification process is utilized to generate electricity and heat. The plant has an installed capacity to generate around 16 MW of electricity, which is supplied to the local power grid. The Okhla plant plays a crucial role in waste management and environmental sustainability. By converting municipal waste into energy, it helps in reducing the volume of waste sent to landfills and mitigating greenhouse gas emissions. The plant is designed to comply with the emission standards set by the Central Pollution Control Board (CPCB) to ensure minimal impact on air quality and environmental pollution.

WTE Plant, Jabalpur: The WTE plant in Jabalpur, Madhya Pradesh, utilizes MSW for the generation of electricity. It helps in reducing the waste volume and generates renewable energy. The Jabalpur MSW project is a pioneering and acclaimed initiative in India's waste management sector. It is currently operating with full capacity. The plant has successfully generated 11.5 MW of energy, which is equivalent to powering approximately 18,000 households. On a daily basis, it efficiently processes 600 tons of un-segregated solid waste, minimizing environmental impact, and promoting sustainable waste management practices. One of the remarkable achievements of the Jabalpur MSW project is the reduction of around 37,000 tons of carbon emissions in Jabalpur. By processing 2,19,000 tons of solid waste annually using **incineration**, the project has also contributed to saving approximately 4.4 hectares of land that would otherwise be required for landfill purposes. These significant outcomes highlight the plant's positive impact on the environment, energy generation, and land preservation. The Jabalpur MSW project serves as a remarkable model for waste management in India, demonstrating the potential to convert waste into a valuable resource while simultaneously addressing environmental challenges. Its success has garnered recognition and serves as an inspiration for similar WTE initiatives across the country.

Indore Bio Energy Park, Indore: In Indore, Bio-CNG (compressed natural gas) plant is known as the Indore Bio Energy Park. The plant is operated by the Indore Municipal Corporation (IMC) and is one of the largest Bio-CNG plants in India. The plant has the capability to process 550 metric tons of organic waste per day. This includes various types of organic waste, such as food waste, agricultural residue, and organic waste from industries. The plant uses anaerobic digestion technology to convert organic waste into biogas, which is further purified and upgraded to produce Bio-CNG. The plant has the capacity to produce a substantial amount of Bio-CNG, which can be used as a clean fuel for vehicles and other applications. The Bio-CNG plant contributes significantly to the generation of renewable energy. The production of Bio-CNG helps reduce dependency on fossil fuels, thereby reducing carbon emissions and promoting environmental sustainability. The operation of the Bio-CNG plant helps in the effective management of organic waste, which would otherwise end up in landfills and contribute to environmental pollution. By converting organic waste into Bio-CNG, the plant reduces greenhouse gas emissions and helps in mitigating climate change. The establishment of the Bio-CNG plant aligns with the goals of the Swachh Bharat Mission, which aims to achieve cleanliness and proper waste management across the country. It serves as a model for other cities and regions to adopt sustainable waste management practices. The plant has been set up on 15 acres of a trenching ground of the Indore Municipal Corporation in Devguradia area with an investment of Rs 150 crore through public–private partnership (PPP) mode. The plant has the capacity to produce 19,000 kg of Bio-CNG (compressed natural gas) per day from 550 tonnes of wet waste (of fruits, vegetables, raw meat, stale food, green leaves, and flowers).

Organic Fraction of Municipal Solid Waste into Bioethanol: Municipal waste is collected and segregated to separate organic waste, which serves as the primary feedstock for bioethanol production. The organic waste undergoes pre-treatment, which may include processes such as shredding, size reduction, temperature applications, and removal of contaminants. This step helps enhance the efficiency of subsequent conversion processes. The pre-treated organic waste is subjected to hydrolysis, where enzymes or acid catalysts are used to break down complex organic molecules into simpler sugars such as glucose. The hydrolyzed sugars are then fermented using specific strains of microorganisms (yeast or bacteria) that convert the sugars into ethanol through the process of anaerobic fermentation. The fermented mixture undergoes distillation to separate ethanol from water and other impurities. Additional purification processes like dehydration and rectification may be employed to obtain high-purity bioethanol.

Bio-ethanol is an eco-friendly and unconventional fuel. Increase in the production of bio-ethanol will considerably decrease the dependency on fossil fuel for energy generation. Search of suitable low-cost feedstock which is available through the year in bulk is a great challenge associated with the production of bio-ethanol. Food crops are used as a feedstock for bio-ethanol production which affects the food chain. In this chapter, municipal organic waste (MOW) is suggested as a suitable feedstock for bio-ethanol production. MOW is a perennial, easily available in bulk, almost free of cost, and readily available feedstock. Authors of the chapter currently working on production of bioethanol from organic fraction of MSW of Indore city. Experiments were conducted for maximum recovery of bio-ethanol from the waste. Results of the research prove suitability and cost-effectiveness of organic waste for production of bio-ethanol. A design of the bio-ethanol reactor is proposed which can be used for fabrication of the reactor for production of bio-ethanol on a large scale.

9.9 CONCLUSION

In conclusion, the management of MSW presents a unique opportunity to generate wealth and value from what was once considered a burden on society. Through various waste-to-wealth initiatives, such as composting, recycling, and WTE projects, municipalities and communities can harness the potential of their waste to create economic, social, and environmental benefits.

By implementing efficient and sustainable waste management and technological practices, we can not only reduce the strain on landfills but also minimize environmental pollution and conserve natural resources. Moreover, the revenue generated from these waste-to-wealth initiatives can be reinvested in further waste management infrastructure and initiatives, creating a circular economy approach.

To fully unlock the wealth potential from MSW, it is crucial to have comprehensive waste management policies, adequate infrastructure, public participation, and awareness campaigns. Collaboration between government bodies, private sector entities, and communities is vital to drive innovation, investment, and the adoption of sustainable waste management practices.

By embracing the concept of wealth from waste, we can transform our perception of MSW from a problem to a valuable resource. The journey toward a circular economy where waste is minimized, resources are conserved, and wealth is generated requires a collective effort, but the rewards are significant in terms of economic prosperity, environmental sustainability, and improved quality of life for all.

REFERENCES

1. Abubakar, I. R., Maniruzzaman, K. M., Dano, U. L., Alshihri, F. S., & Alrawaf, T. I., (2022). Environmental sustainability impacts of solid waste management practices in the global south. *International Journal of Innovative Research and Public Health, 19,* 12717. https://doi.org/10.3390/ijerph191912717

2. Aljaradin, M., & Persson, K. M., (2015). Environmental impact of municipal solid waste landfills in semi-arid climates – case study – Jordan. *The Open Waste Management, 1,* 28–37. https://doi.org/10.2174/1876400201205010028

3. Central Pollution Control Board Delhi (2018–19). *Annual Report on Solid Waste Management.* https://cpcb.nic.in/annual-report.php

4. Central Pollution Control Board Delhi (2020–21). *Annual Report on Solid Waste Management.* https://cpcb.nic.in/annual-report.php

5. Galicia, F. G., Páez, A. L. C., & Tejeida Padilla, R. (2019). A study and factor identification of municipal solid waste management in Mexico City. *Sustainability, 11*(22), 6305. https://doi.org/10.3390/su11226305

6. Kaza, S., Yao, L. C., Bhada-Tata, P., & Van Woerden, F., (2018). *What a Waste 2.0: A Global Snapshot of Solid Waste Management to 2050, Urban Development.* Washington, DC: World Bank. http://hdl.handle.net/10986/30317

7. Moya, D., Aldas, C., Lopez, G., & Kaparaju, P., (2017). Municipal solid waste as a valuable renewable energy resource: A worldwide opportunity of energy recovery by using waste-to-energy technologies. *Energy Procedia, 134,* 286–295. http://doi.org/10.1016/j.egypro.2017.09.618

8. Ministry of Housing and Urban Affairs, Government of India. (2021). *Report on Circular Economy in Municipal Solid and Liquid Waste.* https://mohua.gov.in/pdf/627b8318adf18Circular-Economy-in-waste-management-FINAL.pdf

9. Statista. *Share of Waste Processed in India from Financial Year 2016 to 2022.* www.statista.com/statistics/1147662/india-share-of-waste-processed

10 Plastic Waste and Its Management Strategies
Recent Advances

Aryadeep Roychoudhury, Shreyosi Pahari, and Alen D. Rozario

10.1 INTRODUCTION

Due to their diverse applications, chemical compositions, and properties, plastics are being extensively used around the globe. Use of more plastics corresponds to a huge generation of plastic waste. Both marine and terrestrial life are seriously impacted by the waste generated by plastics. Conventional plastics, which are non-biodegradable, require longer time for degradation, show high resistance to aging, and get converted to small debris that accumulates in the environment (1). The unbound monomers and the hazardous by-products of the plastics ultimately get released into the aquatic environment, both freshwater and marine water. This plastic appears as prey to the marine animals so that they consume it along with their food and these plastics get accumulated in their gut. Once the plastics are present in the gut, they can affect the inner lining of the stomach, blocking the digestive tract (1). Many researchers have reported several cases of the accumulation of plastics in the gut. Fragments of plastics enter the filter-feeding organisms which paves the way for entry into the food chain, thus accumulating in the higher marine organisms. Microorganisms and other diatoms start growing over the surfaces of the plastics, thus resulting in biofouling. Biofouling is defined as the biological accumulation of undesirable microorganisms as well as macro-organisms on surfaces exposed to natural water with other constituents. The plastics generally become heavy as a result of biofouling and eventually settle on the sea floor. As more and more plastics are being accumulated on the sea floor, hypoxia conditions are set up due to inhibition of the gaseous exchange between the pores of the accumulated plastic debris and overlying waters. Hypoxia leads to malfunctioning of the marine ecosystem, ultimately causing death of the marine animals due to suffocation. In the case of mesopelagic fishes, the ingestion of the plastics can increase their buoyancy, thereby making it difficult for them to return to deeper water. The chemicals that are used to produce plastics along with their additives and unbound polymers which are having hazardous effects on biological life and the environment, get released plinto the ecosystem during the life cycle of the plastic products. Bisphenol A (BPA), phthalates, polyhalogenated flame retardants, polyfluorinated compounds, and antibacterial compounds pose serious threats to human health and

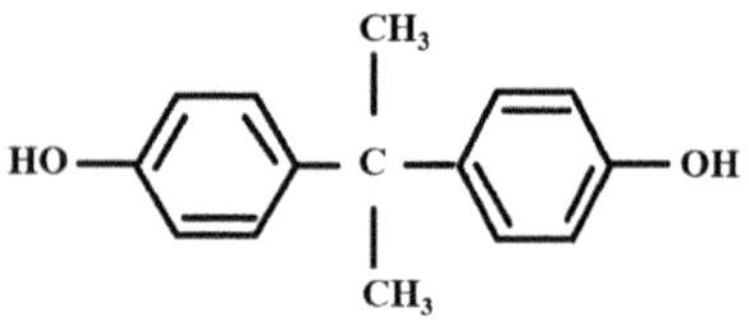

$$HO - \bigcirc - \underset{\underset{CH_3}{|}}{\overset{\overset{CH_3}{|}}{C}} - \bigcirc - OH$$

FIGURE 10.1   Structure of bisphenol A (BPA).

other organisms (1). In some of the electrical and non-electrical plastic production, polybrominated diphenyl ethers (PBDE) are used as fire retardants and are persistent. BPA and phthalates are one of the major concerns due to their biological effects in both humans and animals.

10.2   CHEMICALS INVOLVED IN PLASTICS

10.2.1   BISPHENOL A (BPA)

BPA is a monomer chemical used to produce PVC and polycarbonate plastics. During polymerization, some of the monomers remain unbound which are released and ultimately found in the leachate of the landfills where the plastic waste is being dumped. In the regions receiving leachates, the ecosystems are degraded. BPA is also recognized as a potential endocrine disruptor which results in malfunctioning of the hormones. Other health-related issues are cardiovascular disease, heart attacks, and male sexual dysfunction (Figure 10.1).

10.2.2   PHTHALATES

Phthalates are the diesters of phthalic acid used for the flexibility, softening, and elasticity of rigid plastics. As the phthalates are not covalently bound to the polymer matrix, they are prone to leaching into the environment. Di (2-ethylhexyl) phthalate (DEHP) causes health concerns for humans. Upon entering the human body, DHEP is converted to mono (2-ethylhexyl) phthalate (MEPH) by lipases and gets absorbed by the body. Other phthalate additives such as dibutyl phthalate (DBP) and benzyl butyl phthalate (BBP) produce a toxic metabolite, mono-butyl phthalate. Hormonal disruption, asthma, genital malformations, and the development of some cancers are the other adverse effects related to phthalates (Figure 10.2).

10.3   SOURCES OF THE PLASTIC WASTE

The major contributors to plastic pollution in modern times include the following:

Old trash: From the milk cartons that are lined with plastics to the plastic bottles, to even products that might contain small plastic beads, plastics are found everywhere and every time, all such items are thrown or washed down a sink, and the risk of toxic pollutants entering the environment increases (3).

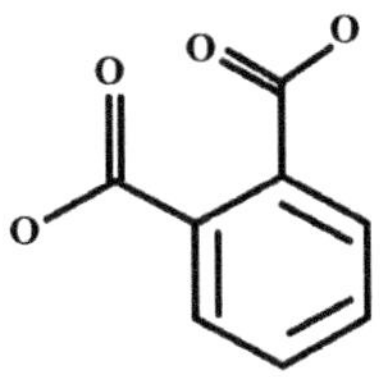

FIGURE 10.2   Structure of phthalate.

Overusing: Plastic is not very expensive. Ever-increasing population growth and rapid urbanization have increased the demand for plastics. As plastics are a very durable and cheap material, they are extensively used in every way possible, ranging from using it as packaging material and in water bottles to straws and carry bags. After use, they are either thrown away but do not get disposed of properly, thus polluting the land and water or they are even burned polluting the air.

Plastics take a longer time to decompose: Plastics are produced by very strong chemical bonds and so require many years to decompose, as many as 400 to 500 years depending on the type. The same properties that make plastics useful, that is their durability and resistance to degradation, make it nearly impossible for nature to completely break plastics down. They do not fully disappear but are broken down into smaller pieces called microplastics, which can enter the human body through inhalation and absorption and accumulate in the organs like liver, spleen, and kidneys (11).

Fishing nets: Fishing is necessitated in many parts of the world where fish are eaten daily. The nets that are used for fishing are made up of plastics which contributes largely to the plastic pollution in the oceans. When these nets spend a lot of time in the submerged water, they leak the toxins in the surrounding water. Plastic waste is also washed to the shores from ships and nets used for fishing. This not only kills marine wildlife but also pollutes the water (3).

Plastic and garbage disposal: When the disposal of plastic is not managed carefully, it results in landfills. Similarly, if plastics are being burned, they can release harmful gases into the atmosphere that can cause serious illness (3). Recycling plastics also should not be encouraged all the time as it essentially uses the existing plastic, albeit in a new form.

Forces of nature: Many a time plastics are transported to the rivers, sewers, and streams, and finally into oceans by wind as they are quite light.

10.4 TYPES OF ENVIRONMENTAL HAZARDS RELATED TO PLASTICS

Land: In human-inhabited areas, plastics are being used in larger quantities. Plastic pollution and related plastic-based products can damage as well as contaminate the terrestrial environment. They can be subsequently transferred to the marine ecosystem (2). It has been seen that the plastic pollutants that are present in the aquatic environment originate almost about 80% from the land (2). Dumping of the plastics on the land leads to biotic and abiotic degradation thus releasing the harmful plastic additives into the environment that can leach and eventually percolate in various aspects of the environment. Reports have shown that microplastics (4) as well as synthetic polymer fibers can be still detectable five years after they have been applied to sewage sludge and soil (5). Chlorinated plastics consist of toxic chemicals that impose the risk of leaching out into the surrounding soil and can also percolate in the underground water, thus adversely affecting the ecosystem.

Water: Approximately 165 million tonnes of plastic waste are estimated to be present in the oceans of the world (6). An average of about 8 million tonnes of plastic is being released annually into the ocean (7). Plastics that are present in the ocean can degrade within a year but not completely. Eighty percent of the waste that is found in the water consists of plastics. Plastic debris found in the ocean can be readily colonized by marine animals. Within the marine ecosystem, plastics can concentrate and absorb the contaminants present in the seawater from other sources. Organic pollutants such as nonylphenol and dichlorodiphe nyldichloroethylene (DDE) have the potential to get accumulated several folds on the plastics, as compared to surrounding water (8). Marine animals such as turtles, seabirds, and fishes generally ingest or are entangled with the plastic debris leading to lesser movement, feeding, reproductive output, and eventually death (9). Rivers and lakes carry the waste generated from plastics, from deep inland to the sea making them contributors to the ocean population (11).

Air: Plastics which are finally decomposed release CO_2 and methane into the surrounding air. Gases such as CO_2 are released when plastics and plastic-based products are burned which are capable of trapping the radiant heat and obstructing it from escaping from the earth, thus leading to global warming (10). Other pollutants released due to the burning of plastics include heavy metals, dioxins, and furans which, when inhaled, can cause respiratory problems (10).

10.5 NATURE OF PLASTICS

Plastics, being a polymeric material, has the capability of being molded or can be shaped by the application of little pressure or adequate heat. This property of plasticity when combined with the properties of low density, electrical conductivity, transparency, and toughness allows plastics to be available in varied forms, such as polyethylene terephthalate (PET), polyvinyl chloride (PVC), insulating food containers, etc. Many of the chemical names of the polymers are better known by their abbreviations or trade names in recent days. For example, foamed polystyrene and polymethyl methacrylate are known as Styrofoam and plexiglass (Perspex), respectively, which are their trademark names. Industrial fabricators of plastic products describe plastics either as specialty resins or commodity resins (12). The plastics that are produced in high volume and low cost for the most common disposable items and durable goods are known as commodity resins. A few examples of commodity resins include polyethylene, polypropylene, and PVC. Specialty resins are the plastics that are produced in low volumes and at higher cost. This group of plastics also includes the so-called engineering plastics that can compete with die-cast metals in plumbing, hardware, and automotive applications. Polyacetal, polyamide (trade name: nylon), polytetrafluoroethylene (Teflon) are some of the engineering resins. Thermoplastic elastomers are another special kind of specialty resins that have the elastic properties of rubber but can still be molded on the application of heat (12).

Plastics can be broadly divided into two categories based on their chemical composition. One category of plastics is made up of polymers which have only aliphatic linear carbon atoms in their backbone chains. This category of plastics consists of all the commodity plastics listed above. The other category of plastics consists primarily of heterochain polymers. This compound contains oxygen, nitrogen, and sulfur in its backbone chains and includes the engineering plastics like polycarbonate.

On a broader view, plastics are classified as thermoplastic resins or thermosetting resins. Thermoplastics such as polyethylene and polystyrene can get molded and remolded easily and repeatedly. The polymer structure associated with thermoplastics is of individual molecules that are separate from one another and flow past one another. Thermoplastic resins may have a low or extensively high molecular weight that may be branched or linear in structure, but the most important feature is that of separability and consequent mobility.

Thermosets, however, cannot be molded or remolded upon heating. During the initial time of their processing, they undergo a chemical reaction that results in an infusible and insoluble network and thus the finished product is like one large molecule (12). For example, an epoxy polymer used in the fiber-reinforced laminate for a golf club undergoes a cross-linking reaction when molded at high temperatures. However, subsequent application of heat does not result in refolding but ultimately causes it to break down.

10.6 PHYSICAL STATES AND MOLECULAR MORPHOLOGIES

The nature of the plastics is also controlled by their morphology characteristics, i.e., the arrangement of the molecules on a large scale. Morphologies can be either amorphous or crystalline. In the case of amorphous structure, molecules are arranged randomly and intertwined but the crystalline structures are arranged closely and in a discernible order. Thermosets belong to amorphous types, while thermoplastics may be amorphous or semi-crystalline, displaying crystalline regions called crystallites within an amorphous matrix.

The thermoplastics retain their molded shape up to a certain temperature set by the glass transition temperature (T_g) of the particular polymer. Below the T_g, the molecules of a polymer material are frozen and are known as glassy states. In a glassy state, there is little or no movement of the molecules past one another. So, the materials appear as stiff and brittle. Above T_g, the amorphous parts enter the rubbery state, where the molecules show increased mobility and the material becomes plastic and elastic. In the case of polystyrene, which is a non-crystalline polymer, if the temperature is raised, it leads to a liquid state. However, for crystalline polymers such as polyethylene, the liquid state is not reached until the T_m (melting temperature) is reached. Beyond this point, the rubbery or liquid polymers can be molded because the crystalline regions are no longer stable.

10.7   PROPERTIES OF PLASTICS

Firstly, the physical state and the morphology greatly influence the mechanical properties of plastics. When a plastic is loaded or stressed in tension, it typically results in its elongation. A glassy polymer such as polystyrene is stiff and shows a high ratio of stress to elongation. The crystalline plastics can be used as films or molded objects because, at room temperature, their amorphous regions are well above their T_g. In PET, which is a semi-crystalline plastic, the crystalline portions exist in a glassy matrix because the T_g of PET is above the room temperature providing stiffness and huge dimensional stability under stress. These are relatively used in beverage bottles and recording tapes. When the plastics are being elongated, they cannot be recovered when the stress is removed which causes the elongation. This situation is called "creep". Creep is very small for the plastics below their T_g and comparatively more above their T_g.

Most of the plastics are poor conductors of heat. Their conductivity can also be significantly reduced when a gas is being incorporated into the material. For example, foamed polystyrene has a thermal conductivity of about one-quarter that of an unfoamed polymer. Besides conductivity, another important property includes the dielectric strength which implies resistance to breakdown at high voltages, and dielectric loss which is referred to as a measure of the energy dissipated as heat when an alternating current is applied. Most of the plastics are also combined with other ingredients and additives during processing and fabrication (12). Plastics are mixed with the additives to assign the plastics the appropriate properties. A few of the known additives of plastics include plasticizers, colorants, and stabilizers. Additives are also used to combat the deterioration of the plastic and significantly improve the life span of the final product. A few of the additives are covered in the below discussion.

10.7.1   COLORANTS

Plastics are being colored for most of the consumer applications. The ease with which colors get incorporated in a molded particle gives it a slight advantage over metals and ceramics. The diverse types of colors depend upon titanium oxide and zinc oxide (white), carbon (black) as also oxides of iron and chromium. Organic compounds can be used to add color, either as dyes or pigments.

10.7.2   STABILIZERS

Stabilizers are usually added to the plastics to have an enhanced lifetime and also to counter the effects of aging. The most common stabilizers are antioxidants as all the carbon-based polymers are subjected to oxidation. Hindered phenols and tertiary amines are used in plastics in low concentrations (12). At processing temperatures, PVC requires the addition of heat stabilizers to reduce dehydrohalogenation. Zinc and calcium soaps and organic phosphites are effective stabilizers.

10.7.3   PLASTICIZERS

Plasticizers are used to change the T_g of a polymer. PVC is often mixed with non-volatile liquids for this reason. For example, a PVC garden hose should remain flexible at 0 °C. A mixture of about 70 parts of PVC and 30 parts of dioctyl phthalate (DOP) has a T_g of around $-10°C$, to use as a garden hose. PVC can retain plasticizers of varying chemical composition and molecular size (12).

Let us now have a broader insight about the other different properties of the plastics (13).

10.8   CHEMICAL RESISTANCE OF PLASTICS

Plastics can offer a greater resistance to chemicals and solvents. During the manufacturing process of the plastics, the degree of resistance is determined. The plastics that are commercially available in the market have greater corrosion resistance and are thus used in place of metals for water-carrying activity.

10.8.1   DIMENSIONAL STABILITY

Thermoplastics can be easily reshaped and can be reused but thermosetting plastics are not possible to be reshaped and thus not applicable to be reused.

10.8.2   DURABILITY OF PLASTICS

Plastics show good durability properties when having sufficient surface hardness. Thermoplastics may be sometimes affected by termites or rodents but that might not impose a serious problem because plastics do not have any nutritional values.

10.8.3   ELECTRIC INSULATION

Plastics are used as linings in electrical cables and electronic goods due to their high and efficient property of insulation.

10.8.4   RESISTANCE TO FIRE

Resistance to fire and varying temperatures depends on the structures of the plastics. As for example, plastics that are being produced from cellulose acetate burn very slowly. PVC-made plastics also do not catch fire easily. Similarly, plastics made from phenol formaldehyde and urea formaldehyde are fireproof.

10.8.5   MELTING POINT

Generally, plastics have very low melting points. However, plastics that have a melting point of around 50°C cannot be used in regions of high temperature. As thermosetting plastics have a higher melting point than thermoplastics, thermosetting plastics cannot be used for recycling purposes.

10.8.6 OPTICAL PROPERTY

Transparent plastics allow light rays to pass through them and do not alter the direction of the light rays. However, the semi-transparent or translucent plastics allow the light rays to pass and, in turn, change the direction of the light rays.

10.8.7 SOUND ABSORPTION

By the saturation of the phenolic resins, it is possible to prepare acoustic boards that have the property of absorbing sound and thus provide sound insulation. Acoustic ceilings can be seen in theatres and seminar halls.

10.8.8 STRENGTH

When fibrous materials are reinforced into plastics, they can significantly improve their strength. If the strength-to-weight ratio of the plastics and the metals are the same, plastics cannot be given enough preference due to certain conditions like poor stiffness, sensitivity against temperature, etc. (13).

10.9 CATEGORIZATION OF PLASTICS

In the current world, there are many diverse types of plastics available having specific sets of unique properties and uses. Some of the categories of the plastics are discussed in the table given below:

CATEGORIES OF PLASTICS	DESCRIPTION	STRUCTURE
Polyethylene terephthalate (PET)	PET is the most abundant synthetic plastic and is also popularly known as thermoplastic resin in the polyester family. PET has a higher chemical resistance to water and organic compounds. It also demonstrates a higher strength-to-weight ratio and is almost shatterproof. This type of plastic is easy to be recycled and is found in everyday items including food and drink containers and also the garment fibers.	Polyethylene Terephthalate

CATEGORIES OF PLASTICS	DESCRIPTION	STRUCTURE
Acrylonitrile butadiene styrene (ABS)	By polymerizing styrene and acrylonitrile in the presence of polybutadiene, ABS can be produced. ABS is glossy, highly processable, flexible, and strong, and is an impact-resistant material that is frequently used in the automotive and refrigeration sectors. Some of the applications of ABS include the production of boxes, protective headgear, as well as luggage and children's toys.	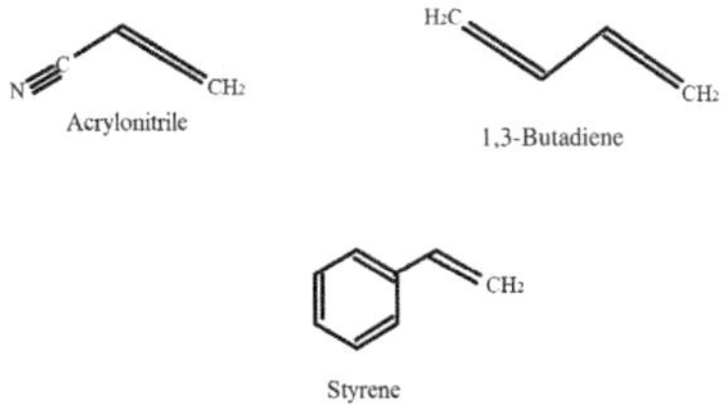
Polyvinyl chloride (PVC)	PVC can be made both flexible and rigid though it is well known for its versatility to be mixed with other materials. Foamed PVC sheets are generally used in the case of store displays and exhibits, whereas rigid PVCs are used extensively in construction materials, bottles, windows, and doors. When PVC is mixed with plasticizers (such as phthalates), softer PVC is possible to be produced which is applied in clothing, plumbing, and medical tubing.	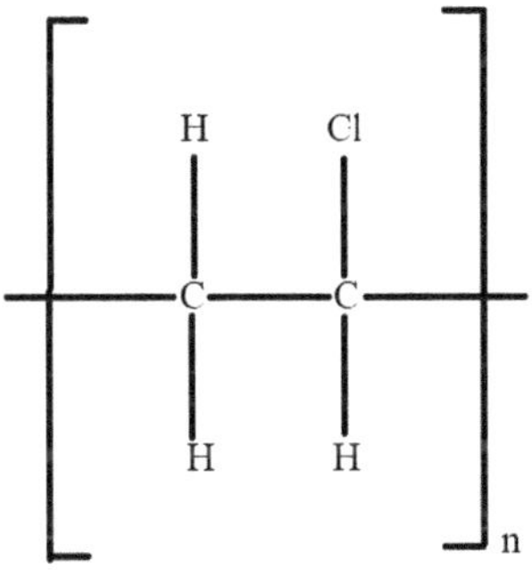
Polypropylene (PP)	PP is the most flexible thermoplastics which is used in the production of food containers, components of automobiles, medical devices, and laboratory equipment. PP, though flexible, is still stronger than PE and is also heat and acid-resistant. It is also easily affordable.	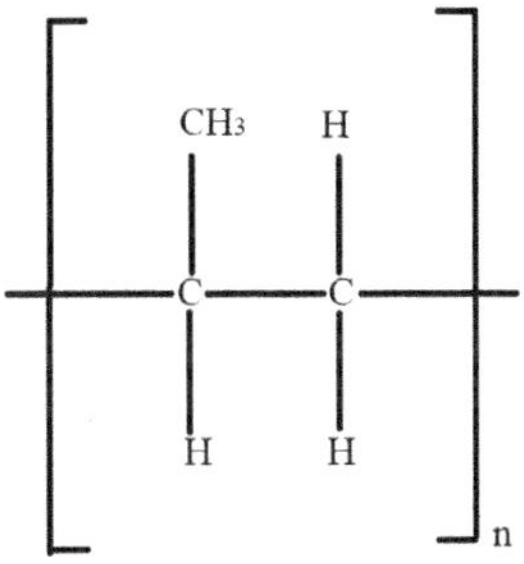

(*continued*)

CATEGORIES OF PLASTICS	DESCRIPTION	STRUCTURE
Polyethylene (PE)	It is the most common type of plastic on the planet. PE comes in varying densities and thus functions for a wide gamut of purposes. According to their density differences, the types of PE are LDPE (low-density polyethylene), MDPE (medium density polyethylene), HDPE (high-density polyethylene), and UHMWPE (ultra-high-molecular-weight polyethylene). MDPE is used for shrink film, gas pipes, screw closures, and carrier bags. UHMWPE is abrasive resistant compared to HDPE as it has unusually longer polymer chains. UHMWPE is used in artificial ice-skating rinks, hydraulic seals, and biomaterials for knee, spine, and hip implants. As LDPE is ductile, it is used in the manufacturing of plastic bags and shopping bags and also in disposable packaging items. HDPE is more robust than both LDPE and MDPE. This is used in plastic bottles, sewage pipe snowboards, and folding chairs.	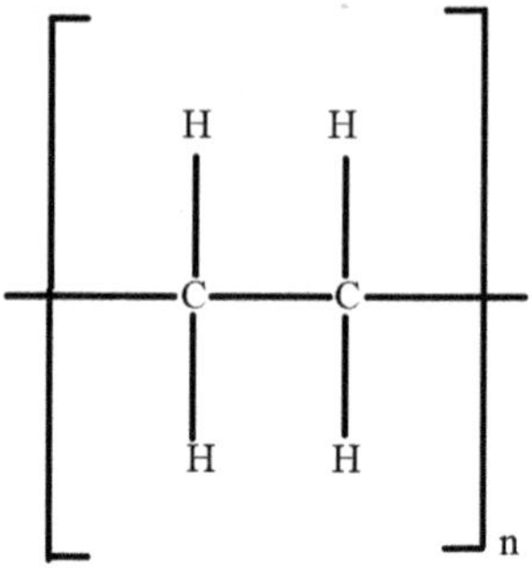 Polyethylene
Polystyrene (PS)	Also known as Styrofoam, it can be both solid and foamed. PS is used in egg cartons, disposable dinnerware as well as drinking cups due to its low cost per unit weight and ease of production. However, it is extremely combustible and when ignited releases toxic compounds.	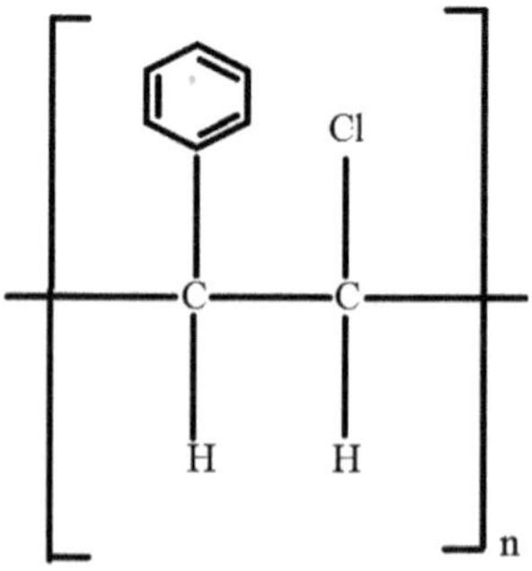 Polystyrene

CATEGORIES OF PLASTICS	DESCRIPTION	STRUCTURE
Polycarbonate (PC)	PC is a good engineering plastic that is strong, stable, and transparent. Its strength-to-weight ratio is 250 times higher than that of steel and has a glass-like clarity. The intrinsic design and flexibility of the PC enable a strong and impact-resistant construction. Unlike glass or acrylic, PC sheets can be readily cut out without needing to be pre-formed and manufactured. Some of the applications of PC lie in the production of CDs, sunglasses, police riot gear, etc.	Polycarbonate
Acrylic or polymethyl methacrylate (PMMA)	PMMA is a particular type of synthetic resin produced from the polymerization of methyl methacrylate (21). Being a lightweight and break-resistant compound, it has applications in optical products and gadgets. PMMA is also resistant to UV rays and static electricity and can also accept bright dyes. It is 17 times more impact-resistant than the glass and PC sheeting used together.	Acrylic or Polymethyl Methacrylate

10.10 MICROPLASTICS

Plastic pollution in the environment leads to the formation of "synthetic solid particles" or "polymeric matrices", called microplastics, which are less than 5 mm in length and generally insoluble in water (15). Ranging from cosmetics to synthetic clothing, to that of plastic bags and bottles, microplastics can be found in a variety of products (14). Microplastics are polymer chains consisting of carbon and hydrogen bonded together. Chemicals such as phthalates and polybrominated diphenyl ethers (PBDEs) are also present in microplastics that can leach out and adversely affect the environment.

Microplastics generally contain two types of chemicals: (i) additives and polymer raw materials, and (ii) chemicals that are absorbed from the surrounding environment. Additives are the specific type of chemicals that are added during the production of plastics which give them specific colors and transparency and also improve resistance against ozone, temperature, light, bacteria, etc. (16). Microplastics are broadly divided into two categories: primary microplastics and secondary microplastics. Some of the basic examples of primary microplastics include the microbeads. The plastic pellets in the industrial manufacturing industries, scrubbers, commercial cleaning abrasives, plastic powder or fluffs, which are used to produce plastic goods along with volatile particulate contaminants such as micro-polyesters, nano Fe_2O_4, and SiO_2 from the printing toner, are the different sources of primary microplastics (17). Primary microplastics can directly enter the environment through diverse ways such as usage of the personal care products which are later washed into wastewater systems from the household or during the abrasion, i.e. laundering of clothing made with synthetic textiles. When the larger-sized plastics are broken down, they form the secondary microplastics. The process of production of the secondary microplastics is initiated when the larger plastics undergo weathering or are exposed to wave action, wind abrasion, and ultraviolet rays. In addition to weathering, the fragmentation process that paves the way for the formation of secondary microplastics also includes the gradual degradation of the plastics in water which principally consists of three processes, namely bio-fragmentation, assimilation, and bio-deterioration (17).

10.11 WHAT ARE NANOPLASTICS?

The particles that range in size between 1 nm to 1 μm and are the omnipresent pollutants of the natural environment, resulting from the fragmentation of the larger plastic debris, are referred to as nanoplastics. The two different types of sources that contribute to the production of the nanoplastics are (i) primary plastic waste and (ii) secondary derivatives (18). Primary nanoplastic waste is disposed of in the environment in nanoscales including the sources of nanomedicines, nanoimaging, nanosensors, and personal care products. The disintegration of the plastics driven by the physical, chemical, and biological forces gives rise to the secondary derivatives of the nanoplastics. The other potential sources of the nanoplastics include the following.

10.11.1 NANOPLASTICS FROM TIRE WEAR

When the vehicles are moving at a considerable speed, the tires are rubbed continuously against the ground and due to constant friction, particles are released into the environment. Approximately 30% of the weight of a tire is emitted into the environment from the use of scrap materials. In addition, an analysis of the airborne particles near a road has shown that the sizes of the particles ranged from 6 to 562 nm and from 30 to 60 nm. This source of nanoplastics should be given proper attention due to its adverse effects and large amounts emitted from tire wear (19).

10.11.2 Nanoplastics from Laundry Wastewater

Nanoplastics from the wastewater generated by the laundries are considered a major source. Plastic fibers are among the most frequently detected nanoplastics from wastewater for household washing purposes. Acrylic, nylon, and polyester microfibers are released from synthetic textiles to laundry wastewater with an average of 7360 fibers $m^{-2} L^{-1}$. The annual emission of microfibers from synthetic and cotton materials from household laundry wastewater is projected to be 565,000 kg each year (18).

Nanomaterials, including the nanoplastics, have a special property, i.e., their surface area is extremely large and active. The size of the nanoplastics corresponds to the size of the cell membranes. Due to its hydrophobic character, it is easily absorbed into the cells (19). This proves the cellular toxicity or cytotoxicity of the nanoplastics. The cellular responses depend on the specific way and the route of the absorption of the nanoplastics and the location in the body where they are stacked. The airway cells and the gastrointestinal tract are the most exposed places to nanoplastics (19).

Through a wide range of entry points, microplastics are able to accumulate in soil and affect its physical, chemical, and biological characteristics in both positive and negative ways (21). The bioavailability of microplastics may be improved by biological factors. When confronted with polystyrene (PS) beads, suspension-feeding bivalve mollusks consumed considerably more of the 100 nm-sized beads. These were added to manually created aggregates that were created in the laboratories by rolling actual seawater. Energy is transferred between pelagic and benthic habitats by the seasonal flocculation of natural particles into sinking aggregates (22). Microplastics may be successively egged after ingestion within feces. Detrivores and suspension feeders may consume these aggregated microplastics. Organisms that live in sediment, like the lugworm *Arenicola marina*, are able to cycle the topmost sediment layers through a process called bioturbation. The sediment may absorb microplastic particles that have accumulated on the benthos, making them available to fauna (23).

10.12 THE VULNERABILITY OF AQUATIC LIFE TO INGESTING MICROPLASTICS

There is evidence of selectivity in the ingestion of natural particles by a variety of species; thus, it is likely that the same selection will also apply to microplastics.

- Amphipods could mistakenly absorb microplastics as natural food sources, making them the main consumers of microplastics (24). Benthic holothurians like *Thyonella gemmate, Holothuria floridana, H. grisea,* and *Cucumaria frondosa* consume microplastic, using a non-selective feeding strategy (23). It has been found that individuals belonging to four species of two orders ingested significantly more plastic than expected, based on plastic-to-sand grain ratios from each sediment treatment (25).
- Plastics contain toxic chemicals such as bisphenol-A (BPA), monomers, flame retardants, oligomers, metal ions, and antibiotics. These toxic chemicals can build up in marine species that unintentionally swallow plastics. Flame

retardants and phthalates, which are included in some plastics, have the potential to be hazardous to fish, mollusks, and mammals (26).

- The manufacture, handling, and disposal of microplastics and their waste have an impact on the aquaculture industry. On both wild and domestic fish stocks, plastic pollution has been shown to have a significant impact. Due to this, there may now be health hazards associated with some harvested goods (27).

- Microplastics can have either physical or chemical impacts on aquaculture fish species. The particle size and shape determine its physical effects, whereas its chemical makeup, the presence of additives, or the presence of other pollutants such as heavy metals determine its consequences. The intake of MPs can have negative effects on the digestive system, such as blockage and mechanical damage, which can lead to famishness due to an incorrect sense of satiety and, in severe cases, fatal stomach and intestine punctures (28).

- Although there is limited proof, it is possible for microplastic to penetrate the food chain. According to Eriksson and Burton (29), plastic fragments discovered in the scat of fur seals (*Arctocephalus* spp.) were consumed by lantern fish (*Electrona subaspera*), which were then consumed by the seals (29). Norway lobsters (*Nephrops norvegicus*) were fed with pieces of fish, and implanted with polypropylene strands, and 24 hours later, the plastic still remained in their stomachs (24).

- Bioaccumulation—The associated physical effects of ingesting microplastics are likely influenced by the propensity of the organism to accumulate microplastics. A 20-mm long chaetognath named *Sagitta elegans* was found in a plankton tow in the coastal water of south New England. Its intestine featured a 0.6 mm diameter spherule. Although it was not established whether this was plastic, the spherule was said to be the same as PS spherules that were also found in the tow. However, this illustrates how comparable particles might gather in marine creatures (30). Microplastic accumulation in marine invertebrates can obstruct the digestive tract, which would reduce feeding because of satiation. Alternately, predation of marine invertebrates polluted with microplastics might provide a route for plastic to go up the food chain (23). In addition to internal buildup, exterior adsorption of MPs may be potentially harmful. The binding of plastic beads (20 nm) to the algal species *Chlorella* and *Scenedesmus* hindered photosynthesis (31) possibly because of the physical obstruction of light and air. Additionally, it boosted the generation of reactive oxygen species, indicating an oxidative stress condition (31).

10.13 IMPACT OF MICROPLASTICS ON TERRESTRIAL ECOSYSTEM

- Similar to human populations, natural populations suffer significant negative effects from consuming or being around the aforementioned harmful substances. Plastic materials, however, have the ability to absorb harmful chemicals that have the potential to bioaccumulate over time. Persistent organic pollutants (POPs), a class of significant hazardous chemicals, are remarkably resistant to biodegradation. POPs include industrial chemicals like polychlorinated biphenyls (PCBs), dioxins, such as polychlorinated dibenzo-p-dioxins (PCDD), and insecticides like DDT, which are organochlorine pesticides (32).

- Complex processes such as advection, dispersion, settling, adsorption, diffusion, degradation, and aggregation are all part of the MP transport mechanism in terrestrial soil. MPs are transported by air, water, and wind before being further deposited in the deeper areas of the terrestrial boundary (33). MP density, contact angle, and movement in soil are all factors that influence their setting and diffusion. Earthworms and plant roots also contribute to MP movement in soil, as decomposed roots leave macropores that facilitate transport (34).

- Microplastics pose a growing hazard to the health of the soil biota and terrestrial ecosystems (35). The presence of MPs in soil has negative effects on the soil microbiome, notably altering the structure and enzymatic activity of the microbial community, which mediate crucial ecosystem functions and are of great interest to the scientific communities. China has been utilizing plastic film mulches for 30 years, and as a result, there are now 50–260 kg/ha worth of plastic fragments in the topsoil (0–20 cm), which may inhibit plant growth (36).

- The findings in a study showed that PVC MPs at 0.1% and 1% levels did not significantly alter the diversity and composition of the bacterial community in soil for over 35 days. However, they significantly decreased or enriched a number of bacterial genera and altered the amount of accessible P in the soil. Further research is needed to better understand the dangers of MP pollution in soil settings (37).

- It has been claimed that MPs modify the microbial makeup of sediment to affect the nitrogen cycling mechanism in soil. While PVC treatment affects both processes in soil, MPs such as polyurethane foam and polylactic acid may alter the microbiota in benthic sediments and increase the de-ammonification and de-nitrification processes (38). By introducing organic matter and boosting soil porosity, earthworms, and microorganisms contribute significantly to soil fertility (39). However, there is no proof that earthworms (*Eisenia fetida*) consume MP or accumulating pesticides. The flourishing of plants thriving in contaminated soil may benefit from the ability of MPs to absorb hazardous pollutants from plant roots (40).

- As MPs are mostly made of carbon, they can promote the growth of microorganisms while also sapping the vital nutrients of the soil, thereby impairing plant growth. MPs may or may not degrade naturally (41). Excess carbon can accelerate microbial growth, but it can also quench nutrients in the soil, leaving plants with little or no nutrition. This could have a negative impact on plant growth and alter the soil carbon-to-nitrogen ratios (42). The movement of MPs inside soil by springtails and earthworms has been observed, with structural changes documented in their burrows and altered gut microbiomes. This has had a negative impact on development and reproduction, as well as altered gut microflora and isotopic signatures (^{15}N and ^{13}C). Even without direct evidence of ingestion, springtails exposed to MPs showed negative impacts on development and reproduction, altered gut microflora, and altered isotopic signatures (43, 44).

10.14 IMPACT OF PLASTICS ON HUMAN HEALTH

- The majority of large plastic particles are not poisonous enough to kill. However, exposure to, ingestion, and uptake of microscopic MPs can be harmful and serve as a long-term environmental stressor. After earthworms were exposed to MPs in their meal, sublethal adverse reactions such as growth loss were noted. It has been suggested that these effects may be partially explained by the histological harm and gene expression brought upon by MP exposure. Additionally, due to the stronger adsorption of metal to high-density polyethylene MP, MPs may operate as a vector of harmful Zn to earthworms under environmental conditions (47, 48, 49).

- The response and sensitivity to polystyrene nanobeads in 5 mM NaCl growth media of *Saccharomyces cerevisiae* yeast cells showed lethal effects (100% mortality). The heterogeneity of cell wall was found to explain the varying sensitivity of filamentous fungi, viz., *Aspergillus oryzae* and *Aspergillus nidulans*. In vertebrates, the response and sensitivity to nanobeads were connected to immunological characteristics of the respiratory system, specifically surface respiratory macrophages of domestic ducks and rabbits (48, 49, 50). Microplastics are persistent and some animals are more sensitive to them, resulting in the selection of species features, which could have an impact on phenotypic, genetic, and functional biodiversity.

- Terrain animals are likely to be exposed to plastic pollution levels or could be exposed to them in the future, which could change the physiological and ecosystem process baselines. This new anthropogenic stressor may already be putting some species under pressure to evolve and pollution from MPs and nanoplastics may have significant effects on the biodiversity of continental systems. Further research is needed to provide accurate data on environmental behavior and ecotoxicological information regarding various MP pollutants in terrestrial ecosystems.

- When MPs are consumed along with related POPs and EDCs, they cause thyroid dysfunction and developmental problems (51). Phthalates can reduce thyroid weight and lead to hyperactivity in children (52, 53).

- TDCs associated with plastic enter the body through the gastrointestinal tract and interfere with thyroid hormone production and metabolism. They form complexes with protein of thyroid hormones and reach the brain, disrupting thyroid health. SCTD is caused by abnormal low or high levels of TSH (56, 57). BPA inhibits T3 binding and transcriptional activity, which affects thyroid functioning (58). Phthalates interfere with metabolic activity and gene expression in the HPT axis, reducing thyroid function through inhibition of T3 protein binding, antagonistic interactions, and disruption of thyroid receptor transcriptional activity (59, 60).

- BPA increases adrenal index, causes vascular obstruction, reduces antioxidant enzyme activity, and reduces vimentin and alpha-smooth muscle actin expression (61). DEHP is associated with lower angiotensin II expression, reducing aldosterone levels (62).

- The pituitary gland is affected by EDCs transmitted by MPs and nanoplastics (61) in two ways: prolactinoma, a non-cancerous pituitary tumor, is induced, and prolactin and TSH are stimulated (62). The regulation mechanism of the HPT axis is disturbed by BPA as an essential addition (63) through altered TSH levels, and it directly impacts the pituitary by changing the way it reacts to TRH, secreted by the hypothalamus (64). By changing the HP thyroid/gonadal axis, mercury bioaccumulates in the pituitary and thyroid glands and produces endocrine poisoning (67). Among the most dangerous EDCs, cadmium and arsenic modify hormone production, which has a negative impact on the endocrine system (68).

10.15 STRATEGIES FOR MANAGING PLASTIC WASTE

At first look, strategies like recycling might appear to save resources, but over time, they cause alteration of consumer behavior, leading them to consume more and reducing the overall resource savings. To avoid the drawbacks of conventional recycling, it is essential to investigate alternative solutions that entail converting plastic into different end products. With the help of a special recycling technique called wood–plastic composites, plastic may be transformed into a form that offers improved material performance, processing efficiency, and user acceptance (69, 70). Other techniques for dealing with plastic trash include pyrolysis, liquefaction, and use of plastic in cement and road construction.

10.15.1 RECYCLING

Recycling is a waste management practice that involves gathering waste materials and turning them into raw materials that may be used again to create other useful products. It is also referred to as "renewing or reusing" to safeguard the environment and society from harm. Due to their carbon-based composition and use of additional polymers, plastics are not biodegradable. It includes bottles and other items that can be melted down and used to make furniture made of plastic, such as tables and chairs. The following six phases are used to complete this process: gathering waste plastics, sorting or categorizing plastics, washing to eliminate pollutants, shredding and resizing, identifying and separating plastics, and compounding (69). When plastic waste is recycled rather than being disposed of in undesirable locations, the world can benefit in a number of ways. One of these benefits is the protection of human life by reducing carbon dioxide and other harmful emissions in the atmosphere, which may result during incineration or combustion of the waste (70). Recycling uses less energy, lessens pollutants across the ecosystem, and promotes the preservation of the environment. It reduces the need for fossil fuel consumption and conserves landfill space, which is quickly running out. Additionally, it supports a sustainable way of life and advances the economy of the country. Recycling provides many advantages for the community, but it also has some drawbacks that are manageable and controllable. Some chemicals are discharged into the environment during the recycling process. Some of these compounds, such as volatile gases produced by the compositions of plastic waste and organic chains of monomers that create plastic chains of organic fumes and ashes, can damage plant structures and have an adverse effect on wildlife when ingested by various animals that reside close to the recycling zone. Plastics must be heated during the process to melt them, which results in the release of sulfur, carbon, and other gases into the atmosphere. These gases can result in acid rain, the greenhouse effect, and global warming, all of which have different negative environmental effects (71). This may cause health problems for people who enter the recycling zone for plastics. Waste is separated to ensure ongoing recycling after the downcycling of plastics, which explains why the plastic is ultimately unsuitable for further recycling. This implies that rather than seeing them as an alternate use of leftover plastic garbage, they end up in a landfill (72). Additionally, it makes significant contributions to waste reduction and the creation of electricity from waste, both of which are crucial to modern industrialization.

10.15.2 INCINERATION

The term "waste incineration" refers to the process of burning waste in oxygen, also referred to chemically as "complete combustion," which emits carbon dioxide and water molecules into the atmosphere (73). A minor amount of hydrochloric acid, ash, and other volatile compounds make up the waste created during incineration. In general, not all plastic garbage is a viable candidate for combustion; some of it is explosive and oxygen-heating resistant. It is not required that all household trash plastics be adequately treated for incineration. To prevent these unforeseen explosive mishaps, we must be cautious when choosing plastics for incineration among non-combustible garbage. Energy, commonly referred to as fuel, can be produced through the combustion of organic molecules. The fuels utilized in the propulsion of cars and airplanes can be found in a variety of physical states, including liquid, solid, and gas. Instead of producing energy, this technique of incineration has numerous positive effects on society. It makes significant contributions to waste reduction and the creation of electricity from waste, both of which are crucial to modern industrialization (74). A key role in the creation of green power from biomass resources has been performed by waste incineration. In addition to being used more than 450 times in Europe, incinerating waste also includes heat recovery (75). In the European Union, recycling and incineration techniques have evolved, with 26% of trash being recycled, 15% composted, 31% being dumped in landfills, and 26% being burned. In comparison, between 1995 and 2013, respectively, 18% and 43% of garbage were recycled and burned (76). The handling of plastic trash through incineration has numerous advantages, including less garbage entering the ecosystem, producing heat and electricity needed for various tasks, lowering air pollution, saving money on waste transport costs, and eliminating dangerous microorganisms and chemicals. Additionally, it can be used in any climate or season and stops the formation of methane gas. Similar to all other biochemical or scientific processes, incineration as a chemical process has both advantages and disadvantages. In comparison to alternative waste management techniques, the incineration process has some drawbacks, such as an expensive setup. It can harm public health and contaminate the environment. It emits waste ash that could be harmful to both people and the environment (77).

10.15.3 LANDFILLS

After being used, plastics are disposed of in various trash cans and end up in landfills. Landfills are any locations or areas where we discard all disposable plastic garbage after use, before burrowing it beneath the surface of the earth. Many safety precautions should be taken throughout this manual disposal process to prevent additional side effects including groundwater pollutants and soil deterioration that can arise from subpar processing (78). To meet the aforementioned goals, landfill arrangements are designed to provide a safer region for the disposal of plastic waste while also protecting aquatic life and airspace. It warrants a lot of work on the part of the community, such as excavating a deep pit or dumping in it, filling it with waste, and allowing it to decay. This procedure is conducted very slowly and may take over a year (79). Each organic molecule goes through biodegradation and decomposition throughout this landfill processing. In landfill processing, plastic bags and other long polymer waste might take between 10 and 100 years to disintegrate (80). Due to their unique biochemical characteristics and environmental elements, including sunlight, wind, and climate change, different plastic waste might take a very long period to degrade (81). To properly dispose of all plastic products, reuse or recycling must be the first option. Due to the emissions of carbon dioxide and methane gas produced during the biodegradation process, landfills make a great energy source. It maintains hygiene, keeps cities clean, and separates hazardous garbage from various types of waste. Additionally, this is a financially sensible way to handle plastic garbage. Even though this approach may be employed to treat plastic trash, it has significant drawbacks, such as contributing to climate change and igniting methane as a combustible gas. Waste incineration has performed an important role in the creation of green power from biomass resources. In addition to being used more than 450 times in Europe, incinerating waste also includes heat recovery (82).

10.15.4 PYROLYSIS

To acquire hydrocarbons and recover crude petrochemicals, a technique known as pyrolysis must be used to transform gases and fatty oils. Even crude petrochemicals are recovered with it, and plastic garbage is used to produce sustainable energy (83). According to how much heat energy is required to destroy plastic connections, the pyrolysis process can be divided into three basic groups. Some media types are based on high, medium, and low temperatures. The range of temperatures utilized to destroy the plastic structure is the basis for the terms "medium" and "high" temperature (83). The corresponding temperature that defines the three pyrolysis states are, respectively, between 600 and 800 °C, and above 800 °C (84). The two primary groups of plastics that are most frequently utilized in human everyday demands are thermosets and thermoplastics, with thermoplastics accounting for around 80% of all used plastics. This is contingent on their simple molecular reformation capability during exposure to heat and their changeability (84). There are two main methods for doing it, and they can be distinguished by the existence or lack of a catalyst. By applying high pressure and temperatures to PW, thermal pyrolysis destroys the compound by a combination of scission, in which the carbon chain splits at its center, along with chain crosslinking and cyclization of linear structures (85). However, catalytic pyrolysis makes use of a catalyst to increase the degradation efficiency and cut back on energy requirements. The outcomes of PW pyrolysis differ depending on the type and quantity of feedstock used as well as the reactor used. These can be divided into three different groups, including char, gaseous, and liquid hydrocarbons. The essential liquid by-products of pyrolysis can be used either directly as fuel or after being mixed with petrol, motor oil, or diesel as long as they maintain the required properties (86). Plastics can be pyrolyzed to produce a variety of products, depending on the reactor type, residence period, plastics used, condensation set up, feeding set up, and temperature used (87, 88). The initial by-products of petroleum refining cuts and petrochemicals are converted to their monomers and other useful components, such as additives and plasters, by catalytic means of chemical processes or heating (89). Tar is an organic compound produced when PW is co-gasified or co-pyrolyzed with other compounds, but its condensable behavior makes its existence in gas-manufacturing facilities difficult due to its proclivity for slagging and inhibiting catalysts (90, 91). Tar is used extensively and contributes value in other industries, despite having a negative impact on gas-producing plants. It has many uses in the coating industry and is used to treat both human and plant ailments (92). Road construction is the most common application for liquid tar (93). In many states, particularly those that are developing, roads are built utilizing a bitumen base and subsequent layers on top of it (94). Over time, these layers are squeezed to create a solid and stable structure. Over time, these layers gradually disintegrate and are replaced. Recently, the technology of coating bitumen with PW has gained prominence (95, 96). The advantages are twofold: non-biodegradable PW may be removed effectively, and PW-based roads have outperformed conventional tar roads (97). One kilometer of road requires over 1 tonne of PW to be paved, which solves the complex problem of waste management and resulting emissions (98). India, the Netherlands, and the United Kingdom are the countries that mostly advocate this method. PW is also being used to build roads in other developing countries like Ghana and Malaysia (99). The addition of PW to bitumen causes it to have a higher melting point and more flexibility, and roads with plastic in them can withstand more rainfall. The stiffening effect of PW is caused by an increase in attraction forces between bitumen and PW, and the bonding forces are further enhanced by oxidizing and linking agents. The PW-modified road can support higher loads over longer distances (100, 101).

10.15.5 CONCRETE PRODUCTION

Because it is so frequently employed in the construction industry, concrete forms the basis for industrial progress (102, 103). Recent studies on concrete have emphasized the use of other components as substitutes for natural aggregates, with a particular emphasis on lighter materials (104, 105). PW might have the ability to be used as an alternative to aggregate in the production of concrete (106, 107, 108, 109). PW incorporation has a negative impact on the strength, and characteristics of the concrete (110). Consequently, these reused plastic aggregates have the capacity to improve several material properties and may be utilized in sound-insulating, heat treatment, and light-weight substances (111). A specific proportion of PW may be employed in structural concrete applications when lower stresses are applied and endurance is not as important (112). According to Gu and Ozbakkaloglu (113), PW is preferred to produce lightweight concrete. Additionally, PW-containing concrete is appropriate for thermal and soundproofing applications because of its improved functional performance (114, 115, 116). It was shown that composites made using PW had a coefficient of thermal conductivity five times lower than typical composites because of the lower conductivity of plastic and the increased porosity of such composites (117). Due to the hydrophobic nature of plastic, thermal conductivity decreased with increasing concentration, causing voids to form in the composites (118). Numerous studies have shown that PW has the potential to be employed as a sound-absorbing substance in concrete because of its porous nature (115). The decreased strength of the concrete is primarily caused by two factors: a higher air content and a poorer ability of plastic particles to connect. Therefore, a greater investigation into the long-term behavior of plastic accumulates in concrete and their effects on the ecosystem and service life is advised. PW has been used in construction for additional purposes such as plastic bottle bricks (119), plastic bottle blocks (120), and plastic fibers in concrete (121).

10.16 USING BIOTECHNOLOGY TO GET RID OF MICROPLASTICS

Extracellular enzymes produced by microbes—mostly bacteria and fungi—help to break down different types of bio and fossil-based polymers (122). These polymers are broken down by bacteria and fungi into CO_2 and H_2O using a variety of metabolic and enzymatic processes. Depending on the microbiological species as well as within the strains, different enzymes have different characteristics and catalytic activities. A variety of enzymes have been found to break down distinct forms of polymer due to this selectivity. For instance, proteases produced by *Bacillus* spp. and *Brevibacillus* spp. are engaged in the degradation of different polymers (123). To catalyze the oxidation of aromatic and non-aromatic chemicals, laccases are typically found in fungi that physiologically break down lignin (124). These microbial enzymes have an effective and environmentally responsible impact on the bio-decay rate of polymers. It has been observed that numerous bacteria and their enzymes are linked to both bio-degradable and non-biodegradable polymers, including PHA, PLA, PET, PHB, PVC, PCL, and PBS (125). Microbes adhering to polymers and then colonizing the surfaces is the main mechanism of plastic biodegradation. Two processes are involved in the hydrolysis of plastics using enzymes. Prior to hydrolytic division, the enzymes first connect to the polymer substrate. Oligomers, dimers, and monomers are examples of polymer degradation products that have significantly lower molecular weight and are eventually mineralized to produce CO_2 and H_2O (126). In an aerobic environment, bacteria employ oxygen as an electron acceptor before synthesizing smaller organic molecules, which results in the production of CO_2 and water by-products (127). Microorganisms break down polymers under anaerobic conditions when there is no oxygen present.

Anaerobic bacteria utilize sulfate, nitrate, iron, carbon dioxide, and manganese as electron acceptors (127). To improve the circumstances under which polymers can be decomposed effectively, novel microbial enzymes and routes must be investigated.

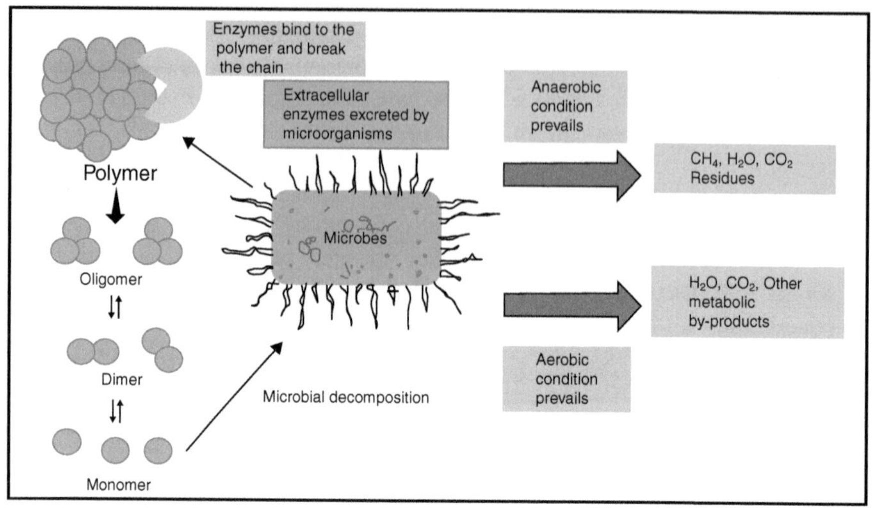

FIGURE 10.3   Plastic biodegradation mechanisms in aerobic and anaerobic environments.

Fossil-based and bio-based polymers that are thought to be non-biodegradable may actually be fully degradable once novel microbial strains and their mechanisms to do so have been fully elucidated. To be used in packaging and the health care sector in a sustainable and environmentally friendly way, new polymers must be added to the category of biodegradable polymers.

Bacillus cereus and *Bacillus gottheilii*, two separate *Bacillus* bacterial strains retrieved from mangrove sediments, were tested to remove various MPs made of polyethylene, polystyrene, polyethylene terephthalate, and polypropylene (128). Electron microscopy and FTIR studies were used to scan the morphological and structural modifications, and the weight loss of the MPs was used to calculate the rate of biodegradation. *B. cereus* isolated with polystyrene MPs provided the fastest mass decrease and lowest half-life of degradation, while *B. gottheilii* on polyethylene provided 0.0016/day and 431.25 days in total. SEM revealed cracks, holes, and erosions, and *B. gottheilii* emerged as a more effective prospective MP degrader (129) (Figure 10.3).

10.16.1   DEGRADATION OF MICROPLASTICS BY ALGAE

The biological degradation of polymeric material can be accomplished by using microalgae, their enzymes, and their toxins (130, 131, 132). When contrasted to the bacterial system, their key benefit is that they do not need a rich supply of carbon for growth and are adaptable to a wide range of habitats, where the majority of the MPs are found (133). In wastewater streams, microalgae are known to colonize plastic surfaces, and this adherence causes the ligninolytic and exopolysaccharide enzymes to start breaking down the plastic. These polymers primarily act as a supply of carbon, increasing cellular proteins and carbohydrates while also speeding up growth.

Scanning electron microscopy has just recently revealed the breakdown or surface deterioration of low-density polyethylene sheets caused by algal colonization (134).

The production of biofilms on the surface of polymers is typically linked to the breakdown of MPs . Several cyanobacterial strains, including those from the genus *Microcystis, Rivularia, Pleurocapsa, Synechococcus, Prochlorothrix, Leptolyngbya, Calothrix,* and *Scytonema,* were also found capable of developing biofilms on the MP polymers (134, 135, 136, 137). Diatoms, which aid in photosynthesis, also thrive in the biofilms in addition to cyanobacterial species (138).

10.16.2 Degradation of Microplastics by Fungi

Aspergillus niger and *Penicillium pinophilum* were responsible for low-density polyethylene degradation, which require physical pretreatment techniques such as thermo-oxidization at 80 °C for 15 days (139). *Aspergillus* spp. and *Lysinibacillus* spp. demonstrated 15.8% and 29.5% biodegradation of non-UV-irradiated and UV-irradiated polymer films, respectively (140).

Fungi produce a wide variety of intracellular and extracellular enzymes that can catalyze a variety of processes and break down polymers made of petroleum. The breakdown and utilization of aliphatic, alicyclic, and aromatic compounds is aided by the epoxidases and transferases of the cytochrome P450 family, which are linked to oxidation and conjugation processes (141, 142). They also use cofactors including heme, $NADPH + H^+$, and FAD to create the spore wall and keep the integrity of the hyphal wall.

Due to their pervasiveness and detrimental effects on both individuals and the ecosystem, MPs are gradually receiving more attention. Due to the extensive usage of single-use personal protective equipment goods after the coronavirus disease pandemic of 2019, MP exposure has multiplied (143, 144, 145). Microplastic exposure can result in health problems like neurotoxicity, endocrine system disruption, carcinogenicity, and metabolic obstacles (146, 147). It can also occur through ingestion, digestion, and dermal absorption. Therefore, effective MP bioremediation is required.

10.16.3 Modern Biotechnology Techniques to Accelerate the Breakdown of Microplastics

Genetic alterations have been introduced to encourage the ability of the bacterial biofilm to trap polyvinyl chloride (148). By genetically modifying *Pseudomonas aeruginosa* and removing the *wspF* gene, it has been possible for *Pseudomonas* to accumulate MPs in its biofilm by increasing the production of sticky exopolymeric compounds. Additionally, the *yhjH* gene was added to the bacterium under the direction of an arabinose-induced promoter. Induced expression of the gene decreased the biofilm formation, sufficient to release the trapped MPs because the function of *yhjH* was to lower the levels of cyclic dimeric guanosine monophosphate. The synthetic "capture and release" technique would make it possible to develop effective MP scavengers for aquatic ecosystem bioremediation (149). Genetic engineering techniques have enabled the modification of the genetic makeup of microorganisms to increase their capacity for biodegradation in the presence of hydrocarbons and toxic metals. However, few studies have been done to develop strains that are better at degrading plastics (150). Enzyme selectivity and affinity toward various MPs are changed using gene editing approaches. The three main procedures are site-directed mutagenesis, antisense RNA technology, and polymerase chain reaction. Gene expression in host cells can be controlled via antisense RNA technology, and the activity of genes involved in the breakdown of MPs can be changed by site-directed mutagenesis. To improve the capacity of *Archaeoglobus fulgidus* to break down polyethylene terephthalate, it has been reported to alter carboxylesterase via in silico site-directed mutagenesis to generate BTA hydrolase (151). A genetically modified strain of *E. coli* and *S. cerevisiae* BY 4741 produces the main enzymes involved in the breakdown of MPs, such as the manganese-dependent peroxidase, and a reconstituted biological strain of *E. coli* BL21 and *P. chrysosporium*-produced laccase enzymes (153, 154). These genetically altered enzymes can degrade polyethylene terephthalate more effectively.

Gene editing techniques have been used to design the genomes of plants, animals, and microbes to express particular genes (153, 154). The manipulation of organisms has become simpler with the development of various gene editing tools, such as zinc finger proteins, transcription activator-like effector nucleases, and clustered regularly interspaced palindromic repeats (CRISPR)/Cas9 (155, 156). The alteration of a gene of significance through genome editing also aids in the elimination and restoration of functional studies that change the expression of several genes. *Streptomyces albogriseolus* LBX-2 has three CRISPR sequences, making it a good candidate for genetic engineering, as oxygenase is the key enzyme involved in the breakdown of polyethylene (157).

10.16.4 METAGENOMIC INVESTIGATION OF MICROBES THAT DEGRADE PLASTIC FOR BIOTECHNOLOGICAL USE

The processing of millions of DNA/RNA fragments and their simultaneous analysis allowed for the examination of a large number of environmental samples because of recent developments in next-generation sequencing technology and bioinformatics tools (158, 159). The metagenomic investigation of bacteria that break down plastic is also possible, and it can move forward by analyzing the structure of the microbial community in the "plastisphere" and discovering new genes or enzymes that break down simple to complex polymers (160, 161). The untapped microbial gene reservoir can be made known in this way for biotechnological application and further exploitation.

The two methods used today for metagenomic study of any local microbial community are structural approach and functional approach (162). In a structural metagenomic technique, environmental sample sequencing was primarily used to reveal the microbial community composition of any specific ecology. It will primarily give a taxonomic identity of microbial population in a culture-independent way, and it can also be used to investigate other properties like the discovery of novel genes and forecasting of gene function with potential applications in intricate metabolic pathways. It sheds light on the microbial population dynamics of the specified ecology at various spatiotemporal scales and determines if the contributions of a member to the evolution of the community structure are modest or significant from a geo-ecological perspective. Conversely, functional metagenomics makes use of the sequence or structural data of genes to identify their function. The prediction of potential desired genes from the metagenome library, followed by their heterologous expression for additional activity-based screening and functional confirmation, begins with the extraction of DNA from ambient sources (163, 164). To help annotate genes from the enormous metagenomic database, the functional metagenomics methodology is utilized as a complement to sequence-based structural metagenomics (165, 166). The "core" microbial population or the particular microbial families, that are remarkably common in "plastispheres" across geographical locations (167, 168, 169) and continually endure for an extended period to carry out an important function of plastic degradation, will also be identified by the comparative metagenomic research in various "plastisphere" of diverse ecology. Moreover, their biological niches (from the marine to the terrestrial), as well as their mechanisms of adaptability, fitness, and survival with the potential for plastic degradation, can all be studied. Therefore, it will be possible to accelerate the degradation process by designing the microbial ecosystem and its metabolic processes in the plastisphere.

10.17 BIOTECHNOLOGICAL IMPUTATION ON METAGENOMIC INFORMATION

Up to this point, substantial strides have been developed in the biotechnological implication of metagenomic data for different microbial ecology, from the human gut to the plant rhizosphere (159, 170), and a variety of goods created from microorganisms are readily available.

10.17.1 Changing the Composition of the Microbial Community

The manipulation of the microbial community makeup in various human, animal, plant, and soil-based habitats is the most traditional and modern technique that is well-established. It consists of several chemical, biological, and molecular techniques for manipulation at a wide scale with increased precision and size (171). Prebiotic (chemical) and probiotic (cellular) techniques can be effectively used on any plastisphere. The prebiotic strategy alters the environment for greater microbial adaptation. For this, a variety of substances, including oligosaccharides and polysaccharides, can alter the microbiome composition and favor the proliferation of plastic degraders specifically. On the plastic surface, chemical substrates such as chitin, cellulose, starch, glycolipids, lipopeptides, etc. operate as biosurfactant and aid in the development of biofilms (172, 173). However, as a biofilm matures, the nutritional content steadily declines because of the increased concentration of carbon, relative to nitrogen, brought about by the breakdown of hydrocarbons. The rate of microbial breakdown is therefore increased by the external supplementation of new resources such as nitrogen and phosphorus (174). Chemical stimulants can be used to promote the emergence of microbes and protect their ecosystem from hazardous pollutants emitted from the deteriorating polymer surface. These stimulants have buffering capabilities to control pH, temperature, and oxygen levels in an appropriate range for the potential activity of bacteria (173, 175, 176, 177).

10.18   ENGINEERING MICROBIOLOGICAL GENETIC ORGANIZATION

The production of genetically modified microorganisms to break down intricate polymers in the environment is becoming practically feasible because of advances in molecular biology and microbial genetics. It is feasible to create programmable biological devices, such as synthetic cells or synthetic life, with precise and unique features in the age of systems biology due to the ease with which biosynthetic units or genetic circuits may be created (178). Using genetic engineering, protein and enzyme engineering, and genome modification technologies, these "synthetic" microbial cells can be produced and then used to modify the microbiome of the plastisphere. Either *ex situ* custom microbe design or *in situ* metagenome alteration can be used to genetically manipulate the microbiome. Complex polymeric plastic waste biodegrades through a variety of oxidation stages to transform into monomers before being assimilated into the TCA cycle (179). No single species could complete all of these diverse biodegradation stages (159). Genetic engineering technologies can be used to supplement several genes from whole plastic metabolic routes or an entire gene circuit into various microbial species to solve the problem. The genetically modified bacteria may be made to produce various enzymes, control quorum sensing mechanisms for biofilm development and other functions. As proof, PETase, a crucial enzyme for the breakdown of PET, may be expressed through genetic engineering in bacteria other than its native host *Ideonella sakaiensis* (181), including *Bacillus subtilis* and *Escherichia coli* (182, 183). The PETase enzyme in *Bacillus subtilis* was 3.8-fold expressed using native signal peptide and Tat-independent secretory pathway, while in *Escherichia coli*, it was produced extracellularly using the Sec system. PET film degradation experiment was used to determine functionality (182).

10.19   EMPLOYING BIOINFORMATICS STRATEGY

The Environmental Contaminant Biotransformation Pathway Resource, MetaCyc database, and BioCyc database, among others, have been created as databases that provide details on the metabolic pathways, microbial enzymes, and genes involved in the biodegradation process. These databases have been used to evaluate the biodegradation process (184, 185, 186, 187).

A novel strategy for the decomposition of plastic can be developed using the databases and computational approaches that assist identification of the enzymes participating in the process of metabolism of interest and forecast the biodegradation routes of hazardous compounds (188).

Despite all these benefits, the main drawback of bioinformatics is the scarcity of field-based data and the validation of that data, both of which are essential for future research. Additionally, there is a significant information gap between the relevant enzymes of different groups of bacteria that break down synthetic polymers. Therefore, a thorough examination is needed to determine the best metabolic routes for the disintegration of polymers and the enzymes that go along with them. In the near future, a combination of methods utilizing metabolic engineering, genetics, molecular and systems biology, as well as bioinformatics tools, may aid in our search for an appropriate and long-term solution for the biodegradation of MPs (189).

10.20 CONCLUSION

Plastics have a favorable influence on our daily lives because they are commonly used in home tasks and commercial packaging for all kinds of end products. Due to their physicochemical makeup, plastics play a variety of significant functions in our lives, but if their disposal after use is not properly managed, they can have negative effects on both human life and the ecosystem. In order for plastic life to be cyclic, we have shown a variety of strategies that can be used to treat waste plastic. The use of these techniques will save money by recycling raw materials and reusing plastics, as well as the lives of people, animals, and the environment (190).

The problem of managing solid waste will be resolved and raw material reserves will be depleted by the use of PW in construction applications. Moreover, it promotes the circular economy sustainability trend by providing a way to utilize these trashes for long-term as opposed to short-term applications. PW can replace every component in cementitious composites with some tolerable negative consequences on the efficiency of the finished composite due to its potential use as a binder, aggregate, and fiber. Utilizing PW to fulfill different construction applications will result in a variety of revenue streams (191).

In conclusion, it is the holistic problem regarding the way we make use of our resources that should worry us because it can only be solved using a "systems thinking" strategy. Given that most people on earth reside in cities and towns, it will be essential to include integrated ('systems') design methodologies in the current urban infrastructure, such as permaculture and the circular economy. It is evident that the only way we can effectively "care for our common home" and chart a viable course for the future of humanity is by switching from the current linear, "take, make, dispose (waste-creation)" model of resource consumption to the systemic, circular alternative of "reduce, reuse, recycle regenerate" (192).

Genetic engineering is still only permissible on a laboratory scale due to inefficiency and environmental safety concerns, but advances made in genome modification and synthetic biology have enabled the programmable change of microbial genomes. These methods can be used for gene overexpression or repression in addition to deletion or gene insertion (193). The improvement of environmental tolerance in microbial species is another benefit of gene control by deactivated Cas (dCas) (194). Genome editing aids in the customization of microbes for the manufacture of innovative enzymes and additional metabolites (195); as a result, altered bacteria are a better probiotic target. The latest methods aim to biologically design or model bacteria at the genome scale to produce desired phenotypes. Broad-spectrum applications therefore necessitate the introduction of engineered regulatory gene networks in living cells (196). Systems biology and gene editing, together with advanced bioinformatics techniques, may provide novel approaches to the plastic problem.

ACKNOWLEDGMENTS

Financial assistance from the Science and Engineering Research Board, Government of India, and the Department of Science and Technology and Biotechnology, Government of West Bengal are gratefully acknowledged.

REFERENCES

1. Yogalakshmi, K.N., Singh, S. (2020). Plastic waste: Environmental hazards, is biodegradation, and challenges. In: Saxena, G., Bharagava, R. (eds) *Bioremediation of Industrial Waste for Environmental Safety*. Springer, Singapore. https://doi.org/10.1007/978-981-13-1891-7_6

2. Alabi, O.A., Ologbonjaye, K.I., Awosolu, O., Alalade, O.E. Public and environmental health effects of plastic wastes disposal: A review. *J. Toxicol. Risk Assess.* 2019;5(021). doi: 10.23937/2572-4061.1510021

3. www.conserve-energy-future.com/causes-effects-solutions-of-plastic-pollution.php

4. Dubaish, F., Liebezeit, G. Suspended microplastics and black carbon particles in the Jade System, Southern North Sea. *Water Air Soil Pollut.* 2013 Jan 17;224(2):1352. doi: 10.1007/s11270-012-1352-9

5. Zubris, K.A., Richards, B.K. Synthetic fibers as an indicator of land application of sludge. *Environ. Pollut.* 2005 Nov;138(2):201–11. PMID: 15967553. doi: 10.1016/j.envpol.2005.04.013

6. Knight, G. (2012) Plastic Pollut. Raintree Publisher, London.

7. Eriksen, M., Lebreton, L.C., Carson, H.S., Thiel, M., Moore, C.J., Borerro, J.C., Galgani, F., Ryan, P.G., Reisser, J. Plastic pollution in the world's oceans: More than 5 trillion plastic pieces weighing over 250,000 tons afloat at sea. *PLoS One.* 2014 Dec 10;9(12):e111913. PMID: 25494041; PMCID: PMC4262196. doi: 10.1371/journal.pone.0111913

8. Mato, Y., Isobe, T., Takada, H., Kanehiro, H., Ohtake, C., Kaminuma, T. Plastic resin pellets as a transport medium for toxic chemicals in the marine environment. *Environ. Sci. Technol.* 2001 Jan 15;35(2):318–24. PMID: 11347604. doi: 10.1021/es0010498

9. Laist, D.W. (1997). Impacts of Marine Debris: Entanglement of Marine Life in Marine Debris Including a Comprehensive List of Species with Entanglement and Ingestion Records. In: Coe, J.M., Rogers, D.B. (eds) *Marine Debris. Springer Series on Environmental Management*. Springer, New York, NY. https://doi.org/10.1007/978-1-4613-8486-1_10

10. Hamlet, C., Matte, T., Mehta, S. Combating Plastic Air Pollution on Earth's Day. Vital Strategies Environmental Health Division. 2018.

11. www.unep.org/interactives/beat-plastic-pollution/#:~:text=Cigarette%20butts%20%E2%80%94%20whose%20filters%20contain,the%20next%20most%20common%20items

12. Rodriguez, F. "Plastic". Encyclopedia Britannica, 9 May 2023. www.britannica.com/science/plastic Accessed 10 June 2023.

13. https://theconstructor.org/building/plastics-construction-material/12438/

14. Rogers, K. "Microplastics". Encyclopedia Britannica, 5 April 2022. www.britannica.com/technology/microplastic Accessed 27 May 2023.

15. Frias, J., Nash, R. Microplastics: Finding a consensus on the definition. Mar. Pollut. Bull. 2018;138:145–147. doi: 10.1016/j.marpolbul.2018.11.022

16. Hahladakis, N.J., Costas, A.V., Weber, R., Iacovidou, E., Purnell, P. An overview of chemical additives present in plastics: Migration, release, fate and environmental

impact during their use, disposal and recycling. *J. Hazard. Mater.* 2018;344:179–199. doi: 10.1016/j.jhazmat.2017.10.014

17. Lamichhane, G., Acharya, A., Marahatha, R. et al. Microplastics in environment: Global concern, challenges, and controlling measures. *Int. J. Environ. Sci. Technol.* 2023;20:4673–4694. https://doi.org/10.1007/s13762-022-04261-1

18. Lai, H., Liu, X., Qu, M. Nanoplastics and human health: Hazard identification and biointerface. *Nanomaterials (Basel).* 2022 April11;12(8):1298. PMID: 35458006; PMCID: PMC9026096. doi: 10.3390/nano12081298

19. Joksimovic, N., Selakovic, D., Jovicic, N., Jankovic, N., Pradeepkumar, P., Eftekhari, A., Rosic, G. Nanoplastics as an invisible threat to humans and the environment. *J. Nanomater.* 2022; 6707819, 15 pages. https://doi.org/10.1155/2022/6707819

20. Britannica, The Editors of Encyclopaedia. "polymethyl methacrylate". Encyclopedia Britannica, 25 May 2023. www.britannica.com/science/polymethyl-methacrylate. Accessed 8 June 2023.

21. Dissanayake, P.D. Soobin kim effects of microplastics on the terrestrial environment: A critical review. *Environ. Res.* 2022 June;209:112734. https://doi.org/10.1016/j.envres.2022.112734

22. Ward, J.E., Kach, D.J. Marine aggregates facilitate ingestion of nanoparticles by suspension-feeding bivalves. *Mar. Environ. Res.* 2009;68(3):137–42. doi: 10.1016/j.marenvres.2009.05.002

23. Wright, S.L., Thompson, R.C., Galloway T.S. The physical impacts of microplastics on marine organisms: a review. *Environ. Pollut.* 2013 Jul;178:483–92. doi: 10.1016/j.envpol.2013.02.031

24. Murray, F., Cowie, P.R. Plastic contamination in the decapod crustacean *Nephrops norvegicus* (Linnaeus, 1758). *Mar. Pollut. Bull.* 2011 Jun;62(6):1207–17. doi: 10.1016/j.marpolbul.2011.03.032

25. Graham, E.R.,. Thompson, J.T. Deposit- and suspension-feeding sea cucumbers (Echinodermata) ingest plastic fragments. *J. Exp. Mar. Biol. Ecol.* 2009;368(1):22–29. ISSN 0022-0981. https://doi.org/10.1016/j.jembe.2008.09.007

26. Lithner, D., Larsson, Å., Dave, G. Environmental and health hazard ranking and assessment of plastic polymers based on chemical composition. Sci. Total Environ. 2011;409(18):3309–24, ISSN 0048-9697. https://doi.org/10.1016/j.scitotenv.2011.04.038

27. www.was.org/articles/Plastics-in-Aquaculture-The-WAS-IMarEST-Roundtables. aspx#.YkaASKgo9LM

28. Farrell, P., Nelson, K. Trophic level transfer of microplastic: *Mytilus edulis* (L.) to *Carcinusmaenas* (L.). *Environ. Pollut.* 2013;177:1–3. ISSN 0269-7491. https://doi.org/10.1016/j.envpol.2013.01.046

29. Eriksson C, Burton H. Origins and biological accumulation of small plastic particles in fur seals from Macquarie Island. *Ambio.* 2003 Sep;32(6):380–4. PMID: 14627365. doi: 10.1579/0044-7447-32.6.380

30. Carpenter, E.J., Smith, K.L., Jr. Plastics on the Sargasso sea surface. *Science.* 1972 Mar 17;175(4027):1240–1. doi: 10.1126/science.175.4027.1240

31. Bhattacharya, P., Lin, S., Turner, J.P., Ke, P.C. *J. Phys. Chem. C.* 2010;114(39):16556–61. doi: 10.1021/jp1054759

32. Hirai, H., Takada, H., Ogata, Y., Yamashita, R., Mizukawa, K., Saha, M., Kwan, C., Moore, C., Gray, H., Laursen, D., Zettler, E.R., Farrington, J.W., Reddy, K.M., Peacock, E.E., Ward, M.W. Organic micropollutants in marine plastics debris from the open ocean and remote and urban beaches. *Mar. Pollut. Bull.* 2011;62(8):1683–92. ISSN 0025-326X. https://doi.org/10.1016/j.marpolbul.2011.06.004

33. Hitchcock, J.N. Storm events as key moments of microplastic contamination in aquatic ecosystems. *Sci. Total Environ.* 2020;734:139436. ISSN 0048-9697. https://doi.org/10.1016/j.scitotenv.2020.139436

34. Kumar, A., Mishra, S., Pandey, R., Yu, Z.G., Kumar, M., Khoo, K.S., Thakur, T.K., Show, P.L., Microplastics in terrestrial ecosystems: Un-ignorable impacts on soil characterises, nutrient storage and its cycling. *TrAC Trends Anal. Chem.* 2023;158:116869. ISSN 0165-9936. https://doi.org/10.1016/j.trac.2022.116869

35. Khalid, N., Aqeel, M., Noman, A. Microplastics could be a threat to plants in terrestrial systems directly or indirectly. *Environ. Pollut.* 2020;267, , 115653. ISSN 0269-7491. https://doi.org/10.1016/j.envpol.2020.115653

36. Yin Liu, Qing Huang, Wen Hu, Jiemin Qin, Yingrui Zheng, Junfeng Wang, Qingqing Wang, Yuxin Xu, Genmao Guo, Shan Hu, Li Xu, Effects of plastic mulch film residues on soil-microbe-plant systems under different soil pH conditions. *Chemosphere* 2021;267:128901. ISSN 0045-6535. https://doi.org/10.1016/j.chemosphere.2020.128901

37. Yan, Y., Chen, Z., Zhu, F. et al. Effect of polyvinyl chloride microplastics on bacterial community and nutrient status in two agricultural soils. *Bull. Environ. Contam. Toxicol.* 2021;107:602–9.

38. Seeley, M.E., Song, B., Passie, R., Hale, R.C. Microplastics affect sedimentary microbial communities and nitrogen cycling. *Nat. Commun.* 2020 May 12;11(1):2372. PMID: 32398678; PMCID: PMC7217880. doi: 10.1038/s41467-020-16235-3

39. Lwanga, E.H., Vega, J.M., Ku Quej, V., et al. Field evidence for transfer of plastic debris along a terrestrial food chain. *Sci. Rep.* 2017;7:14071. https://doi.org/10.1038/s41598-017-14588-2

40. Andrés, R.-S., Bruna, S., Eduardo, F.da.S., Anabela, C., Ruth, P. Low-density polyethylene microplastics as a source and carriers of agrochemicals to soil and earthworms. *Environ. Chem.* 2019;16:8–17. https://doi.org/10.1071/EN18162

41. de Souza Machado, A.A., Kloas, W., Zarfl, C., Hempel, S., Rillig, M.C. Microplastics as an emerging threat to terrestrial ecosystems. *Glob. Change Biol.* 2018;24:1405–16. https://doi.org/10.1111/gcb.14020

42. Qi, Y., Beriot, N., Gort, G., Lwanga, E.H., Gooren, H., Yang, X., Geissen, V. Impact of plastic mulch film debris on soil physicochemical and hydrological properties. *Environ. Pollut.* 2020;266(Part 3):115097. ISSN 0269-7491. https://doi.org/10.1016/j.envpol.2020.115097

43. Zhu, D., Chen, Q.-L., An, X.-L., Yang, X.-R., Christie, P., Ke, X., Wu, L.-H., Zhu, Y.-G. Exposure of soil collembolans to microplastics perturbs their gut microbiota and alters their isotopic composition. *Soil Biol. Biochem.* 2018;116:302–10. ISSN 0038-0717. https://doi.org/10.1016/j.soilbio.2017.10.027

44. Lwanga, E.H., Gertsen, H., Gooren, H., Peters, P., Salánki, T., van der Ploeg, M., Besseling, E., Koelmans, A.A., Geissen, V. Incorporation of microplastics from litter into burrows of Lumbricusterrestris. *Environ. Pollut.* 2017;220(Part A):523–31. ISSN 0269-7491. https://doi.org/10.1016/j.envpol.2016.09.096

45. Lwanga, E.H., Gertsen, H., Gooren, H., Peters, P., Salánki, T., van der Ploeg, M., Besseling, E.,Koelmans, A.A., Geissen, V. *Environ. Sci. Technol.* 2016;50(5), 2685–91 doi: 10.1021/acs.est.5b05478

46. Rodriguez-Seijo, A., Lourenco, J., Rocha-Santos, TaP., Da Costa, J., Duarte, A. C., Vala, H., & Pereira, R. Histopathological and molecular effects of microplastics in Eisenia andrei Bouche. *Environ. Pollut.* 2017;220:495–503. https://doi.org/10.1016/j.envpol.2016.09.092

47. Hodson, M.E., Duffus-Hodson, C.A., Clark, A., Prendergast-Miller, M.T., & Thorpe, K.L. Plastic bag derived-microplastics as a vector for metal exposure in terrestrial invertebrates. *Environ. Sci. Technol.* 2017;51:4714–21. https://doi.org/10.1021/acs.est.7b00635

48. Miyazaki, J., Kuriyama, Y., Miyamoto, A., Tokumoto, H., Konishi, Y., Nomura, T. Adhesion and internalization of functionalized polystyrene latex nanoparticles toward the yeast *Saccharomyces cerevisiae. Adv. Powder Technol.* 2014;25:1394–1397. https://doi.org/10.1016/j.apt.2014.06.014

49. Nomura, T., Tani, S., Yamamoto, M., Nakagawa, T., Toyoda, S., Fujisawa, E., ... Konishi, Y. Cytotoxicity and colloidal behavior of polystyrene latex nanoparticles toward filamentous fungi in isotonic solutions. *Chemosphere* 2016;149:84–90. https://doi.org/10.1016/j.chemosphere.2016.01.09

50. Mutua, P.M., Gicheru, M.M., Makanya, A.N., Kiama, S.G. Comparative quantitative and qualitative attributes of the surface respiratory macrophages in the domestic duck and the rabbit. *Intl J. Morphol.* 29:353–62. https://doi.org/10.4067/S0717-950220 11000200008

51. Gallo, F., Fossi, C., Weber, R., Santillo, D., Sousa, J., Ingram, I., et al. Marine litter plastics and microplastics and their toxic chemicals components: The need for urgent preventive measures. *Environ. Sci. Europe* 2018;30(1):1–14. doi: 10.1186/s12302-018-0139-z

52. Boas, M., Frederiksen, H., Feldt-Rasmussen, U., Skakkebæk, N.E., Hegedüs, L., Hilsted, L., et al. Childhood exposure to phthalates: Associations with thyroid function, insulin-like growth factor I, and growth. *Environ. Health Perspect.* 2010;118(10):1458–64. doi: 10.1289/ehp.0901331

53. Mathieu-Denoncourt, J., Wallace, S.J., de Solla, S.R., Langlois, V.S. Plasticizer endocrine disruption: Highlighting developmental and reproductive effects in mammals and non-mammalian aquatic species. *Gen. Comp. Endocrinol.* 2015;219:74–88. doi: 10.1016/j.ygcen.2014.11.003

54. Andra, S.S., Makris, K.C. Thyroid disrupting chemicals in plastic additives and thyroid health. *J. Environ. Sci. Health. Part C* 2012;30(2):107–51. doi: 10.1080/10590501.2012.681487

55. Ayala, A.R., Danese, M.D., Ladenson, P.W. When to treat mild hypothyroidism. *Endocrinol. Metab. Clinics* 2000;29(2):399–415. doi: 10.1016/S0889-8529(05)70139-0

56. Moriyama, K., Tagami, T., Akamizu, T., Usui, T., Saijo, M., Kanamoto, N., et al. Thyroid hormone action is disrupted by bisphenol a as an antagonist. *J. Clin. Endocrinol. Metab.* 2002;87(11):5185–90. doi: 10.1210/jc.2002-020209

57. Ishihara, A., Sawatsubashi, S., Yamauchi, K. Endocrine disrupting chemicals: Interference of thyroid hormone binding to transthyretins and to thyroid hormone receptors. Mol. Cell. Endocrinol. 2003;199(1–2):105–17. doi: 10.1016/S0303-7207(02)00302-7

58. Derakhshan, A., Shu, H., Broeren, M.A.C., Lindh, C.H., Peeters, R.P., Kortenkamp, A., et al. Association of phthalate exposure with thyroid function during pregnancy. *Environ. Int.* 2021;157:106795. doi: 10.1016/j.envint.2021.106795

59. Olukole, S.G., Lanipekun, D.O., Ola-Davies, E.O., Bankole, O.O. Melatonin attenuates bisphenol A—induced toxicity of the adrenal gland of wistar rats. *Environ. Sci. Pollut. Res.* 2019;26(6):5971–82. doi: 10.1007/s11356-018-4024-5

60. Campioli E, Martinez-Arguelles DB, Papadopoulos V. In utero exposure to the endocrine disruptor di-(2-ethylhexyl) phthalate promotes local adipose and

systemic inflammation in adult male offspring. Nutr. Diabetes 2014;4(5):e115–e115. doi: 10.1038/nutd.2014.13

61. Laskar, N., Upendra, K. Plastics and microplastics: A threat to environment. *Environ. Technol. Innovation.* 2019;14:100352. doi: 10.1016/j.eti.2019.100352

62. Fujimoto, N. Effects of endocrine disruptors on the pituitary gland. *J. Toxicol. Pathol.* 2001;14(1):65–5. doi: 10.1293/tox.14.65

63. Graceli, J.B., Dettogni, R.S., Merlo, E., Niño, O., da Costa, C.S., Zanol, J.F., et al. The impact of endocrine-disrupting chemical exposure in the mammalian hypothalamic–pituitary axis. *Mol. Cell Endocrinol.* 2020;518:110997. doi: 10.1016/j.mce.2020.110997

64. Fernandez, M.O., Bourguignon, N.S., Arocena, P., Rosa, M., Libertun, C., Lux-Lantos, V. Neonatal exposure to bisphenol A alters the hypothalamic–pituitary–thyroid axis in female rats. Toxicol. Lett. 2018;285:81–6. doi: 10.1016/j.toxlet.2017.12.029

65. Bjørklund, G., Chirumbolo, S., Dadar, M., Pivina, L., Lindh, U., Butnariu, M., et al. Mercury exposure and its effects on fertility and pregnancy outcome. *Basic Clin. Pharmacol. Toxicol.* 2019;125(4):317–27. doi: 10.1111/bcpt.13264

66. Sabir, S., Hamid Akash, M.S., Fiayyaz, F., Saleem, U., Mehmood, M.H., Rehman, K. Role of cadmium and arsenic as endocrine disruptors in the metabolism of carbohydrates: Inserting the association into perspectives. Biomed. Pharmacother. 2019;114:108802. doi: 10.1016/j.biopha.2019.108802

67. Adhikary, K.B., Shusheng, P., Staiger, M.P. Dimensional stability and mechanical behaviour of wood–plastic composites based on recycled and virgin high-density polyethylene (HDPE). *Composites Part B: Engineering* 2008;39(5):807–15. ISSN 1359-8368. https://doi.org/10.1016/j.compositesb.2007.10.005

68. Martinez Lopez, Y., Paes, J.B., Gustave, D., Gonçalves, F.G., Méndez, F.C., Theodoro Nantet, A.C. Production of wood-plastic composites using cedrela odorata sawdust waste and recycled thermoplastics mixture from post-consumer products—a sustainable approach for cleaner production in Cuba. *J. Clean. Product.* 2020;244: 118723. doi: 10.1016/j.jclepro.2019.118723

69. Szostak, E., et al. Characteristics of plastic waste processing in the modern recycling plant operating in Poland. Energies 2020;14(1):35.

70. Vollmer, I., et al. Beyond mechanical recycling: Giving new life to plastic waste. Angew. Chemie Int. Ed. 2020;59(36):15402–23.

71. Shen, L., Worrell, E. Chapter 13—Plastic recycling. In: Ernst Worrell, M.A. (ed) *Reuter, Handbook of Recycling.* Elsevier, 2014:179–190. ISBN 9780123964595. https://doi.org/10.1016/B978-0-12-396459-5.00013-1

72. Zheng, Y., Yanful, E.K., Bassi, A.S. A review of plastic waste biodegradation. *Crit. Rev. Biotechnol.* 2005;25(4):243–50. doi: 10.1080/07388550500346359

73. Shome, R. Role of microbial enzymes in bioremediation. *ELifePress* 2020;1:15–20.

74. Weerdt, L.D., Sasao, T., Compernolle, T., Passel, S.V., De Jaeger, S. The effect of waste incineration taxation on industrial plastic waste generation: A panel analysis. *Resour. Conser. Recycl.* 2020;157:104717. ISSN 0921-3449. https://doi.org/10.1016/j.resconrec.2020.104717

75. European Commission. Green Paper on a European Strategy on Plastic Waste in the Environment. European Commission, Brussels, Belgium, 2013.

76. Block, C., Van Caneghem, J., Van Brecht, A., et al. Incineration of hazardous waste: A sustainable process? *Waste Biomass Valor* 2015;6:137–145. https://doi.org/10.1007/s12649-014-9334-3

77. Netzer, F.P., Noguera, P. *'Introduction', Oxide Thin Films and Nanostructures* (online ed.). Oxford Academic, Oxford, 22 Apr. 2021. https://doi.org/10.1093/oso/978019 8834618.003.0001

78. Zheng, Y., Yanful, E.K., Bassi, A.S., A review of plastic waste biodegradation. *Crit. Rev. Biotechnol.* 2005;25(4):243–50. doi: 10.1080/07388550500346359

79. Liang, Y., Tan, Q., Song, Q., Li, J. An analysis of the plastic waste trade and management in Asia. *Waste Manag.* 2021;119:242–53. ISSN 0956-053X. https://doi.org/10.1016/j.wasman.2020.09.049

80. Thiounn, T., Smith, R.C. Advances and approaches for chemical recycling of plastic waste. *J. Polym. Sci.* 2020;58:1347–1364. https://doi.org/10.1002/pol.20190261

81. Zhou, H., Fang, W., Xu, W., Cao, A., Wang, R. Characteristics and the recovery potential of plastic wastes obtained from landfill mining. *J. Clean. Product.* 2014;80:80–6. ISSN 0959-6526. https://doi.org/10.1016/j.jclepro.2014.05.083

82. Kedzierski, M., Frère, D., Le Maguer, G., Bruzaud, S. Why is there plastic packaging in the natural environment? Understanding the roots of our individual plastic waste management behaviours. *Sci. Total Environ.* 2020;740: 139985. ISSN 0048-9697. https://doi.org/10.1016/j.scitotenv.2020.139985

83. Sharuddin, S.D.A., et al. A review on pyrolysis of plastic wastes. *Energy Conver. Manag.* 2016;115:308–26.

84. Nabi, G., et al. Smooth growth, characterization and optical properties of Cu_2SnS_3 thin film via spray pyrolysis method. *Phys. B Conden. Mat.* 2021;602:412498.

85. Panda, A.K., Singh, R.K., Mishra, D.K. Thermolysis of waste plastics to liquid fuel: A suitable method for plastic waste management and manufacture of value added products—a world prospective. *Renew. Sustain. Energy Rev.* 2010;14:233–48. doi: 10.1016/j.rser.2009.07.005

86. Sharma, B.K., Moser, B.R., Vermillion, K.E., Doll, K.M., Rajagopalan, N. Production, characterization and fuel properties of alternative diesel fuel from pyrolysis of waste plastic grocery bags. *Fuel Process. Technol.* 2014;122:79–90. doi: 10.1016/j.fuproc.2014.01.019

87. Miandad, R., Barakat, M.A., Aburiazaiza, A.S., Rehan, M., Nizami, A.S. Catalytic pyrolysis of plastic waste: A review. *Proc. Safety Environ. Protect.* 2016;102:822–38. ISSN 0957-5820. https://doi.org/10.1016/j.psep.2016.06.022

88. Qureshi, M.S., Oasmaa, A., Pihkola, H., Deviatkin, I., Tenhunen, A., Mannila, J., Minkkinen, H., Pohjakallio, M., Laine-Ylijoki, J. Pyrolysis of plastic waste: Opportunities and challenges. *J. Anal. Appl. Pyrol.* 2020;152:104804. ISSN 0165-2370. https://doi.org/10.1016/j.jaap.2020.104804

89. Shent, H., Pugh, R.J., Forssberg, E. A review of plastics waste recycling and the flotation of plastics. *Resour. Conser. Recycl.* 1999;25(2):85–109. ISSN 0921-3449. https://doi.org/10.1016/S0921-3449(98)00017-2

90. Arena, U., Zaccariello, L., Mastellone, M.L. Tar removal during the fluidized bed gasification of plastic waste. *Waste Manag.* 2009;29:783–91. doi: 10.1016/j.wasman.2008.05.010

91. Mastellone, M.L., Arena, U. Olivine as a tar removal catalyst during fluidized bed gasification of plastic waste. *AIChE J.* 2008;54:1656–67. doi: 10.1002/aic.11497

92. Kurt, Y., Isik, K. Comparison of tar produced by traditional and laboratory methods. *Stud. Ethno-Med.* 2012;6:77–83. doi: 10.1080/09735070.2012.11886423

93. Limantara, A.D., Gardjito, E., Ridwan, A., Subiyanto, B., Raharjo, D., Santoso, A., Heryanto, B., Sudarmanto, H.L. Comparative study of bio-asphalt, coconut shell distillation tar, and plastic road in terms of construction, economical, and regulatory aspects. *J. Phys. Conf. Ser.* 2019;1364:012058. doi: 10.1088/1742-6596/1364/1/012058

94. Siham, K., Fabrice, B., Edine, A.N., Patrick, D. Marine dredged sediments as new materials resource for road construction. *Waste Manag.* 2008;28:919–28. doi: 10.1016/j.wasman.2007.03.027

95. Chavan, M.A.J. Use of plastic waste in flexible pavements. *Int. J. Appl. Innov. Eng. Manag.* 2013;2:540–52.

96. Caputo, P., Porto, M., Loise, V., Teltayev, B., Rossi, C.O. Analysis of mechanical performance of bitumen modified with waste plastic and rubber additives by rheology and self diffusion NMR experiments. *Eurasian Chem.-Technol. J.* 2019;21:235–239. doi: 10.18321/ectj864

97. Manju, R., Sathya, S., Sheema, K. Use of plastic waste in bituminous pavement. *Int. J. Chem. Tech. Res.* 2017;10:804–811.

98. Vasudevan, R., Sekar, A.R.C., Sundarakannan, B., Velkennedy R. A technique to dispose waste plastics in an ecofriendly way—application in construction of flexible pavements. *Constr. Build. Mater.* 2012;28:311–320. doi: 10.1016/j.conbuildmat.2011.08.031

99. Gopinath, K.P., Nagarajan, V.M., Krishnan, A., Malolan, R. A critical review on the influence of energy, environmental and economic factors on various processes used to handle and recycle plastic wastes: Development of a comprehensive index. *J. Clean. Prod.* 2020;274:123031. doi: 10.1016/j.jclepro.2020.123031

100. Jafar, J.J. Utilisation of waste plastic in bituminous mix for improved performance of roads. *KSCE J. Civ. Eng.* 2016;20:243–9. doi: 10.1007/s12205-015-0511-0

101. Behl, A., Sharma, G., Kumar, G. A sustainable approach: Utilization of waste PVC in asphalting of roads. *Constr. Build. Mater.* 2014;54:113–117. doi: 10.1016/j.conbuildmat.2013.12.050

102. Ahmad, W., Farooq, S.H., Usman, M., Khan, M., Ahmad, A., Aslam, F., Yousef, R.A., Abduljabbar, H.A., Sufian, M. Effect of coconut fiber length and content on properties of high strength concrete. *Materials.* 2020;13:1075. doi: 10.3390/ma13051075

103. Khan, M., Ali, M. Improvement in concrete behavior with fly ash, silica-fume and coconut fibres. Constr. Build. Mater. 2019;203:174–87. doi: 10.1016/j.conbuildmat.2019.01.103

104. Saikia, N., De Brito, J. Use of plastic waste as aggregate in cement mortar and concrete preparation: A review. *Constr. Build. Mater.* 2012;34:385–401. doi: 10.1016/j.conbuildmat.2012.02.066

105. Ismail, Z.Z., Al-Hashmi, E.A. Use of waste plastic in concrete mixture as aggregate replacement. *Waste Manag.* 2008;28:2041–7. doi: 10.1016/j.wasman.2007.08.023

106. Mohammed A.A., Mohammed I.I., Mohammed S.A. Some properties of concrete with plastic aggregate derived from shredded PVC sheets. *Constr. Build. Mater.* 2019;201:232–45. doi: 10.1016/j.conbuildmat.2018.12.145

107. Manjunath B.T.A. Partial replacement of E-plastic waste as coarse-aggregate in concrete. *Procedia Environ. Sci.* 2016;35:731–9. doi: 10.1016/j.proenv.2016.07.079

108. Basha, S.I., Ali, M.R., Al-Dulaijan, S.U., Maslehuddin, M. Mechanical and thermal properties of lightweight recycled plastic aggregate concrete. J. Build. Eng. 2020;32:101710. doi: 10.1016/j.jobe.2020.101710

109. Alaloul, W.S., John, V.O., Musarat, M.A. Mechanical and thermal properties of interlocking bricks utilizing wasted polyethylene terephthalate. *Int. J. Concr. Struct. Mater.* 2020;14:1–11. doi: 10.1186/s40069-020-00399-9

110. Liguori, B., Iucolano, F., Capasso, I., Lavorgna, M., Verdolotti, L. The effect of recycled plastic aggregate on chemico-physical and functional properties of composite mortars. *Mater. Des.* 2014;57:578–584. doi: 10.1016/j.matdes.2014.01.006

111. Alyousef, R., Ahmad, W., Ahmad, A., Aslam, F., Joyklad, P., Alabduljabbar, H. Potential use of recycled plastic and rubber aggregate in cementitious materials for sustainable construction: A review. *J. Clean. Prod.* 2021;329:129736. doi: 10.1016/j.jclepro.2021.129736

112   Babafemi, A.J., Šavija, B., Paul, S.C., Anggraini, V. Engineering properties of concrete with waste recycled plastic: A review. *Sustainability* 2018;10:3875. doi: 10.3390/su10113875

113.   Gu L., Ozbakkaloglu T. Use of recycled plastics in concrete: A critical review. *Waste Manag.* 2016;51:19–42. doi: 10.1016/j.wasman.2016.03.005

114.   Yesilata, B., Isıker, Y., Turgut, P. Thermal insulation enhancement in concretes by adding waste PET and rubber pieces. *Constr. Build. Mater.* 2009;23:1878–82. doi: 10.1016/j.conbuildmat.2008.09.014

115.   Rahman M.M., Islam M.A., Ahmed M., Salam M.A. Recycled polymer materials as aggregates for concrete and blocks. *J. Chem. Eng.* 2012;27:53–57. doi: 10.3329/jce.v27i1.15859.

116.   Rahmani E., Dehestani M., Beygi M., Allahyari H., Nikbin I. On the mechanical properties of concrete containing waste PET particles. *Constr. Build. Mater.* 2013;47:1302–1308. doi: 10.1016/j.conbuildmat.2013.06.041.

117.   Murugan, D., Varughese, S., Swaminathan, T. Recycled polyolefin-based plastic wastes for sound absorption. *Polym.-Plast. Technol. Eng.* 2006;45:885–8. doi: 10.1080/03602550600611818

118.   Wang, R., Meyer, C. Performance of cement mortar made with recycled high impact polystyrene. *Cem. Concr. Compos.* 2012;34:975–81. doi: 10.1016/j.cemconcomp.2012.06.014

119.   Safinia, S., Alkalbani, A. Use of recycled plastic water bottles in concrete blocks. *Procedia Eng.* 2016;164:214–21. doi: 10.1016/j.proeng.2016.11.612

120.   Mokhtar, M., Sahat, S., Hamid, B., Kaamin, M., Kesot, M.J., Wen, L.C., Xin, L.Y., Ling, N.P., Lei, V.S.J. Application of plastic bottle as a wall structure for green house. *ARPN J. Eng. Appl. Sci.* 2016;11:7617–21.

121.   Sharma, R., Bansal, P.P. Use of different forms of waste plastic in concrete—a review. *J. Clean. Prod.* 2016;112:473–82. doi: 10.1016/j.jclepro.2015.08.042

122.   Shah, A.A., Kato, S., Shintani, N., Kamini, N.R., Nakajima-Kambe, T. Microbial degradation of aliphatic and aliphatic-aromatic co-polyesters. *Appl. Microbiol. Biotechnol.* 2014;98(8):3437–47. https://doi.org/ 10.1007/s00253-014-5558-1

123.   Sivan, A. New perspectives in plastic biodegradation. *Curr. Opin. Biotechnol.* 2011;22(3):422–426. https://doi.org/10.1016/j.copbio.2011. 01.013

124.   Muhamad, W.N.A.W., Othman, R., Shaharuddin, R.I., Irani, M.S. Microorganism as plastic biodegradation agent towards sustainable environment. Adv. Environ. Biol. 2015;9:8–14.

125.   Mayer, A.M., Staples, R.C. Laccase: New functions for an old enzyme. *Phytochemistry* 2002;60(6):551–65. https://doi.org/10.1016/ S0031-9422(02)00171-1

126.   Tokiwa, Y., Calabia, B.P., Ugwu, C.U., Aiba, S. Biodegradability of plastics. Int. J. Mol. Sci. 2009;10(9):3722–42. https://doi.org/10.3390/ijms10093722

127.   Priyanka, N., Archana, T. Biodegradability of polythene and plastic by the help of microorganism: A way for brighter future. *J. Environ. Anal. Toxicol.* 2012;1:12–15.

128.   Auta, H., Emenike, C., Fauziah, S. Screening of *Bacillus* strains isolated from mangrove ecosystems in Peninsular Malaysia for microplastic degradation. *Environ. Pollut.* 2017;231:1552–9. https://doi.org/10.1016/j.envpol.2017.09.043

129.   Sowmya, H., Ramalingappa, M., Thippeswamy, B. Biodegradation of polyethylene by *Bacillus cereus. Adv. Polym. Sci. Technol. Int. J.* 2014;4(2):28–32. https://doi.org/ 10.1016/j.protcy.2016.05.031

130.   Moog, D., Schmitt, J., Senger, J., Zarzycki, J., Rexer, K.H., Linne, U., Erb, T., Maier, U.G. Using a marine microalga as a chassis for polyethylene terephthalate (PET) degradation. *Microb. Cell Fact.* 2019;18(1):1–15. doi: 10.1186/s12934-019-1220-7

131. Chia, W.Y., Tang, D.Y.Y., Khoo, K.S., Lup, A.N.K., Chew, K.W. Nature's fight against plastic pollution: Algae for plastic biodegradation and bioplastics production. *Environ. Sci. Ecotechnol.* 2020;4:100065. doi: 10.1016/j.ese.2020.100065

132. Manzi, H.P., Abou-Shanab, R.A., Jeon, B.H., Wang, J., Salama, E.S. Algae: A frontline photosynthetic organism in the microplastic catastrophe. *Trends Plant Sci.* 2022;S1360–1385(22):157–161. doi: 10.1016/j.tplants.2022.06.005

133. Sanniyasi, E., Gopal, R.K., Gunasekar, D.K., Raj, P.P. Biodegradation of low-density polyethylene (LDPE) sheet by microalga, Uronema Africanum Borge. *Sci. Rep.* 2021;11(1):1–33. doi: 10.1038/s41598-021-96315-6

134. Bryant, J.A., Clemente, T.M., Viviani, D.A., Fong, A.A., Thomas, K.A., Kemp, P., Karl, D.M., White, A.E., DeLong, E.F. Diversity and activity of communities inhabiting plastic debris in the North Pacific Gyre. *mSystems* 2016;1:1–19. doi: 10.1128/mSystems.00024-16

135. Debroas, D., Mone, A., Ter Halle, A. Plastics in the North Atlantic garbage patch: A boat-microbe for hitchhikers and plastic degraders. *Sci. Total Environ.* 2017;599–600:1222–32. doi: 10.1016/j.scitotenv.2017.05.059

136. Dussud, C., Meistertzheim, A.L., Conan, P., Pujo-Pay, M., George, M., Fabre, P., Coudane, J., Higgs, P., Elineau, A., Pedrotti, M.L., Gorsky, G., Ghiglione, J.F. Evidence of niche partitioning among bacteria living on plastics, organic particles and surrounding seawaters. *Environ. Pollut.* 2018;236:807–16. doi: 10.1016/j.envpol.2017.12.027

137. Muthukrishnan, T., Al Khaburi, M., Abed, R.M.M. Fouling microbial communities on plastics compared with wood and steel: Are they substrate- or location-specific? *Microb. Ecol.* 2019;78:361–74. doi: 10.1007/s00248-018-1303-0

138. Amaral-Zettler, L.A., Zettler, E.R., Mincer, T.J. Ecology of the plastisphere. *Nat Rev Microbiol.* 2020;18(3):139–51. doi: 10.1038/s41579-019-0308-0.

139. Volke-Sepúlveda, T., Saucedo-Castañeda, G., Gutiérrez-Rojas, M., Manzur, A., Favela-Torres, E. Thermally treated low density polyethylene biodegradation by *Penicillium pinophilum* and *Aspergillus niger. J. Appl. Polym. Sci.* 2002;83(2):305–14. doi: 10.1002/app.2245

140. Esmaeili, A., Pourbabaee, A.A., Alikhani, H.A., Shabani, F., Esmaeili, E. Biodegradation of low-density polyethylene (LDPE) by mixed culture of *Lysinibacillus xylanilyticus* and *Aspergillus niger* in soil. *PLoS One.* 2013;8(9):e71720. doi: 10.1371/journal.pone.0071720

141. Schwartz, M., Perrot, T., Aubert, E., Dumarçay, S., Favier, F., Gérardin, P., Gelhaye, E. Molecular recognition of wood polyphenols by phase II detoxification enzymes of the white rot *Trametes versicolor. Sci. Rep.* 2018;8(1):1–11. doi: 10.1038/s41598-018-26601-3

142. Shin, J., Kim, J.E., Lee, Y.W., Son, H. Fungal cytochrome P450s and the P450 complement (CYPome) of *Fusarium graminearum. Toxins.* 2018;10(3):112. doi: 10.3390/toxins10030112

143. Anand, U., Li, X., Sunita, K., Lokhandwala, S., Gautam, P., Suresh, S., Sarma, H., Vellingiri, B., Dey, A., Bontempi, E., Jiang, G. SARS-CoV-2 and other pathogens in municipal wastewater, landfill leachate, and solid waste: A review about virus surveillance, infectivity, and inactivation. *Environ. Res.* 2022;203:111839. doi: 10.1016/j.envres.2021.111839

144. De-la-Torre, G.E., Aragaw, T.A. What we need to know about PPE associated with the COVID-19 pandemic in the marine environment. *Mar. Pollut. Bull.* 2021;163:111879. doi: 10.1016/j.marpolbul.2020.111879

145. Yang, S., Cheng, Y., Liu, T. et al. Impact of waste of COVID-19 protective equipment on the environment, animals and human health: a review. *Environ. Chem. Lett.* 20, 2951–2970 (2022). https://doi.org/10.1007/s10311-022-01462-5

146. Naqash, N., Prakash, S., Kapoor, D., Singh, R. Interaction of freshwater microplastics with biota and heavy metals: A review. *Environ. Chem. Lett.* 2020;18(6):1813–24. doi: 10.1007/s10311-020-01044-3

147. Rahman, A., Sarkar, A., Yadav, O.P., Achari, G., Slobodnik, J. Potential human health risks due to environmental exposure to nano- and microplastics and knowledge gaps: A scoping review. *Sci. Total Environ.* 2021;757:143872. doi: 10.1016/j.scitotenv.2020.143872

148. Liu, S.Y., Leung, M.M.L., Fang, J.K.H., Chua, S.L. Engineering a microbial 'trap and release' mechanism for microplastics removal. *Chem. Eng. J.* 2021;404:127079. doi: 10.1016/j.cej.2020.127079

149. Anand, U., Dey, S., Bontempi, E., Ducoli, S., Vethaak, A.D., Dey, A., Federici, S. Biotechnological methods to remove microplastics: A review. *Environ. Chem. Lett.* 2023;21(3):1787–810. Epub 2023 Feb 8. PMID: 36785620; PMCID: PMC9907217. doi: 10.1007/s10311-022-01552-4

150. Kumar, M., Xiong, X., He, M., Tsang, D.C., Gupta, J., Khan, E., Bolan, N.S. Microplastics as pollutants in agricultural soils. *Environ. Pollut.* 2020;265:114980. doi: 10.1016/j.envpol.2020.114980.

151. Lameh, F., Baseer, A.Q., Ashiru, A.G. Comparative molecular docking and molecular-dynamic simulation of wild-type- and mutant carboxylesterase with BTA-hydrolase for enhanced binding to plastic. *Eng. Life Sci.* 2022;22(1):13–29. doi: 10.1002/elsc.202100083

152. Sharma, B., Dangi, A.K., Shukla, P. Contemporary enzyme based technologies for bioremediation: A review. *J. Environ. Manage.* 2018;210:10–22. doi: 10.1016/j.jenvman.2017.12.075

153. Paço, A., Jacinto, J., da Costa, J.P., Santos, P.S., Vitorino, R., Duarte, A.C., Rocha-Santos, T. Biotechnological tools for the effective management of plastics in the environment. *Crit. Rev. Environ. Sci. Technol.* 2019;49(5):410–41. doi: 10.1080/10643389.2018.1548862

154. Bhattacharyya, N., Anand, U., Kumar, R., Ghorai, M., Aftab, T., Jha, N. K., Rajapaksha, A. U., Bundschuh, J., Bontempi, E., & Dey, A. (2023). Phytoremediation and sequestration of soil metals using the CRISPR/Cas9 technology to modify plants: a review. *Environ. Chem. Lett.* 21(1), 429–445. https://doi.org/10.1007/s10311-022-01474-1

155. Jiang, W., Bikard, D., Cox, D., Zhang, F., Marraffini, L.A. RNA-guided editing of bacterial genomes using CRISPR-Cas systems. *Nat. Biotechnol.* 2013;31(3):233–9. doi: 10.1038/nbt.2508

156. Gaj, T., Gersbach, C.A., Barbas, C.F., III. ZFN, TALEN, and CRISPR/Cas-based methods for genome engineering. *Trends Biotechnol.* 2013;31:397–405. doi: 10.1016/j.tibtech.2013.04.004

157. Shao, H., Chen, M., Fei, X., Zhang, R., Zhong, Y., Ni, W., Tan, X. Complete genome sequence and characterization of a polyethylene biodegradation strain, *Streptomyces albogriseolus* LBX-2. *Microorganisms.* 2019;7(10):379. doi: 10.3390/microorganisms7100379

158. Danso, D., Chow, J., Streit, W.R. Plastics: environmental and biotechnological perspectives on microbial degradation. *Appl. Environ. Microbiol.* 2019;85(19):e01095–e19. doi: 10.1128/AEM.01095-19

159. Jaiswal, S., Sharma, B., Shukla, P. Integrated approaches in microbial degradation of plastics. *Environ. Technol. Inno.* 2020;17:100567. doi: 10.1016/j.eti.2019.100567

160. Kirstein, I.V., Wichels, A., Gullans, E., Krohne, G., Gerdts, G. The plastisphere—uncovering tightly attached plastic "specific" microorganisms. *PLoS One.* 2019;14(4):e0215859. doi: 10.1371/journal.pone.0215859

161. Jeffries, T.C., Rayu, S., Nielsen, U.N., Lai, K., Ijaz, A., Nazaries, L., Singh B.K. Metagenomic functional potential predicts degradation rates of a model organophosphorus xenobiotic in pesticide contaminated soils. *Front. Microbiol.* 2018;9:147. doi: 10.3389/fmicb.2018.00147

162. Alves, L.F., Westmann, C.A., Lovate, G.L., de Siqueira, G.M.V., Borelli, T.C., Guazzaroni, M.E. Metagenomic approaches for understanding new concepts in microbial science. *Int. J. Genomics.* 2018;2018:2312987. doi: 10.1155/2018/2312987

163. Pinnell, L.J., Turner, J.W. Shotgun metagenomics reveals the benthic microbial community response to plastic and bioplastic in a coastal marine environment. *Front. Microbiol.* 2019;10:1252. doi: 10.3389/fmicb.2019.01252

164. Lam, K.N., Cheng, J., Engel, K., Neufeld, J.D., Charles, T.C. Current and future resources for functional metagenomics. *Front. Microbiol.* 2015;6:1196. doi: 10.3389/fmicb.2015.01196

165. Lam, K.N., Cheng, J., Engel, K., Neufeld, J.D., Charles, T.C. Current and future resources for functional metagenomics. *Front. Microbiol.* 2015;6:1196. doi: 10.3389/fmicb.2015.01196.

166. Amaral-Zettler, L.A., Zettler, E.R., Slikas, B., Boyd, G.D., Melvin, D.W., Morrall, C.E., Proskurowski, G., Mincer, T.J. The biogeography of the Plastisphere: Implications for policy. *Front. Ecol. Environ.* 2015;13(10):541–6. doi: 10.1890/150017

167. Quero, G.M., Luna, G.M. Surfing and dining on the "plastisphere": Microbial life on plastic marine debris. Adv. Oceanol. Limnol. 2017;8(2):199–207. doi: 10.4081/aiol.2017.7211

168. Jacquin, J., Cheng, J., Odobel, C., Pandin, C., Conan, P., Pujo-Pay, M., Barbe, V., Meistertzheim, A.L., Ghiglione, J.F. Microbial ecotoxicology of marine plastic debris: A review on colonization and biodegradation by the "plastisphere". *Front. Microbiol.* 2019;10:865. doi: 10.3389/fmicb.2019.00865

169. Roager, L., Sonnenschein, E.C. Bacterial candidates for colonization and degradation of marine plastic debris. Environ. Sci. Technol. 2019;53(20):11636–43. doi: 10.1021/acs.est.9b02212

170. Ronda, C., Chen, S.P., Cabral, V., Yaung, S.J., Wang, H.H. Metagenomic engineering of the mammalian gut microbiome in situ. *Nat. Methods.* 2019;16(2):167–170. doi: 10.1038/s41592-018-0301-y

171. Sheth, R.U., Cabral, V., Chen, S.P., Wang, H.H. Manipulating bacterial communities by in situ microbiome engineering. *Trends Genet.* 2016;32(4):189–200. doi: 10.1016/j.tig.2016.01.005

172. Kaczorek, E., Pacholak, A., Zdarta, A., Smułek, W. The impact of biosurfactants on microbial cell properties leading to hydrocarbon bioavailability increase. *Coll. Interf.* 2018;2(3):35. doi: 10.3390/colloids2030035

173. Arkatkar, A., Juwarkar, A.A., Bhaduri, S., Uppara, P.V., Doble, M. Growth of *Pseudomonas* and *Bacillus* biofilms on pretreated polypropylene surface. *Int. Biodeter. Biodegr.* 2016;4(6):530–6. doi: 10.1016/j.ibiod.2010.06.002

174. Webb, H.K., Arnott, J., Crawford, R.J., Ivanova, E.P. Plastic degradation and its environmental implications with special reference to poly(ethylene terephthalate). *Polymers (Basel)* 2013;5(1):1–18. doi: 10.3390/polym5010001

175. Ding, T., Lin, K., Yang, M., Bao, L., Li, J., Yang, B., Gan, J. Biodegradation of triclosan in diatom *Navicula* sp.: Kinetics, transformation products, toxicity evaluation and the effects of pH and potassium permanganate. *J. Hazard. Mater.* 2018;344:200–9. doi: 10.1016/j.jhazmat.2017.09.033

176. Fritz, J., Sandhofer, M., Stacher, C., Braun, R. Strategies for detecting ecotoxicological effects of biodegradable polymers in agricultural applications. *Macromol. Symp.* 2003;197:397–410. doi: 10.1002/masy.200350734

177. Haider, T.P., Völker, C., Kramm, J., Landfester, K., Wurm, F.R. Plastics of the future? The impact of biodegradable polymers on the environment and on society. *Angew. Chem. Int. Ed. Engl.* 2019;58(1):50–62. doi: 10.1002/anie.201805766

178. Trosset, J.Y., Carbonell, P. Synthetic biology for pharmaceutical drug discovery. *Drug Des. Devel. Ther.* 2015;9:6285–6302. doi: 10.2147/DDDT.S58049

179. Jacquin, J., Cheng, J., Odobel, C., Pandin, C., Conan, P., Pujo-Pay, M., Barbe, V., Meistertzheim, A.L., Ghiglione, J.F. Microbial ecotoxicology of marine plastic debris: A review on colonization and biodegradation by the "plastisphere". Front. Microbiol. 2019;10:865. doi: 10.3389/fmicb.2019.00865

180. Ghosh, S., Qureshi, A., Purohit H.J. Microbial degradation of plastics: Biofilms and degradation pathways. In: Kumar, V.; Kumar, R.; Singh, J.; Kumar, P. Media A.E. (eds) *Contaminants in Agriculture and Environment: Health Risks and Remediation.* Haridwar, India: 2019:184–99.

181. Swapnil, K.K., Deshmukh, A.G., Dudhare, M.S., Patil, V.B. Microbial degradation of plastic: A review. *J. Biochem. Technol.* 2015;6(2):952–61.

182. Huang, X., Cao, L., Qin, Z., Li, S., Kong, W., Liu, Y. Tat-independent secretion of polyethylene terephthalate hydrolase PETase in *Bacillus subtilis* 168 mediated by its native signal peptide. *J. Agric. Food Chem.* 2018;66(50):13217–13227. doi: 10.1021/acs.jafc.8b05038

183. Seo, H., Kim, S., Son, H.F., Sagong, H.Y., Joo, S., Kim, K.J. Production of extracellular PETase from Ideonellasakaiensis using sec-dependent signal peptides in *E. coli. Biochem. Biophys. Res. Commun.* 2019;508(1):250–255. doi: 10.1016/j.bbrc.2018.11.087

184. Gao, J., Ellis, L.B., Wackett, L.P. The University of Minnesota biocatalysis/biodegradation database: improving public access. *Nucleic Acids Res.* 2010;38(suppl_1):D488–91. doi: 10.1093/nar/gkp771

185. Wicker, J., Lorsbach, T., Gütlein, M., Schmid, E., Latino, D., Kramer, S., Fenner, K. enviPath—the environmental contaminant biotransformation pathway resource. *Nucleic Acids Res.* 2016;44(D1):D502–8. doi: 10.1093/nar/gkv1229

186. Karp, P.D., Billington, R., Caspi, R., Fulcher, C.A., Latendresse, M., Kothari, A., Keseler, I.M., Krummenacker, M., Midford, P.E., Ong, Q., Ong, W.K. The BioCyc collection of microbial genomes and metabolic pathways. *Brief Bioinform.* 2019;20(4):1085–1093. doi: 10.1093/bib/bbx085

187. Caspi, R., Billington, R., Keseler, I.M., Kothari, A., Krummenacker, M., Midford, P.E., Ong, W.K., Paley, S., Subhraveti, P., Karp, P.D. The MetaCyc database of metabolic pathways and enzymes—a 2019 update. *Nucleic Acids Res.* 2020;48(D1):D445–53. doi: 10.1093/nar/gkz862

188. Ali, S.S., Elsamahy, T., Koutra, E., Kornaros, M., El-Sheekh, M., Abdelkarim, E.A., Sun, J. Degradation of conventional plastic wastes in the environment: A review on current status of knowledge and future perspectives of disposal. *Sci. Total Environ.* 2021;771:144719. doi: 10.1016/j.scitotenv.2020.144719

189. Anand, U., Dey, S., Bontempi, E., Ducoli, S., Vethaak, A.D., Dey, A., Federici, S. Biotechnological methods to remove microplastics: A review. *Environ. Chem. Lett.*

2023;21(3):1787–1810. Epub 2023 Feb 8. PMID: 36785620; PMCID: PMC9907217. doi: 10.1007/s10311-022-01552-4

190. Evode, N., Qamar, S.A., Bilal, M., Barceló, D., Iqbal, H.M.N. Plastic waste and its management strategies for environmental sustainability. *Case Stud. Chem. Environ. Eng.* 2021;4:100142. ISSN 2666-0164. https://doi.org/10.1016/j.cscee.2021.100142

191. Awoyera, P.O., Adesina, A., Plastic wastes to construction products: Status, limitations and future perspective. *Case Stud. Construct. Mater.* 2020:12:e00330. ISSN 2214-5095. https://doi.org/10.1016/j.cscm.2020.e00330

192. Rhodes, C.J. Plastic pollution and potential solutions. *Sci. Progr.* 2018;101(3):207–60. doi: 10.3184/003685018X15294876706211

193. Chattopadhyay, A., Purohit, J., Tiwari, K.K., Deshmukh, R. Targeting transcription factors for plant disease resistance: Shifting paradigm. *Curr. Sci.* 2019;117(1):1598–607. doi: 10.18520/cs/v117/i10/1598-1607

194. Otoupal, P.B., Chatterjee, A. CRISPR gene perturbations provide insights for improving bacterial biofuel tolerance. *Front. Bioeng. Biotechnol.* 2018;6:122. doi: 10.3389/fbioe.2018.00122

195. Mohammad, T., Hassan, M.I. Modern approaches in synthetic biology: Genome editing, quorum sensing, and microbiome engineering. In: Singh S. (ed.) *Synthetic Biology*. Springer, Singapore; 2018:189–205.

196. Jusiak, B., Cleto, S., Perez-Piñera, P., Lu, T.K. Engineering synthetic gene circuits in living cells with CRISPR technology. *Trends Biotechnol.* 2016;34(7):535–47. doi: 10.1016/j.tibtech.2015.12.014

11 Nanowaste and Its Management

New Waste Management Paradigm

Ekta Tiwari, Nitin Khandelwal, and Nisha Singh

11.1 INTRODUCTION TO NANOWASTE

Advancement in science and technology has began a new era of small miracles – nanotechnology or nanoscience. The ability to manipulate matter at the atomic scale and its characterization at that scale has brought about a revolution in the world of material engineering. As of now, nano-based materials have entered almost every aspect of production and consumer life cycle due to their smaller size (1–100 nm) and increased efficacy compared to bulk counterparts. For example, gold (Au) in its bulk form is generally inert; however, once brought down to the nano-size range, Au becomes chemically reactive at least when illuminated with light (Cortie, 2004). Au nanoparticles have excellent photocatalytic properties and are typically used for organic degradation and oxidation reactions (Liu et al., 2017). Similarly, Ag nanoparticles demonstrate antibacterial properties and therefore may inhibit the natural degradation of contaminants through the inactivation of organic degraders (Varier et al., 2019). Demand for nanomaterials is growing rapidly in various sectors, for example, personal care, healthcare and life science, energy, electronics and consumer goods, paint and coating, adhesive and sealants, etc. (Bundschuh et al., 2018). Besides intentional use or industrial application, nanomaterials can be of natural origin. Asbestos, which is a part of a naturally occurring mineral, contains nano-sized fibrils that can be directly released into the environment and have potentially harmful environmental impacts (Wang et al., 2017).

Nanomaterials have been explored and utilized because of their extraordinarily useful properties, but their other side also needs attention when it comes to their disposal. Nanowaste can be simply defined as any nano-sized particle or a by-product of nanomaterials discarded in the environment having nano-specific properties. While nanotechnology helps us to bring down the big wonders in small packets, at the same time it is challenging to manage the nanoscale by-products released from these miraculous solutions. The size of nanowaste is the foremost parameter that makes it difficult to segregate it from regular waste. Being highly chemically reactive due to increased surface–volume ratio, nanomaterials act differently when compared to their bulk counterpart and hence cannot be treated similarly. The high reactivity and

DOI: 10.1201/9781003543176-11

smaller size of nanomaterials can cause serious threats to the environment and human health when released untreated. Studies support the easy uptake of nanomaterials by plants and animals causing various detrimental health issues due to increased bioavailability. Apart from the nanomaterial's properties, different environmental parameters also govern their mobility and toxicity in the environment. Numerous studies focused on understanding the environmental fate, stability, and transport of nanoparticles, suggesting that dominant chemical entities in different soils and water bodies, i.e., bicarbonates and organic matter, help in the stabilization of most of the nanomaterials and thereby facilitate their transport in various environmental compartments and enhance bioavailability. Hazards caused by the nano-specific properties of nanomaterials have not been well documented in the scientific community which makes it more challenging to frame proper regulatory protocols for the management and disposal of nanowaste.

Estimates suggest that a substantial percentage of globally produced engineered nanomaterials end up in landfills. Despite early regulations on waste containing micron- and nano-sized materials, a global consensus on nanowaste management remains elusive. This poses challenges, particularly for research institutions, where a significant portion of nanomaterial production takes place. In the series of exploring and designing newer nanomaterials, research laboratories generate more than 20% of total nanowaste around the world (Schwab et al., 2023). Many of the novel nanomaterials being researched lack regulatory testing and safety data sheets required for risk assessments. However, this fraction may represent the highest risk as most of these nanomaterials tend to have very little existing information on their toxic impacts. Industrial-scale nanomaterial manufacturing also presents challenges related to handling large quantities of nanomaterials and complying with prescribed industrial standards. Efforts to address these issues and establish effective nanowaste management practices are crucial for the responsible development of nanotechnology. In some cases, chemical industries generally provide safety data sheets comparable to their bulk, which is not a true representation of potential risks. Therefore, proper nanowaste management and disposal guidelines can help in minimizing waste generation and release.

This chapter develops an understanding of various emerging nanowaste in the environment, their sources, potential classification, associated toxicities, limitations of conventional waste management techniques in managing nanowaste, and emerging solutions being explored.

11.2 NANOWASTE CLASSIFICATION AND SOURCES

Nanomaterials, whether of natural or synthetic origin, may be introduced into the environment through various sources. Based on the nature of its origin nanowaste sources can be divided into two broad categories: (i) primary sources and (ii) secondary sources.

11.2.1 PRIMARY SOURCES

Primary sources include the processing unit of raw materials and nano-enabled products, release during direct use of nanomaterials in different industries, release of newly synthesized nanomaterials from research laboratories, and disposal of used nanomaterials in water from household activity, waste treatment plants, and landfills.

11.2.2 Secondary Sources

There can be indirect sources as well that lead to the accumulation of nanomaterials in different environmental compartments such as via leaching from landfills (Dulger et al., 2016, Mitrano et al., 2017), application of sewage sludge to the soils (Kim et al., 2010), and unintended release of nano-sized particulates from vehicular and traffic emissions.

Similarly, in a wider context, nanomaterials in the waste can be further categorized based on composition and chemical constituents as (i) metallic nanoparticles, (ii) metal-oxides, (iii) carbon-based nanomaterials, (iv) metallo-organic nanostructures, (v) porous organic polymers, (vi) nanoplastics (vii) nanocomposites, (viii) nanoceramics, (ix) thin films and nanomembranes, and (x) nanofibers.

As presented in Table 11.1, different nanomaterials pose various health hazards to the environment and human health. Based on their reactivity and toxicity, nanowaste can be further divided into classes (Younis, El-Fawal et al., 2018) as shown in Figure 11.1.

Class 1: Non-hazardous or Class 1 nanowaste does not pose any potential harm to human and environmental health upon exposure. Non-hazardous nanowaste has low to medium levels of exposure and they do not have any special requirement for disposal.

Class 2: Class 2 waste can be described as the discarded bulk materials containing some form of nanomaterials, for example, carbon nanotubes (CNTs), and possess a low to medium risk to the human and environmental health upon exposure. These discarded materials have the potential to release nanoparticles in the environment if not managed properly.

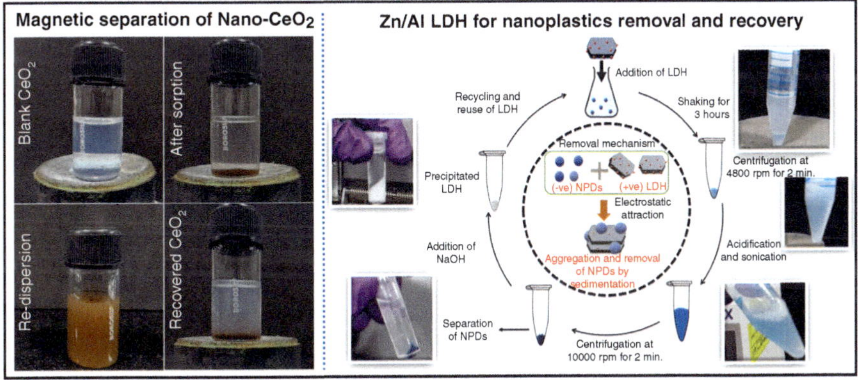

FIGURE 11.1 Classification of nanowaste based on associated risk and toxicity.

TABLE 11.1

Uses of different types of nanomaterials and their health and environmental hazards (Puri et al. 2021; Gupta and Bhandari 2022; Khan, Malik et al. 2022)

Nanomaterial classes	Examples	Use in daily-life products	Health hazards	Environmental hazards
Metallic nanoparticles	Ag, Cu, Au, Fe nanoparticles	Electronics, clothing (antimicrobial properties), cosmetics (pigments), food packaging (antimicrobial coatings), and automotive coatings (scratch resistance and improved appearance)	Exposure to metallic nanoparticles through inhalation, ingestion, or dermal contact can lead to adverse health effects, including respiratory issues, inflammation, and potential toxicity to organs.	Metallic nanoparticles, like silver and copper, can accumulate in the environment, potentially impacting ecosystems and disrupting natural processes. They may harm aquatic organisms and other wildlife.
Nano metal-oxides	TiO_2, ZnO, CuO, Fe_3O_4 nanoparticles	Sunscreens and cosmetics (UV protection), paints and coatings (pigments, UV resistance), and electronics (semiconductors, sensors)	Inhalation or ingestion of metal-oxide nanoparticles can pose health risks, including respiratory problems and potential toxicity to organs. Skin contact may cause irritation or sensitization.	Metal-oxide nanoparticles, such as titanium dioxide and zinc oxide, can accumulate in soil and water, affecting soil quality and disrupting aquatic ecosystems. They may also impact plant growth and aquatic organisms.
Carbon-based nanomaterials	Carbon nanotubes (CNTs), activated charcoal nanoparticles, graphene nanoparticles, fullerene nanoparticles	Electronics, energy storage, and composite materials. Charcoal is used for water filtration and air purification. Graphene is utilized in electronics, batteries, and sensors. Fullerene has applications in pharmaceuticals and solar cells	Improper disposal or release of carbon-based nanomaterials can lead to environmental contamination. CNTs, for example, may accumulate in soil or water and affect microbial communities and aquatic organisms	Inhalation of carbon-based nanomaterials, particularly CNTs, can potentially cause respiratory issues, such as inflammation and fibrosis. The toxicity of other carbon-based nanomaterials is still being studied.

Quantum dots	Cadmium-based quantum dots (CdSe, CdTe), lead-based quantum dots (PbS, PbSe).	Quantum dots are semiconductor nanocrystals with unique optical properties. They are used in displays (QLED TVs), lighting (LEDs), solar cells, and biological imaging (fluorescent probes).	Quantum dots contain heavy metals, such as cadmium and lead, which can be released into the environment. They may pose risks to aquatic organisms and ecosystems.	Quantum dots' heavy metal content raises concerns regarding potential human exposure and toxicity, particularly if they are r eleased or accumulate in the body.
Two-dimensional nanomaterials	Titanium carbide MXene (Ti_3C_2), graphene oxide (GO), molybdenum disulfide (MoS_2).	Applications in energy storage (batteries, supercapacitors), water purification (membranes), sensors, and electronics.	Limited information is available on the specific human health hazards of MXenes and other two-dimensional nanomaterials. Further research is warranted to assess their potential risks.	The environmental hazards of MXenes and other two-dimensional nanomaterials are still being explored. Their potential impacts on ecosystems and organisms have not been extensively studied.
Porous organic polymers	Metal-organic frameworks (MOFs), covalent organic frameworks (COFs), zeolitic imidazolate frameworks (ZIFs).	Gas storage and separation, catalysis, sensing, and drug delivery systems	Nanoporous organic polymers may have limited environmental hazards due to their inherent stability. However, their long-term effects on ecosystems and organisms are not well understood.	The potential human health hazards of nanoporous organic polymers are still under investigation, and more research is warranted to assess any associated risks.

(continued)

TABLE 11.1 (Continued)
Uses of different types of nanomaterials and their health and environmental hazards (Puri et al. 2021; Gupta and Bhandari 2022; Khan, Malik et al. 2022)

Nanomaterial classes	Examples	Use in daily-life products	Health hazards	Environmental hazards
Nanoplastics	Nanosized fragments or particles derived from the degradation of larger plastic materials, such as polyethylene, polystyrene latex beads, or polypropylene	Nanoplastics with dimensions in the nanoscale range. They are used in consumer products such as personal care items, textiles, and food packaging.	The potential health impacts of nanoplastics on humans are still being investigated. Ingestion or inhalation of nanoplastics could lead to localized or systemic effects, but more research is warranted.	Nanoplastics, formed from the degradation or fragmentation of larger plastic particles, can contaminate ecosystems, affect aquatic life, and contribute to the microplastic pollution problem.
Other complex nanostructures				
Nanoceramics	Nano-sized ceramic particles such as titanium nitride (TiN), aluminum oxide (Al_2O_3), and silicon carbide (SiC).	Coatings, catalysts, biomedical implants, and electronic components	Inhalation of nanoceramic particles may pose risks to respiratory health, similar to other airborne particulate matter. Specific health effects depend on factors such as particle size, shape, and surface characteristics.	Nanoceramic particles released into the environment may accumulate in soil, water, and organisms, potentially impacting ecosystems and biological processes.
Nanofibers	Electrospun nanofibers made of polymers like polyethylene, polycaprolactone, or polylactic acid.	Textiles, filtration membranes, tissue engineering scaffolds, and electronics	Inhalation of nanofibers can potentially lead to respiratory issues, as their small size and high aspect ratio may cause lung damage. There may also be concerns regarding skin exposure or ingestion.	Nanofibers released into the environment can accumulate in soil and water systems, affecting aquatic organisms and potentially disrupting ecological balance.

Nanocomposites	Polymer-based nanocomposites containing nanofillers such as clay nanoparticles, CNTs, or metal nanoparticles.	Combination of nanoparticles or nanofillers dispersed in a matrix material. They offer enhanced mechanical, electrical, or thermal properties and find applications in automotive parts, aerospace components, and sporting goods	The release of nanoparticles from nanocomposites during their lifecycle could pose health risks if inhaled, ingested, or exposed to the skin. The toxicity will depend on the specific nanoparticles and their characteristics.	Nanoparticle release from nanocomposites, particularly during manufacturing, use, or disposal, can lead to the introduction of nanoscale pollutants into the environment, potentially affecting ecosystems and wildlife.
Nanomembranes	Thin nanoscale membranes made of materials like graphene, CNTs, or polymer nanocomposites	Are thin films with nanoscale thickness. They are used in filtration, separation processes, and sensors for their selective permeability and high surface-to-volume ratio.	Exposure to nanomembranes can occur through inhalation, dermal contact, or ingestion. Health effects depend on the specific material and potential chemical interactions.	Disposal of nanomembranes and their associated materials may introduce nanoscale pollutants into the environment, potentially affecting ecosystems and organisms.

Class 3: It can be considered as toxic to moderately toxic to the environment and human health upon exposure and possess medium to high risk. For example, nanoparticles are used in packaging and additives for food. Class 3 nanowaste requires a standard protocol for its disposal and risk management.

Class 4: They fall in the category of toxic to extremely toxic and pose a high risk to humans and environmental health. Class 4 requires specially designated sites to be disposed of as hazardous nanowaste streams.

Class 5: Any material containing a colloidal suspension of nanoparticles can be categorized as Class 5 nanowaste and severely toxic similar to Class 4. They need special disposal sites for their disposal and management because improper management may lead to their long-range transport in the environmental matrices due to their high stability and mobility.

11.3 ENVIRONMENTAL AND HUMAN HEALTH RISKS OF NANOWASTE

The increasing market potential of nanomaterials on a global scale has raised a major challenge for conventional waste treatment facilities that are inadequate to manage the mixture of resulting nanomaterials. The eventual release of nanomaterials and lack of data on the generation of nanowaste processes in different sectors including industrial and commercial from end-of-life results in an underestimation of the associated ecological and human health risks. For example, aluminum-based nanoparticles are being used in many areas such as fuel cells, paints, coatings, biomaterials, textiles, etc., and have been reported to disturb cell viability, alter mitochondrial function, increase oxidative stress, and alter tight junction protein expression of the blood–brain barrier in the rat (Chen et al., 2008). Ag nanoparticles are another commonly used nanoparticle and have been reported as toxic due to their antimicrobial properties (Ferdous and Nemmar, 2020). Bioaccumulation and biotransformation of CuO nanoparticles (20 – 40 nm) were reported in *Zea mays L.*, which inhibited the growth of the seedling (Wang et al., 2012). The potential risk associated with nanowaste is governed by a number of physicochemical properties as compared to their bulk counterparts including their composition, size, shape, surface heterogeneity, and reaction rate in given physiological and environmental conditions (Wiesner, Lowry et al., 2006; Bhagat, Nishimura et al., 2021; Singh, Bhagat et al., 2021). Similarly, large aggregated TiO_2 nanoparticles showed less cytotoxicity than the smaller particles, whereas rod-shaped TiO_2 was found more toxic (Kose et al., 2020). Similarly, polyvinylpyrrolidone (PVP)-coated Ag nanoparticles were found less toxic compared to uncoated Ag nanoparticles due to differences in surface functionality (Geppert et al., 2021). Direct uptake and transport of Au nanoparticles (15–50 nm) have been reported in poplar roots, stems, and leaves (Zhai et al., 2014).

Few nanomaterials regarded as inert toward human health and the environment due to their chemical stability were eventually identified to be hazardous including, Au, Fe, and TiO_2 (Umair, Javed et al., 2016). Elevated levels of pigment-grade (powdered) and ultrafine TiO_2 are associated with the development of respiratory tract cancer in rats (Lee, Trochimowicz et al., 1985) and have been included in the 'Group 2B: Carcinogen' class by the International Agency for Research on Cancer (IARC) (Skocaj, Filipic et al., 2011). Cationic nanoparticles are documented to be hazardous as they can readily infiltrate cells and tissues, leading to organ dysfunction, tissue injury, and destabilization of membranes (Gupta and Xie 2018). Silver nanoparticles with antibacterial effects negatively hamper cell membranes and DNA and are shown to be toxic for micro-biota even at very low concentrations. Nanoplastics are identified to cross the blood–brain barrier and placenta, the organ that nourishes an unborn child in humans (Prüst, Meijer et al. 2020). The degree of toxicity is influenced by the size of TiO_2 nanoparticles; smaller particle sizes trigger a higher number of intracellular reactive oxygen species (ROS) in Daphnia Magna (Kim, Klaine et al. 2010). Although studies are in progress to understand the impact on humans, nanoplastics are known to cause oxidative stress in zebra fish along with altering the toxicity of other contaminants (Singh, Bhagat et al. 2021). Numerous studies have focused on the ecotoxicity arising from certain nanomaterials such as CNTs, and metal oxide nanoparticles; there is a great deal of data scarcity on polymer nanoparticles.

Waste stabilization plays a significant role in the fate and behavior of nanowaste, but there is a lack of thumb rule due to the high diversity and complexity of nanoproducts. Nanowaste can enter the environment at different stages of nanomaterial life including manufacturing, distribution, utilization, or disposal (Gupta and Xie 2018). However, these processes are studied with respect to exposure assessment during the application of nanoproducts though their importance in the environmental cycle largely remains unknown. Nanowaste once released can undergo processes leading to an unidentified fate and transport in the environment. These processes include absorption, biodegradation and biotransformation, combustion, dissolution, oxidation, reduction, precipitation and sedimentation, photochemical reaction, physical and mechanical aberration, etc. (Lead, Batley et al., 2018).

The identification of the zone of accumulation of nanowaste in the environment is yet to be investigated. However, 63–91% of 10 major nanomaterials are buried in landfills and therefore soil is stipulated to be the major sink of nanowaste (Lead, Batley et al., 2018). Conversely, surface water may receive a huge amount of nanowaste due to soil treatment by sludge. Several studies suggest that dominant chemical entities in different soils and water bodies, i.e., bicarbonates and organic matter, help in the stabilization of most of the nanomaterials, easing their transport in various environmental compartments and enhancing bioavailability. A study by Singh et al. on the fate of polymeric nanowaste's aggregation behavior in the environment demonstrates the impact of the varying environmental parameters (Singh, Tiwari et al., 2019). The higher electrolyte concentration of the aqueous system supported the aggregation of otherwise stable polystyrene nanoparticles where the divalent cationic species exerted a higher agglomeration in the system by shielding the electrostatic repulsion between the nanoparticles. Further, the presence of dissolved organic matter had a diverse influence depending upon the background ionic species for example polymeric particles were found to be more stable due to steric repulsion but in the presence of divalent cationic species a faster aggregation was reported because of cationic bridging. A similar observation has been observed with metal and metal oxide nanoparticles. Furthermore, several environmental processes such as humidity, temperature, photochemical interactions, and turbulence influence the distribution of nanowaste concentration in the atmosphere (Abbas, Yousaf et al., 2020). The minute size of nanowaste allows prolonged suspension in the atmosphere, thus facilitating dispersion and subsequently raising respiratory and cardiovascular concerns in humans. The body's clearance mechanisms, including mucociliary escalation and macrophage activity, are influenced following such exposure. Besides, the detection of nanowaste in different environmental matrices, such as air, soil, water, or biota, is highly jeopardized due to underdeveloped analytical techniques that do not allow differentiation of nanowaste from naturally occurring nanoparticles. This also results in a significant gap in the fate and transport of nanowaste across complex environmental systems raising concern about the nanowaste risk due to the lack of regulations.

11.4   CHALLENGES OF NANOWASTE MANAGEMENT

Nanowaste management is of crucial need due to its unique properties and potential environmental risks. Current toxicity assessment methods, which focus on mass-based measurements, fall short of capturing the true dangers posed by nanomaterials' shape, scale, surface characteristics, and reactivity. To address this challenge, we need innovative techniques to detect and monitor nano-residues in real environmental matrices, considering their physicochemical properties. These methods will enable the creation of tailored toxicity indices, leading to better policies and laws for controlling nano-pollution. Furthermore, we must establish clear permissible concentration limits for nanomaterials, requiring multiple indexes and units to express them accurately. Research efforts should prioritize understanding the accumulation of nanomaterials in ecosystems and evaluate existing measurement methods to adapt them to nano-residues. By taking these steps, we can effectively mitigate the environmental impact of nanomaterials and ensure a safer future. For example, Ag nanoparticles are routinely used in several of the daily-use industrial products and therefore currently have higher release potential in the environment. Nanosilver waste presents a substantial threat to the environment, as it can undergo transformations that have the potential to disrupt ecosystems. The soil biodiversity is particularly vulnerable, as nanosilver can harm various organisms, including bacteria, fungi, and essential species like earthworms owing to its property to damage DNA. Moreover, its interactions with other toxic substances can exacerbate its environmental toxicity, affecting groundwater quality and bioaccumulating in the food chain. The consequences extend to wastewater treatment systems, where the presence of nanosilver poses significant challenges. Heterotrophic microorganisms responsible for organic and nutrient removal, as well as autotrophic microorganisms crucial for nitrification, are particularly susceptible to the inhibitory effects of silver nanoparticles. This jeopardizes the overall efficiency and effectiveness of wastewater treatment processes.

Wide variations in composition, surface properties, environmental stability, and therefore associated toxicity of different laboratory-designed or industrially produced nanoparticles, make it difficult to manage released nanowaste. Even a specific kind of nanomaterials designed using different synthetic pathways or having variations in shape, size, surface area, or functionality can lead to drastic variations in their toxicity to humans and other organisms (Ispas et al., 2009). In addition, research findings suggest drastic variations in the toxicities of nanoparticles in the presence or absence of other known environmental contaminants (Vineeth Kumar et al., 2022; Singh et al., 2021a). Some of the sources, health, and environmental impacts of this group of nanowaste are presented in Table 11.1.

Some nanomaterials such as metal nanoparticles can be dissolved in acids, resulting in conventional metal waste to deal with. Similarly, soft organic nanomaterials or polymeric substances that can be dissolved in specific organic solvents may be converted to conventional organic waste and disposed of in well-labeled containers (Qin and Wu, 2012; de Moura Souza et al., 2019; Bossert et al., 2019). However, several nanomaterials have strong frameworks and environmental rigidness such as MXenes, carbon nanomaterials, covalent organic frameworks, etc. (Khandelwal and Darbha, 2021; Huang et al., 2017). Such nanomaterials represent the risk of long-term persistence, bio-availability, and trophic transfer, and therefore management of such nanowaste becomes more challenging.

11.5 CONVENTIONAL MANAGEMENT METHODS AND NANOWASTE

General waste management practices include landfilling, recycling, composting, anaerobic digestion, incineration, open dumping, etc. (Vinti et al., 2021). These may not be suitable for nanowaste and may cause several health and environmental concerns. For example, the disposal of nanowaste in landfills may cause their direct leaching into soils and groundwater (Melchor-Martínez et al., 2021). Most nanomaterials are highly reactive and may undergo dissolution or degradation and generate by-products that may leach across the environmental matrix (Ray et al., 2009). In addition, liners are used in landfills to prevent the leaching of chemicals but are not tested for nanomaterials, and if the lining is compromised, it may release nanowaste and other contaminants in other environmental compartments (Mitrano et al., 2017). Most importantly, exposure to nanowaste can also create occupational health concerns for workers in and around landfills. Similarly, incineration also involves potential risks such as the release of toxic fumes and nanoparticulate in the air, potential respiratory health risks, air pollution, incomplete combustion of organic nanoparticles that may result in hazardous by-products, and ash may contain super concentrated nanomaterials requiring specific disposal procedure (Jones and Harrison, 2016). Incineration decreases the amount of waste, eradicates harmful organic waste, and harnesses the energy contained within the waste. Limited information on the fate and transport of the engineered nanomaterial during incineration presents the challenge of disposing of these materials directly into the conventional systems. Waste combustion results in numerous hazardous by-products at different steps of the process that encompasses acid gasses (HCl, SO_2, NO_x), heavy metals, polycyclic aromatic hydrocarbons, particulate matter, and semi-volatile compounds. With varying conditions in different combustion zones, there are a number of possibilities for the nanowaste when incinerated, including: (1) undergoing complete combustion; (2) transformation due to reaction with other substances at high temperature (chlorides, oxides), finer fractions, or agglomerated clusters of particles; and (3) remaining recalcitrant to any chemical or physical alteration and end up into the ash, slag, or filter residues (Walser, Limbach et al., 2012). For example, CNTs demonstrate a complete decomposition (94%) into CO_2, whereas Ag, SiO_2, TiO_2, and ZnO are resistant to incineration. For example, experiments have shown that even with modern filters and separation systems in place, nanoparticles such as cerium oxide particles were not effectively removed in an incineration plant (Walser et al., 2012). This raises concerns about the potential release of nanoparticles into the environment during incineration, both in controlled scenarios and worst-case scenarios with massive nanoparticle release. Therefore, nanowaste recycling can be considered a safer alternative than disposal in landfills or incineration.

11.5.1 Physical, Chemical, and Biological Recycling Approaches

Various processes, such as centrifugation, sedimentation, filtration, crystallization, and evaporation, are used to physically separate nanomaterials from compounds in solution during recycling. This allows for the reuse of the solution and prevents the release of nanomaterials into the environment. These are well-known traditional approaches that may not be viable for large-scale applications due to high time and energy consumption along with some other limitations (Dericiler et al., 2022).

For example, in tire recycling, granular material is used for various applications, but the presence of nanomaterials in tires may lead to the leaching of nanomaterials from recycled products. Even mechanical abrasion and processing can lead to the release of nano-sized plastic debris from tires. Exposure to nanoparticles during recycling can occur in different forms:

1. **Exposure to nano-objects in liquid media:** This occurs during the cleaning of recycling equipment or products before mechanical recycling, where nanoobjects present in water or solvents may pose a risk.
2. **Exposure to nanoobjects from thermal processes:** Nanoparticles can be generated in combustion gases or ambient air during thermal processes such as pyrolysis, heating, and soldering, leading to potential exposure risks.
3. **Exposure to dust-containing nanoobjects:** Fine or ultrafine dust containing loose nanoobjects can be emitted during transportation, sorting, and shredding of recyclable materials, posing a potential risk to individuals.

Therefore, careful consideration is necessary to ensure the safe handling and disposal of nanowaste during recycling processes to minimize potential environmental contamination.

11.5.2 Bioremediation Approaches

Researchers have observed that both plants and microorganisms can uptake nanoparticles and therefore bioremediation can be another possible pathway to eliminate nanowaste (Singh et al., 2019; Taylor et al., 2014). However, most of the nanoparticles have antibacterial properties, and therefore nanowaste can compromise microbial activity in waste treatment plants, leading to untreated nanowaste along with other contaminants (Wang et al., 2017).

Conversely, bioinspired materials can be a promising solution. For example, researchers showed that jellyfish mucus, rich in glycoproteins and glycans, interacts favorably with nanoparticles. The study reveals that the strong interactions between nanoparticle surfaces and the biomolecules present in the mucus lead to the accumulation and aggregation of nanoparticles. The strong affinity between nanoparticles and biomolecules is attributed to hydrogen bonding, ionic interactions, and dehydration of polar groups (Ben-David et al., 2023; Patwa et al., 2015).

However, bioremediation includes several challenges such as bioaccumulation and transfer across the food chain, toxicity in microorganisms, etc. Similarly, the bloom of jellyfish or any other animal for nanowaste remediation can impart negative effects on nutrient cycling and other species (Patwa et al., 2015). These observations highlight the need for more advanced approaches for nanowaste remediation.

11.6 NANOWASTE MANAGEMENT PARADIGM: EMERGING APPROACHES AND CHALLENGES

Considering the challenges associated with the utilization of conventional waste management approaches for handling nanowaste, researchers have started exploring other emerging approaches as listed below.

11.6.1 MAGNETIC SEPARATION

Nanoparticles and nanocomposites that have magnetic properties are easy to separate and can be eliminated or recovered from the waste by applying a strong magnetic field (Powell et al., 2020). Recovered magnetic particles can be given further secondary treatment for their activation and reuse in a similar work process or can be explored for their use in other applications. Nevertheless, magnetic nanoparticles such as iron oxy-hydroxides also act as flocculants and are found to be effective in aggregating non-magnetic nanomaterials leading to their sedimentation or easy magnetic separation (Mandel et al., 2013). Therefore, magnetic properties can be used as a sustainable pathway to recover both magnetic and non-magnetic nanoparticles.

For example, Gupta et al. demonstrated that 25 nm-sized cerium oxide nanoparticles can be removed from different complex aqueous matrices by using magnetite nanoparticles (Figure 11.2) designed through a solvothermal approach. The sorption capacity was found to be around 680 mg/g and sorbed nanoparticles were recovered through a simple sonication approach which resulted in >80% recovery (Gupta et al., 2019).

Similarly, Singh et al. showed possible remediation of different-size nanoplastics with varying surface functionality using biochar-supported magnetite nanocomposite. Electrostatic attraction and surface complexation were found to be the major attachment mechanisms (Singh et al., 2021b).

Besides that, researchers utilizing or exploring the applications of magnetic nanoparticles or composites can make it a routine practice to separate particles from

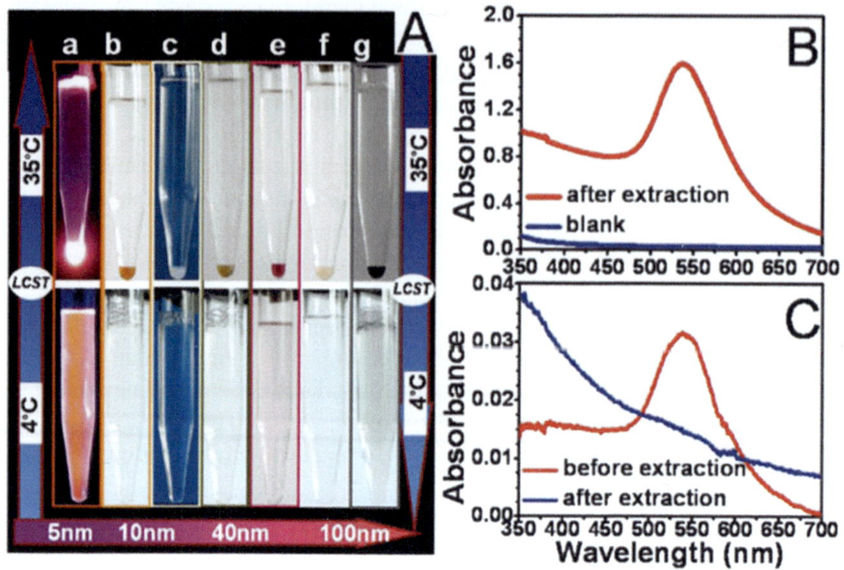

FIGURE 11.2   Nano–CeO$_2$ separation using magnetite nanoparticles (left) and removal and recovery of nanoplastics using layered double hydroxides (right).

Source: (Left) modified from Gupta et al. (2019); (Right) Tiwari et al. (2020).

generated waste for their possible reuse or alternate applications. This will minimize at least one of the laboratory sources for nanowaste generation and release.

11.6.2   ADSORPTION-BASED APPROACH FOR REMOVAL AND RECOVERY OF NANOPARTICLES

Some of the metal-containing inorganic nanomaterials are soluble in acidic medium, while carbon-based nanomaterials, nanoplastics, covalent organic frameworks, and some metallic compounds (e.g., lead chromate) show strong resistance to acids. This allows possible sorption-assisted removal of nanoparticles from complex matrices using specific nanomaterials.

For example, Tiwari et al. used Zn/Al-layered double hydroxide (Figure 11.2) for strong electrostatic attachment and removal of 50 nm size polystyrene nanoplastics from different aqueous solutions. Nanoplastics were recovered by dissolving LDH in an acidic solution, while LDH was again recovered by increasing the solution pH to alkaline (Tiwari et al., 2020).

In another experiment, Khandelwal et al. removed chromate ions through strong Pb–Cr complexation in Pb–Mg/Al LDH. As formed PbCrO$_4$ was not soluble in an acidic solution while Mg/Al LDH can dissolve, authors recovered pure monocrystalline rods of PbCrO$_4$ which can have possible applications in electric industries while the waste acidic solution can be treated again using a base to obtain LDH (Khandelwal et al., 2022). Similarly, organic nanomaterials having solubility in specific organic solvents and other nanoparticles such as mesoporous silica which is soluble in the basic medium can be used to recover other nanomaterials having reverse properties.

11.6.3 Thermal Processes

Researchers have explored various thermal processes for the recovery of nanoparticles from different waste materials. For example, Wang et al. developed a method to recover Ag nanoparticles by thermally decomposing $AgNO_3$ in octadecyl amine (ODA). The ODA, known for its high boiling point and stability, can be recycled and reused for subsequent synthesis of Ag nanoparticles, making the process economical and scalable (Wang et al., 2008).

Thermal evaporation under vacuum and selective thermal transformation was employed to recover lead (Pb) and copper–tin (Cu–Sn) nanoparticles from waste-printed circuit boards (Shokri et al., 2017). In the case of incinerated organic solar cells, silver recovery was achieved by fusion and agglomeration of nanomaterials during the high-temperature incineration process (Søndergaard et al., 2016). CNTs were recovered from waste plastics through pyrolysis, utilizing different systems such as autoclaves, quartz tubes, crucibles, and muffle furnaces (Bazargan and McKay, 2012). These thermal processes offer efficient and environmentally friendly approaches to recycling and recovering valuable nanoparticles from various waste sources. However, the focus should be kept on the safe methods of pyrolysis as released gases may contain toxic products and by-products including smaller nanoparticles in the air.

11.6.4 Antisolvent Technique by Using CO_2

Antisolvent CO_2 precipitation involves supercritical carbon dioxide (CO_2) precipitating solutes from a solution (Reverchon et al., 1998). This is because compressed CO_2 solubility in organic solvents is a function of pressure and therefore solvent properties can be tuned by varying pressure. Researchers have utilized CO_2 as an antisolvent to recover nanoparticles from reverse micelles, such as ZnS, TiO_2, Ag, organic nanoparticles, etc. (Zhang et al., 2002, Baranowska-Wójcik, 2021, Zhang et al., 2001, Liu et al., 2003). This technique offers the advantage of preventing thermal degradation of materials while operating at near-room temperature, but pressure-related damages may still occur depending on the nanoparticle properties.

11.6.5 Microemulsion-Based Phase Separation Techniques

The microemulsion method, which involves thermodynamically stable dispersions of immiscible liquids with an emulsifier or surfactant, offers a promising approach to recycling and reusing nanoparticles. By creating aqueous two-phase systems or adding excess water to water/oil microemulsions, researchers have success-fully separated nanoparticles from impurities or other solute components (Mdlovu et al., 2018; Donga et al., 2019). The hydrophobic nanoparticles remain dispersed in the upper oil-rich phase, while the surfactant and hydrophilic reactants partition into the aqueous-rich phase. This method enables the purification and recovery of nanoparticles while utilizing water which is a green solvent (Hollamby et al., 2010). Recently, researchers have also utilized non-adsorbing polymers that can be added to the microemulsion to separate nanoparticles. These separated nanoparticles can be recovered or dispersed again by the addition of extra microemulsion (Myakonkaya et al., 2011).

11.6.6 NANOPARTICLES RECOVERY USING ELECTRODEPOSITION

Electrodeposition is a technique that utilizes voltage to drive chemical reactions in aqueous solutions which is also a process to synthesize nanoparticles (Tonelli et al., 2019). In recent years, there has been a rise in green synthesis approaches using biocompatible materials like plants and microorganisms. Electrokinetic processes have also been employed for the recovery of nanoparticles. For instance, Cu_2O/TiO_2 photocatalysts have been successfully recovered using electrokinetic methods (Xiu and Zhang, 2009). Moreover, the combination of supercritical water oxidation and electrokinetic processes has facilitated the creation and recovery of Cu_2O nanoparticles of varying sizes from waste-printed circuit boards (Xiu and Zhang, 2012). These techniques offer efficient and sustainable routes for nanoparticle synthesis and recovery, contributing to the field of nanomaterial recycling.

11.6.7 SLUDGE TREATMENT BEFORE INCINERATION

Sludge treatment processes play a crucial role in addressing nanomaterials in wastewater. The disposal of sewage sludge involves incineration or agricultural application but concerns about nanomaterial accumulation and their impact on soil processes have prompted discussions about discontinuing sludge use in farming.

Liu et al. developed a method to recover nano-SnO_2 from tin plate electroplating sludge. The process involves analyzing the sludge, mineralization processing, and acid treatment, yielding high-purity nano-SnO_2 powders with around 90% purity. However, the method has limitations, such as lower Sn recovery efficiency and challenges associated with hazardous waste disposal (Liu et al., 2019). The authors also performed a pilot scale test by treating 2.3 kg of sludge and recovered 90 g SnO_2 nanoparticles (Zhuang et al., 2012). Results suggest that a proper understanding of nanoparticles under varying physicochemical conditions may help in deriving selective remedial or recovery strategies. Certain industrial sludges, like $Mg(OH)_2$ nanowaste from chlorate industries, contain adsorbed heavy metals like carcinogenic Cr(VI) (Liu et al., 2008). Using a suitable mineralizer during hydrothermal coarsening, $Mg(OH)_2$ nanoparticles can be transformed into compounds with reduced surface adsorption properties, facilitating efficient Cr(VI) removal and recycling of the mineralizer as essential reagents for industrial processes (Liu et al., 2008). These approaches underscore the importance of effective sludge treatment and recycling methods for nanomaterials, promoting the sustainable management of nanowaste.

11.6.8 USE OF NANOMEMBRANES

Nanoporous materials and membranes have the potential to separate nanomaterials from contaminated water. Recent laboratory studies have shown promising results in separating nanoparticles including size-based fractionation.

Dey et al. have designed a centimeter-scale covalent organic framework thin film with uniform pores and high permeability and showed its application in the size-selective separation of gold nanoparticles. Tp-Azo COF-based thin film had a pore size of around 2.7 nm which resulted in the selective passing of 2 nm gold nanoparticles through the film while no infiltration of bigger size nanoparticles (Dey et al., 2020). Several COF materials are highly stable under diverse environmental conditions and therefore such films can be helpful in recovering nanoparticles from wastes or mixtures of nanoparticles (Kandambeth et al., 2012). In another work, Walden and Zhang demonstrated the removal of metal oxide nanoparticles using activated sludge and biofilms. Biofilms were found to be effective in removing nanoparticles. However, factors such as colloidal stability, aqueous matrix composition, and surface functionalization of the nanoparticles influenced the process (Walden and Zhang, 2016). Similarly, the presence of environmental complexities, including other colloidal particles such as swelling clays, can lead to film or membrane fouling and may face challenges in upscaling.

11.6.9 USE OF IONIC LIQUIDS

Ionic liquids (ILs) have emerged as promising solvents for the recovery of nanoparticles, providing efficient solvation properties and preventing nanoparticle aggregation and agglomeration through electrostatic repulsion and steric hindrance (Jain et al., 2011). They have found applications in various fields, including the extraction of pesticide residues in fruits and vegetables, the recovery of non-steroidal anti-inflammatory drugs, and the synthesis of nanomaterials (da Silva and Zimnoch dos Santos, 2021). ILs have demonstrated the ability to improve the synthesis procedure, control nanoparticles morphology, and facilitate the recycling of metallic and metal oxide nanoparticles. Additionally, ILs have shown immense potential for the recovery and recycling of cellulose from different sources (Glińska et al., 2020). Huang et al. demonstrated the utilization of room-temperature ionic liquid in recovering CuO nanoparticles with 80–95% recovery within 2 min of contact time (Huang et al., 2006).

However, a major limitation of ILs is their high cost and limited scalability, hindering their widespread use in large-scale industrial applications. Further research is warranted to address these challenges and explore the full potential of ILs for nanomaterial recycling.

11.6.10 CLOUD POINT EXTRACTION

In this approach, environmentally benign surfactants can be used which can form clouds and separate nanoparticles from aqueous dispersions at their critical temperatures (Nazar et al., 2011). For example, Liu et al. have shown the possible extraction of a diversified array of nanoparticles (CdSe/ZnS, Fe_3O_4, TiO_2, Ag, Au, C60, single-walled carbon nanotube (SWCNT)) using a low-cost surfactant (Triton X-114) (Liu et al., 2009b). Considering the complex water matrices, researchers showed the application of the same approach in recovering silver nanoparticles from various complex water matrices such as wastewater, lake, and river water (Liu et al., 2009a).

Although most of these emerging nanowaste remediation or nanoparticle recovery strategies show great potential under a controlled laboratory environment, they are yet to be tested for the justification of SE3 criteria, i.e., a technique should be scalable, efficient, economical, and environmentally friendly. Further research should focus on minimizing the entrance of nanowaste in the environment which can be assured by separate nanowaste collection and handling at the point of generation. Research should also focus on developing general protocols for understanding the fate and transport behavior of different nanoparticles under varying environmental geochemical conditions to find out the competition between dominating environmental parameters and nanoparticle properties in determining the fate and toxicity of nanoparticles.

11.7 CONCLUSION

Nanotechnology has revolutionized various industries, but the rapid growth in nanomaterial usage has resulted in the generation of nanowaste, which poses environmental and health risks. However, there is a lack of clear regulations and guidelines for the handling, disposal, and recycling of nanowaste. This gap hinders the commercial utilization of nanotechnology and limits its potential benefits. International organizations and developed countries are working toward designing comprehensive guidelines and standards to address nanowaste management and ensure the safety of humans and the environment. Additionally, the recycling and recovery of nanomaterials from waste sources are gaining attention as a sustainable approach. Researchers are developing innovative methods to recover nanomaterials from various waste streams, including carbon-, ceramic-, and metal-based nanomaterials. The potential applications of recycled nanomaterials are vast, spanning catalysis, energy devices, biomedicine, and more. However, further research is warranted to understand the potential risks and performance of products incorporating recycled nanomaterials. The responsible management of nanowaste and the development of efficient recycling technologies are essential for the continued advancement of nanotechnology while minimizing its environmental footprint.

Efforts are underway to establish suitable regulations and policies specifically targeting consumer products containing nanomaterials. Our reliance on the current guidelines to store, handle, and dispose of nanomaterials with a knowledge gap on long-term safety, curbing nanomaterial spills, etc., may yield unforeseen ramifications for human health and the environment. Although the threats of nanowaste demonstrated by a number of studies and the industrialization of nanomaterial production have skyrocketed, generalized legislative and policy frameworks are yet to be followed. Establishing global consensus on nanowaste management and implementing comprehensive regulations and, more importantly, safe nanowaste handling and labeled disposal guidelines for research laboratories will pave the way for the safe and sustainable utilization of nanomaterials worldwide. Frameworks like Cradle-to-Cradle Assessment, Extended Producer Responsibility, and Green Chemistry should be encouraged as they support advanced control during the design phase to conscientiously address the fate of nano-by-products.

ACKNOWLEDGMENTS

NK acknowledges the faculty initiation grant (FIG-101017) of IIT Roorkee. Nisha Singh is grateful to Ryota Nakajima and the JAMSTEC Young Researcher Fellowship for their support.

REFERENCES

Abbas, Q., Yousaf, B., Amina, Ali, M. U., Munir, M. A. M., El-Naggar, A., Rinklebe, J. & Naushad, M. 2020. Transformation pathways and fate of engineered nanoparticles (ENPs) in distinct interactive environmental compartments: A review. *Environment International*, 138, 105646.

Baranowska-Wójcik, E. 2021. Factors conditioning the potential effects TiO_2 NPs exposure on human microbiota: A mini-review. *Biological Trace Element Research*, 199, 4458–4465.

Bazargan, A. & Mckay, G. 2012. A review – synthesis of carbon nanotubes from plastic wastes. *Chemical Engineering Journal*, 195–196, 377–391.

Ben-David, E. A., Habibi, M., Haddad, E., Sammar, M., Angel, D. L., Dror, H., Lahovitski, H., Booth, A. M. & Sabbah, I. 2023. Mechanism of nanoplastics capture by jellyfish mucin and its potential as a sustainable water treatment technology. *Science of the Total Environment*, 869, 161824.

Bhagat, J., Nishimura, N. & Y. Shimada (2021). Toxicological interactions of microplastics/nanoplastics and environmental contaminants: Current knowledge and future perspectives. *Journal of Hazardous Materials*, 405, 123913.

Bossert, D., Urban, D. A., Maceroni, M., Ackermann-Hirschi, L., Haeni, L., Yajan, P., Spuch-Calvar, M., Rothen-Rutishauser, B., Rodriguez-Lorenzo, L., Petri-Fink, A. & Schwab, F. 2019. A hydrofluoric acid-free method to dissolve and quantify silica nanoparticles in aqueous and solid matrices. *Scientific Reports*, 9, 7938.

Bundschuh, M., Filser, J., Lüderwald, S., McKee, M. S., Metreveli, G., Schaumann, G. E., Schulz, R. & Wagner, S. 2018. Nanoparticles in the environment: where do we come from, where do we go to? *Environmental Sciences Europe*, 30(1), 6. doi: 10.1186/s12302-018-0132-6

Chen, L., Yokel, R. A., Hennig, B. & Toborek, M. 2008. Manufactured aluminum oxide nanoparticles decrease expression of tight junction proteins in brain vasculature. *Journal of Neuroimmune Pharmacology*, 3, 286–295.

Cortie, M. B. 2004. The weird world of nanoscale gold. *Gold Bulletin*, 37, 12–19. https://doi.org/10.1007/BF03215512

Da Silva, W. L. & Zimnoch Dos Santos, J. H. 2021. Chapter 2 – Applications of ionic liquids in environmental remediation. In: Inamuddin, B. R. & Asiri, A. M. (eds.) *Green Sustainable Process for Chemical and Environmental Engineering and Science* (pp. 15–21). Elsevier.

De Moura Souza, F., Pollo Paniz, F., Pedron, T., Coelho Dos Santos, M. & Lemos Batista, B. 2019. A high-throughput analytical tool for quantification of 15 metallic nanoparticles supported on carbon black. *Heliyon*, 5, e01308.

Dericiler, K., Berktas, I., Dogan, S., Menceloglu, Y. Z. & Saner Okan, B. 2022. Chapter 8 – General techniques for recovery of nanomaterials from wastes. In: Rai, M. & Nguyen, T. A. (eds.) *Nanomaterials Recycling* (pp. 147–174). Elsevier.

Dey, K., Kunjattu H. S., Chahande, A. M. & Banerjee, R. 2020. Nanoparticle size-fractionation through self-standing porous covalent organic framework films. *Angewandte Chemie International Edition*, 59, 1161–1165.

Donga, C., Mabape, K. I. S., Mishra, S. B. & Mishra, A. K. 2019. Chapter 8 – Polymer-based engineering materials for removal of nanowastes from water. In: Mishra, A. K., Anawar, H. M. D. & Drouiche, N. (eds.) *Emerging and Nanomaterial Contaminants in Wastewater* (pp. 217–243). Elsevier.

Dulger, M., Sakallioglu, T., Temizel, I., Demirel, B., Copty, N. K., Onay, T. T., Uyguner-Demirel, C. S. & Karanfil, T. 2016. Leaching potential of nano-scale titanium dioxide in fresh municipal solid waste. *Chemosphere*, 144, 1567–1572. https://doi.org/10.1016/j.chemosphere.2015.10.037

Ferdous, Z. & Nemmar, A. 2020. Health impact of silver nanoparticles: A review of the biodistribution and toxicity following various routes of exposure. *International Journal of Molecular Science*, 21(7), 2375. doi: 10.3390/ijms21072375.

Geppert, M., Sigg, L. & Schirmer, K. 2021. Toxicity and translocation of Ag, CuO, ZnO and TiO_2 nanoparticles upon exposure to fish intestinal epithelial cells. *Environmental Science: Nano*, 8, 2249–2260.

Glińska, K., Stüber, F., Fabregat, A., Giralt, J., Font, J., Mateo-Sanz, J. M., Torrens, E. & Bengoa, C. 2020. Moving municipal WWTP towards circular economy: Cellulose recovery from primary sludge with ionic liquid. *Resources, Conservation and Recycling*, 154, 104626.

Gupta, K., Khandelwal, N. & Darbha, G. K. 2019. Removal and recovery of toxic nanosized cerium oxide using eco-friendly iron oxide nanoparticles. *Frontiers of Environmental Science & Engineering*, 14, 15.

Gupta, P. & Bhandari, S. 2022. Chapter 3 – Classification and sources of nanowastes. In: Rai, M. & Nguyen, T. A. (eds.) *Nanomaterials Recycling*. Elsevier, pp. 37–60.

Gupta, R. & Xie, H. 2018. Nanoparticles in daily life: Applications, toxicity and regulations. *Journal of Environmental Pathology, Toxicology, and Oncology*, 37(3), 209–230.

Hollamby, M. J., Eastoe, J., Chemelli, A., Glatter, O., Rogers, S., Heenan, R. K. & Grillo, I. 2010. Separation and purification of nanoparticles in a single step. *Langmuir*, 26, 6989–6994.

Huang, H.-L., Wang, H. P., Wei, G.-T., Sun, I. W., Huang, J.-F. & Yang, Y. W. 2006. Extraction of nanosize copper pollutants with an ionic liquid. *Environmental Science & Technology*, 40, 4761–4764.

Huang, N., Zhai, L., Xu, H. & Jiang, D. 2017. Stable covalent organic frameworks for exceptional mercury removal from aqueous solutions. *Journal of the American Chemical Society*, 139, 2428–2434.

Ispas, C., Andreescu, D., Patel, A., Goia, D. V., Andreescu, S. & Wallace, K. N. 2009. Toxicity and developmental defects of different sizes and shape nickel nanoparticles in zebrafish. *Environmental Science & Technology*, 43(16), 6349–6356. doi: 10.1021/es9010543. PMID: 19746736

Iain, N., Zhang, X., Hawkett, B. S. & Warr, G. G. 2011. Stable and water-tolerant ionic liquid ferrofluids. *ACS Applied Materials & Interfaces*, 3, 662–667.

Jones, A. M. & Harrison, R. M. 2016. Emission of ultrafine particles from the incineration of municipal solid waste: A review. *Atmospheric Environment*, 140, 519–528.

Kandambeth, S., Mallick, A., Lukose, B., Mane, M. V., Heine, T. & Banerjee, R. 2012. Construction of crystalline 2D covalent organic frameworks with remarkable chemical (acid/base) stability via a combined reversible and irreversible route. *Journal of the American Chemical Society*, 134, 19524–19527.

Khan, A., Malik, S., Ali, N., Nguyen, T. A. & Bilal, M. 2022. 20 – Nanoadsorbents as a green approach for removal of environmental pollutants. In: Iqbal, H. M. N., Bilal, M. and Nguyen, T. A. (eds.) *Nano-Bioremediation: Fundamentals and Applications.* Elsevier, pp. 435–454.

Khandelwal, N. & Darbha, G. K. 2021. A decade of exploring MXenes as aquatic cleaners: Covering a broad range of contaminants, current challenges and future trends. *Chemosphere*, 279, 130587.

Khandelwal, N., Singh, N., Tiwari, E. & Krishna Darbha, G. 2022. Achieving strong Pb–Cr complexation in Mg/Al LDHs for ultrafast chromate ions separation and chrome recovery from complex water matrices. *Environmental Nanotechnology, Monitoring & Management*, 18, 100754.

Kim, K. T., Klaine, S. J., Cho, J., Kim, S.-H. & Kim, S. D. 2010. Oxidative stress responses of *Daphnia magna* exposed to TiO_2 nanoparticles according to size fraction. *Science of the Total Environment*, 408(10), 2268–2272.

Kose, O., Tomatis, M., Leclerc, L., Belblidia, N.-B., Hochepied, J.-F., Turci, F., Pourchez, J. & Forest, V. 2020. Impact of the physicochemical features of TiO_2 nanoparticles on their in vitro toxicity. *Chemical Research in Toxicology*, 33, 2324–2337.

Lead, J. R., Batley, G. E., Alvarez, P. J. J., Croteau, M.-N., Handy, R. D., McLaughlin, M. J., Judy, J. D. & Schirmer, K. 2018. Nanomaterials in the environment: Behavior, fate, bioavailability, and effects – an updated review. *Environmental Toxicology and Chemistry*, 37(8), 2029–2063.

Lee, K. P., Trochimowicz, H. J. & Reinhardt, C. F. 1985. Pulmonary response of rats exposed to titanium dioxide (TiO_2) by inhalation for two years. *Toxicology and Applied Pharmacology*, 79(2), 179–192.

Liu, D., Zhang, J., Han, B., Chen, J., Li, Z., Shen, D. & Yang, G. 2003. Recovery of TiO_2 nanoparticles synthesized in reverse micelles by antisolvent CO_2. *Colloids and Surfaces A: Physicochemical and Engineering Aspects*, 227, 45–48.

Liu, J.-F., Chao, J.-B., Liu, R., Tan, Z.-Q., Yin, Y.-G., Wu, Y. & Jiang, G.-B. 2009a. Cloud point extraction as an dvantageous preconcentration approach for analysis of trace silver nanoparticles in environmental waters. *Analytical Chemistry*, 81, 6496–6502.

Liu, J.-F., Liu, R., Yin, Y.-G. & Jiang, G.-B. 2009b. Triton X-114 based cloud point extraction: A thermoreversible approach for separation/concentration and dispersion of nanomaterials in the aqueous phase. *Chemical Communications*, (12), 1514–1516.

Liu, L., Zhang, X., Yang, L., Ren, L., Wang, D. & Ye, J. 2017. Metal nanoparticles induced photocatalysis. *National Science Review*, 4(5), 761–780. https://doi.org/10.1093/nsr/nwx019

Liu, W., Huang, F., Liao, Y., Zhang, J., Ren, G., Zhuang, Z., Zhen, J., Lin, Z. & Wang, C. 2008. Treatment of CrVI-containing $Mg(OH)_2$ nanowaste. *Angewandte Chemie International Edition*, 47, 5619–5622.

Liu, W., Weng, C., Zheng, J., Peng, X., Zhang, J. & Lin, Z. 2019. Emerging investigator series: Treatment and recycling of heavy metals from nanosludge. *Environmental Science: Nano*, 6, 1657–1673.

Mandel, K., Hutter, F., Gellermann, C. & Sextl, G. 2013. Reusable superparamagnetic nanocomposite particles for magnetic separation of iron hydroxide precipitates to remove and recover heavy metal ions from aqueous solutions. *Separation and Purification Technology*, 109, 144–147.

Mdlovu, N. V., Chiang, C.-L., Lin, K.-S. & Jeng, R.-C. 2018. Recycling copper nanoparticles from printed circuit board waste etchants via a microemulsion process. *Journal of Cleaner Production*, 185, 781–796.

Melchor-Martínez, E. M., Macias-Garbett, R., Malacara-Becerra, A., Iqbal, H. M. N., Sosa-Hernández, J. E. & Parra-Saldívar, R. 2021. Environmental impact of emerging contaminants from battery waste: A mini review. *Case Studies in Chemical and Environmental Engineering*, 3, 100104.

Mitrano, D. M., Mehrabi, K., Dasilva, Y. A. R. & Nowack, B. 2017. Mobility of metallic (nano) particles in leachates from landfills containing waste incineration residues. *Environmental Science: Nano*, 4, 480–492.

Myakonkaya, O., Eastoe, J., Mutch, K. J. & Grillo, I. 2011. Polymer-induced recovery of nanoparticles from microemulsions. *Physical Chemistry Chemical Physics*, 13, 3059–3063.

Nazar, M. F., Shah, S. S., Eastoe, J., Khan, A. M. & Shah, A. 2011. Separation and recycling of nanoparticles using cloud point extraction with non-ionic surfactant mixtures. *Journal of Colloid and Interface Science*, 363, 490–496.

Patwa, A., Thiéry, A., Lombard, F., Lilley, M. K. S., Boisset, C., Bramard, J.-F., Bottero, J.-Y. & Barthélémy, P. 2015. Accumulation of nanoparticles in "jellyfish" mucus: A bio-inspired route to decontamination of nano-waste. *Scientific Reports*, 5, 11387.

Powell, C. D., Atkinson, A. J., Ma, Y., Marcos-Hernandez, M., Villagran, D., Westerhoff, P. & Wong, M. S. 2020. Magnetic nanoparticle recovery device (MagNERD) enables application of iron oxide nanoparticles for water treatment. *Journal of Nanoparticle Research*, 22, 48.

Prüst, M., Meijer, J. & Westerink, R. H. S. 2020. The plastic brain: Neurotoxicity of micro- and nanoplastics. *Particle and Fibre Toxicology* 17(1), 24.

Puri, N., Gupta, A. & Mishra, A. 2021. Recent advances on nano-adsorbents and nanomembranes for the remediation of water. *Journal of Cleaner Production*, 322, 129051.

Qin, X. & Wu, D. 2012. Effect of different solvents on poly(caprolactone) (PCL) electrospun nonwoven membranes. *Journal of Thermal Analysis and Calorimetry*, 107, 1007–1013.

Ray, P. C., Yu, H. & Fu, P. P. 2009. Toxicity and environmental risks of nanomaterials: Challenges and future needs. *Journal of Environmental Science and Health. Part C, Environmental Carcinogenesis & Ecotoxicology Reviews*, 27, 1–35.

Reverchon, E., Della Porta, G., Di Trolio, A. & Pace, S. 1998. Supercritical antisolvent precipitation of nanoparticles of superconductor precursors. *Industrial & Engineering Chemistry Research*, 37, 952–958.

Schwab, F., Rothen-Rutishauser, B., Scherz, A., Meyer, T., Karakoçak, B. B. & Petri-Fink, A. 2023. The need for awareness and action in managing nanowaste. *Nature Nanotechnology*, 18(4), 317–321. doi: 10.1038/s41565-023-01331-4.

Shokri, A., Pahlevani, F., Cole, I. & Sahajwalla, V. 2017. Selective thermal transformation of old computer printed circuit boards to Cu–Sn based alloy. *Journal of Environmental Management*, 199, 7–12.

Singh, J., Vishwakarma, K., Ramawat, N., Rai, P., Singh, V. K., Mishra, R. K., Kumar, V., Tripathi, D. K. & Sharma, S. 2019. Nanomaterials and microbes' interactions: A contemporary overview. *3 Biotech*, 9, 68.

Singh, N., Bhagat, J., Tiwari, E., Khandelwal, N., Darbha, G. K. & Shyama, S. K. 2021a. Metal oxide nanoparticles and polycyclic aromatic hydrocarbons alter nanoplastic's stability and toxicity to zebrafish. *Journal of Hazardous Materials*, 407, 124382.

Singh, N., Tiwari, E., Khandelwal, N. & Darbha, G. K. 2019. Understanding the stability of nanoplastics in aqueous environments: effect of ionic strength, temperature, dissolved organic matter, clay, and heavy metals. *Environmental Science: Nano* 6(10), 2968–2976.

Singh, N., Bhagat, J., Tiwari, E., Khandelwal, N., Darbha, G. K. & Shyama, S. K. 2021a. Metal oxide nanoparticles and polycyclic aromatic hydrocarbons alter nanoplastic's stability and toxicity to zebrafish. *Journal of Hazardous Materials*, 407, 124382.

Singh, N., Khandelwal, N., Ganie, Z. A., Tiwari, E. & Darbha, G. K. 2021b. Eco-friendly magnetic biochar: An effective trap for nanoplastics of varying surface functionality and size in the aqueous environment. *Chemical Engineering Journal*, 418, 129405.

Skocaj, M., Filipic, M., Petkovic, J. & Novak, S. 2011. Titanium dioxide in our everyday life; is it safe? *Radiology and Oncology*, 45(4), 227–247.

Søndergaard, R. R., Zimmermann, Y.-S., Espinosa, N., Lenz, M. & Krebs, F. 2016. Incineration of organic solar cells: Efficient end of life management by quantitative silver recovery. *Energy & Environmental Science*, 9, 857–861.

Taylor, A. F., Rylott, E. L., Anderson, C. W. N. & Bruce, N. C. 2014. Investigating the toxicity, uptake, nanoparticle formation and genetic response of plants to gold. *PLoS One*, 9, e93793.

Tiwari, E., Singh, N., Khandelwal, N., Monikh, F. A. & Darbha, G. K. 2020. Application of Zn/Al layered double hydroxides for the removal of nano-scale plastic debris from aqueous systems. *Journal of Hazardous Materials*, 397, 122769.

Tonelli, D., Scavetta, E. & Gualandi, I. 2019. Electrochemical deposition of nanomaterials for electrochemical sensing. *Sensors (Basel)*, 19(5), 1186.

Umair, M., Javed, I., Rehman, M., Madni, A., Javeed, A., Ghafoor, A. & Ashraf, M. 2016. Nanotoxicity of inert materials: The case of gold, silver and iron. *Journal of Pharmacy & Pharmaceutical Sciences*, 19(2), 161–180.

Varier, K. M., Gudeppu, M., Chinnasamy, A., Thangarajan, S., Balasubramanian, J., Li, Y. & Gajendran, B. 2019. Nanoparticles: Antimicrobial applications and its prospects. *Advanced Nanostructured Materials for Environmental Remediation*, 25, 321–355. doi: 10.1007/978-3-030-04477-0_12

Vineeth Kumar, C. M., Karthick, V., Kumar, V. G., Inbakandan, D., Rene, E. R., Suganya, K. S. U., Embrandiri, A., Dhas, T. S., Ravi, M. & Sowmiya, P. 2022. The impact of engineered nanomaterials on the environment: Release mechanism, toxicity, transformation, and remediation. *Environmental Research*, 212, 113202.

Vinti, G., Bauza, V., Clasen, T., Medlicott, K., Tudor, T., Zurbrügg, C. & Vaccari, M. 2021. Municipal solid waste management and adverse health outcomes: A systematic review. *International Journal of Environmental Research and Public Health*, 18(8), 4331.

Walden, C. & Zhang, W. 2016. Biofilms versus activated sludge: Considerations in metal and metal oxide nanoparticle removal from wastewater. *Environmental Science & Technology*, 50, 8417–8431.

Walser, T., Limbach, L. K., Brogioli, R., Erismann, E., Flamigni, L., Hattendorf, B., Juchli, M., Krumeich, F., Ludwig, C., Prikopsky, K., Rossier, M., Saner, D., Sigg, A., Hellweg, S., Günther, D. & Stark, W. J. 2012. Persistence of engineered nanoparticles in a municipal solid-waste incineration plant. *Nature Nanotechnology*, 7(8), 520–524.

Wang, D., Xie, T., Peng, Q. & Li, Y. 2008. Ag, Ag_2S, and Ag_2Se nanocrystals: Synthesis, assembly, and construction of mesoporous structures. *Journal of the American Chemical Society*, 130, 4016–4022.

Wang, J., Schlagenhauf, L. & Setyan, A. 2017. Transformation of the released asbestos, carbon fibers and carbon nanotubes from composite materials and the changes of their potential health impacts. *Journal of Nanobiotechnology*, 15, 15. https://doi.org/10.1186/s12951-017-0248-7

Wang, L., Hu, C. & Shao, L. 2017. The antimicrobial activity of nanoparticles: Present situation and prospects for the future. *International Journal of Nanomedicine*, 12, 1227–1249.

Wang, Z. Y., Xie, X. Y., Zhao, J., Liu, X. Y., Feng, W. Q., White, J. C. & Xing, B. S. 2012. Xylem- and phloem-based transport of CuO nanoparticles in maize (*Zea mays* L.). *Environmental Science & Technology*, 46, 4434–4441.

Wiesner, M. R., Lowry, G. V., Alvarez, P., Dionysiou, D. & Biswas, P. 2006. Assessing the risks of manufactured nanomaterials. *Environmental Science & Technology*, 40(14), 4336–4345.

Xiu, F.-R. & Zhang, F.-S. 2009. Preparation of nano-Cu_2O/TiO_2 photocatalyst from waste printed circuit boards by electrokinetic process. *Journal of Hazardous Materials*, 172, 1458–1463.

Xiu, F. R. & Zhang, F. S. 2012. Size-controlled preparation of Cu_2O nanoparticles from waste printed circuit boards by supercritical water combined with electrokinetic process. *Journal of Hazardous Materials*, 233–234, 200–206.

Younis, S. A., El-Fawal, E. M. & Serp, P. 2018. Nano-wastes and the environment: Potential challenges and opportunities of nano-waste management paradigm for greener nanotechnologies. In: Hussain, C. M. (ed.) *Handbook of Environmental Materials Management*. Springer International Publishing, pp. 1–72.

Zhai, G. S., Walters, K. S., Peate, D. W., Alvarez, P. J. J. & Schnoor, J. L. 2014. Transport of gold nanoparticles through plasmodesmata and precipitation of gold ions in woody poplar. *Environmental Science & Technology Letters*, 1, 146–151.

Zhang, J., Han, B., Liu, J., Zhang, X., He, J., Liu, Z., Jiang, T. & Yang, G. 2002. Recovery of silver nanoparticles synthesized in AOT/C12E4 mixed reverse micelles by antisolvent CO_2. *Chemistry – A European Journal*, 8, 3879–3883.

Zhang, J., Han, B., Liu, J., Zhang, X., Liu, Z. & He, J. 2001. A new method to recover the nanoparticles from reverse micelles: Recovery of ZnS nanoparticles synthesized in reverse micelles by compressed CO_2. *Chemical Communications*, 2724(24), 2724–2725.

Zhuang, Z., Xu, X., Wang, Y., Wang, Y., Huang, F. & Lin, Z. 2012. Treatment of nanowaste via fast crystal growth: With recycling of nano-SnO_2 from electroplating sludge as a study case. *Journal of Hazardous Materials*, 211–212, 414–419.

12 Algal Nanoparticles as a Green Approach for Bioremediation of Heavy Metals from Wastewater

Sujata, Anuj Sharma, Bansal Deepak, Rachna Bhateria, and Sharma Mona

12.1 INTRODUCTION

In recent times, environment and water resources have been affected by rapid industrialization, advanced practices in agriculture, urbanization, and increasing population (Alcamo et al, 2017). As water use increases and the water quality deteriorates, concerns about clean water are growing for human use and developing countries' sustainable development (Alcamo et al, 2017). For the resolution of water catastrophes, some measures are needed urgently, including water conservation, proper wastewater treatment creating a water-saving society, and improved utilization of water use (Hlongwane, 2019). Reuse of wastewater has a future for pollution reduction and is helpful in fulfilling the crisis of water resources (Lyu et al, 2016). For the reuse of wastewater, proper treatment of water is an essential step of wastewater treatment. Many types of complex pollutants are generated by continuously developing industrial and agricultural effluents that affect the treated wastewater also (Li et al, 2016). According to the World Health Organization (WHO), millions of people die every day due to contaminated water, and billions of people are affected by waterborne health diseases. Table 12.1 describes the various contaminants and their effect. Different industries like tannery, textile, leather, and petroleum refining primarily release heavy metals (lead, zinc, chromium, copper, mercury, platinum, cadmium, nickel, silver, etc.) in wastewater (Mukherjee et al, 2021). Nowadays, a number of conventional methods like chemical, mechanical, activated carbon, activated sludge, oxidation, reverse osmosis, and nanofiltration methods are used for wastewater treatment (Amin et al, 2014). However, all these are not able to treat complicated/complex pollutants like pharmaceutical products, personal care products, industrial additives, and surfactants (Amin et al, 2014). Therefore, the industry requires a long-term, reliable, accurate,

TABLE 12.1

Different water contaminants with major source and different human diseases

Sr No	Water contaminants	Major source	Waterborne disease/ water toxicity	References
1	Heavy metals	Textile, mining, petroleum, chemical industry, agricultural industries.	Cardiovascular disease, cancer, diabetes, hypertension, mellitus, skeletal malformation, mutagenicity/ genetic disease	Hlongwane, 2019; EPA, 2002
2	Dye	Textile industries, printing industries, cosmetic industries	Bladder cancer, disorder of central nervous system, irritation of skin and eyes, dermatitis, allergic conjunctivitis, rhinitis, asthma	Amin et al, 2014; Hunger 2007; Tang et al., 2012; Gu et al., 2016
4	Herbicides/ Pesticides	Agricultural waste, household waste	Cardiovascular disease, muscles, and retinal degenerations, cancer, irreversible cell membrane damage, liver damage, anemia	EPA, 2002; Saleh et al, 2020
5	Virus/bacteria	Fecal matter, soil pollution	Gastroenteritis, typhoid, pneumonia, hepatitis, respiratory infection, meningitis	Symonds et al, 2014; Chahal et al., 2016

safe, and cost-friendly specific green technology. Nanotechnology is a reliable industrial application with low energy consumption as well as a contaminant-free environment because the unique properties of nanomaterial (nanoscale, highly reactive, strong mechanical properties, porosity characteristics, mobility, hydrophilicity) are the best methods for wastewater treatment (Hlongwane, 2019). Many types of toxic metals, organic and inorganic contaminants, and harmful microorganisms have been removed using different nanomaterials. Different studies say that nanocatalysts are useful for treating because of their unique qualities, like low size, high reactivity, and transferrable atomicity.

12.2   SYNTHESIS OF NANOPARTICLES

To determine the characteristics and thin films of nanoparticles, synthesis approaches play an important role. There are several methods for the synthesis of nanoparticles and thin films. The main aim behind exploring numerous methods is to be cost-effective and reliable in terms of utilization. The nucleation, stabilization of nanoparticles pH, pressure growth in size, shape and size distribution, reaction time, and exposure time are the main factors (Singh et al, 2020; Mukherjee et al, 2021). Bottom-up and top-down are the two main types of ways that are used for synthesizing nanoparticles (Khanna et al, 2019). In the top-down approach, large material breaks down into smaller nanosized particles. Top-down approaches are used for the division of bulk material into desired structures with appropriate properties, while bottom-up approach has the potential to produce less waste by using molecule by molecule or atom by atom from the bottom (Singh et al, 2020).

Algae, bacteria, and fungi (biological organisms) are used for synthesizing nanomaterials. For the efficient synthesis of nanoparticles, the extract of algae acts as a natural capping and reducing agent. Ecologically and economically algae are one of the important groups of the important group of photosynthetic organisms (Khan et al, 2022). They are living in freshwater, hot water, frozen water, marine water, and terrestrial region. Algal biomass can be harvested many times in one year without any addition of fertilizers and chemicals (Mukherjee et al, 2021). Algae play a vital role in the aquatic body as the primary producer and being a primary producer; they produce oxygen and organic matter in the water body and purify the water (Sharma et al, 2023). Algae are microscopic plants with various advantages, such as bioactive compounds and a high growth rate for the production of nanoparticles in comparison to higher plants (Jacob et al, 2021). They have many applications in medicine, foods, cosmetics, agriculture, biofuels, and natural dye (LewisOscar et al, 2016). Algae have a high potential for the treatment of wastewater because they can grow easily in the water helping in the bioremediation process (Sharma et al, 2023). Chlorophyceae, Rhodophyceae, Cyanophyceae, and Phaeophyceae are classes of algae by which the production of nanoparticles occurs through the intracellular and extracellular synthesis of gold, silver, and other metallic compounds (Khanna et al, 2019). Heavy metals extracted from wastewater are updated by algal cells and are used as nutrient sources to produce algal biomass by synchronizing their metabolic processes (Khan et al, 2022). During the last few years, algae have been used as a model organism for algae-mediated nanoparticles (Ponnuchamy and Jacob, 2016). Depending on size and age, different concentrations of proteins, carbohydrates, vitamins, pigments, and antioxidants are found in algae (Mukherjee et al, 2021). These bioactive compounds are described as reducing and stabilization agents for nanoparticle synthesis (Ponnuchamy and Jacob, 2016). Intracellular or extracellular methods can be used to synthesize nanomaterials, depending on the species of algae. Different types of algal species used for the synthesis of metallic nanoparticles are clearly described in Table 12.2.

12.3 CONVENTIONAL PRACTICES FOR REMEDIATION OF HEAVY METALS

Heavy metals live for a long time in the environment because heavy metals with high mobility easily leach out and spread ubiquitously in different media and are also absorbed by living organisms (Arora and Khosla, 2021). Heavy metals are non-biodegradable so they can affect every level of food chain organisms followed by biomagnification (Wu et al, 2018). Aggregation of metals in the organs such as the kidney, liver, brain, and liver occurs through the skin, inhalation, and ingestion

TABLE 12.2
Different types of algal species used for the synthesis of metallic nanoparticles

Sr no.	Algal species	Metallic nanoparticles (NPs)	References
1	*Chlorella vulgaris, Ulva fasciata, Spirulina platenis, Lyngbya majuscule Calothrix,* and *Leptolyngbya, Plectonema boryanum, Calothrix pulvinate*	Ag	Chakraborty et al, 2009; Rajesh et al, 2012; Brayner et al, 2007 Lewis Oscar et al, 2016, Mukherjee et al, 2021
2	*Diadesmis gallica, Calothrix, Leptolyngbya Navicula atomus, Plectonema boryanum, Calothrix pulvinate*	Au	Schrofel et al, 2011; Lewis Oscar et al; Brayner et al, 2007, 2016; Mukherjee et al, 2021
3	*Plectonema boryanum, Calothrix pulvinate, Calothrix, Leptolyngbya*	Pt	Mukherjee et al, 2021; Brayner et al, 2007
4	*Bifurcaria bifurcate*	CuO	Abboud et al., 2014
5	*Sargassum plagiophyllum*	AgCl	Gonzalacz-Ballesteros and Rodriguez-Arguelles, 2020
6	*Ulva fasciata*	ZnO	Alsaggal et al, 2021

where they can cause severe disease (Esmaeilzadeh et al, 2019; Adriano et al, 2001). Heavy metals contaminate the soil also inhibit plant growth, enter the food chain, and cause health problems in living organisms (Arora and Khosla, 2021). So, these heavy metals are toxic and carcinogenic. Proper treatment of wastewater is required for the remediation of heavy metals. There are many conventional strategies for the extraction of heavy metals. Many types of methods like physiochemical, electrochemical, and adsorption are described in Figure 12.1.

12.3.1 ELECTROCHEMICAL PROCESS

Electric current passes through the aqueous metal solution in an electrochemical process by utilizing voltage between an insoluble anode and a cathode plate (Tran et al, 2017). In this process, an electrochemical cell in which direct current voltage is applied to the sacrificial electrode, commonly iron or aluminum. In this process, metals are precipitated as hydroxide in an acidic medium or neutralized the electrolyte (Razzak et al, 2022).

12.3.2 ELECTRO-FLOTATION PROCESS

In this process, metals are suspended and become a solution in a liquid phase. Due to the adhesion of small foam of hydrogen and oxygen on both, electrodes move upward

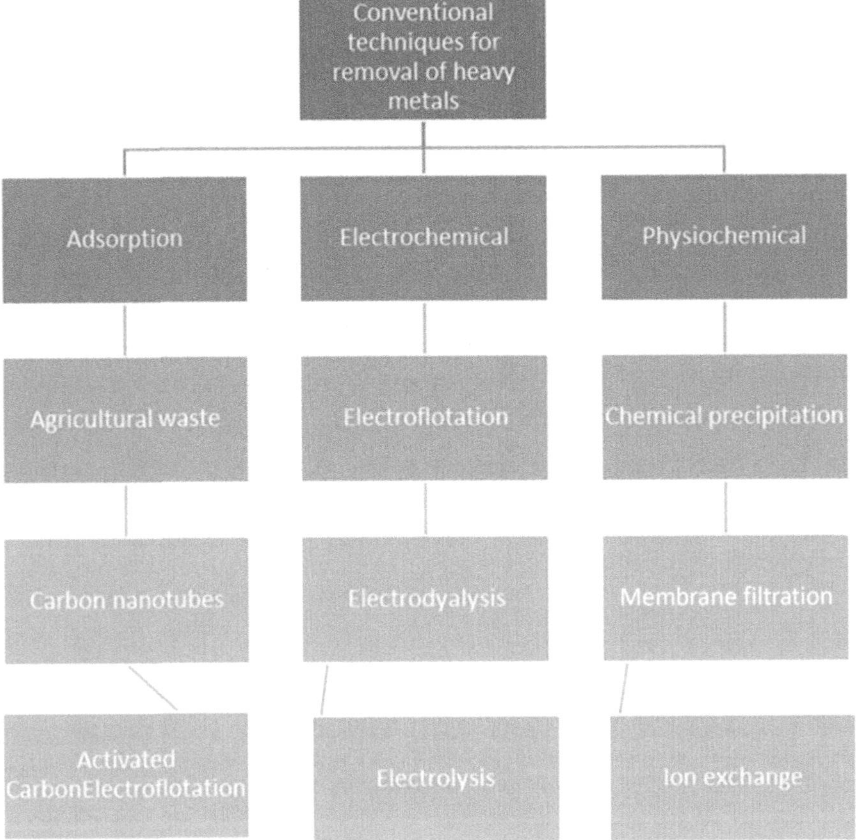

FIGURE 12.1 Conventional techniques for the remediation of heavy metals from wastewater.

in the flotation cell (da Mota et al, 2015). Electro-flotation methods generally consist of three stages (da Mota et al, 2015). First is oxidation of electrodes to form coagulant agents and then the occurrence of electrolysis of water (Al Aji et al, 2012). The second stage is destabilization that causes electrochemical metal reduction of cation on the cathode surface. The formation of hydroxide ions at the cathode increases the pH of the wastewater (Emamjomeh et al, 2009). Further to this, precipitation of metal ions occurs. The third stage contains an aggregation of destabilized phase to form flakes (Parga et al, 2002).

12.3.3 COAGULATION/FLOCCULATION

This method involves the addition of coagulants (such as alum or ferric chloride) to destabilize particles and form larger clusters, which can then be easily separated through sedimentation or filtration (Haung et al, 2015).

12.3.4 Activated Carbon Adsorption

High porosity and surface area are primary features because of which activated carbon is widely utilized in the adsorption of contaminants. It effectively removes a variety of pollutants, including pesticides, pharmaceuticals, and industrial chemicals (Ma et al, 2021).

12.3.5 Biological Treatment

In the biological methods, microorganisms are used to degrade or transform pollutants in wastewater. Many processes like activated sludge, biofilters, and constructed wetlands employ bacteria, fungi, and plants to break down organic matter and remove nutrients (Liu et al, 2022).

12.3.6 Membrane Filtration

Membrane-based technologies, including reverse osmosis, ultrafiltration, and nanofiltration, rely on semi-permeable membranes to separate pollutants from water. These processes effectively remove suspended solids, microorganisms, and dissolved contaminants (Wang et al, 2022).

12.3.7 Oxidation Processes

Advanced oxidation processes (AOPs), such as ozonation, UV/H_2O_2, and Fenton reaction, generate highly reactive hydroxyl radicals to degrade recalcitrant pollutants. These methods are important in removing organic contaminants and emerging pollutants (Zhu et al, 2022). It is to be noted that the selection of a remediation practice depends on the specific contaminant in wastewater, their concentrations, and the desired effluent quality. Ongoing research continues to improve these conventional methods and explore new approaches for efficient and sustainable pollutant remediation.

12.4 SYNTHESIS OF MICROALGAL NANOPARTICLES

Particles having a size <100 nm are classified as nano-adsorbents (Sharma et al, 2023, LewisOscar et al, 2016). Nano-adsorbents are believed to be tremendously efficient in the removal of contaminants, because of having numerous features such as a high surface-to-volume ratio, non-linear optics, and tolerance to high temperatures. Because of having a large surface-to-volume ratio, nanoparticles have more binding sites that help in the remediation of contaminants efficiently (Bhukal et al, 2022). Nanoparticles are synthesized utilizing numerous chemical and physical methodologies like thermolysis, vapor deposition, high-energy ball processing, sol–gel strategy, and chemical fume testimony. Both these physical and chemical processes lead to devastating impacts on the environment and they are time-consuming, expensive, and difficult to handle (Guleri et al, 2020; Sathishkumar et al, 2019; Ting et al, 2017). Synthesis of nanoparticles using biological means using plants, microbes, and algae is preferred by the researchers because of their eco-friendly synthesis. Synthesis of

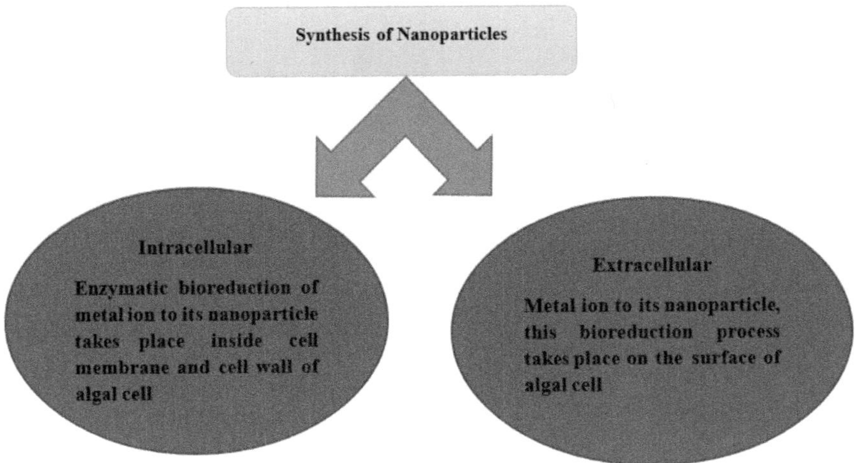

FIGURE 12.2 Intracellular and extracellular mode of algal nanoparticle synthesis.

nanomaterials using biological means is a quite simple and economical procedure as compared to chemical and physical strategies (Thunugunta et al, 2015; Agarwal et al, 2019). The present focus of the researcher is to find an eco-friendly, commercially viable, and competent procedure. Phyco-nanotechnology in which algae is used to synthesize nanomaterials is one of the emerging techniques to synthesize nanoparticles (Khanna et al, 2019). Biomass of algae and its nano-adsorbents both can remove contaminants but algae-mediated nanoparticles have a large surface-to-volume ratio and thus are more efficient. The presence of compounds in algae such as flavonoids helps in the reduction process during nanomaterial synthesis (Molazadeh et al, 2019). Microalgae can be used in various forms during synthesizing nanomaterials such as (1) biomolecules extracted from the microalgae-disrupted cell, (2) cell-free supernatant from microalgae, (3) entire cell of microalgae, and (4) living cell of microalgae (Sharma et al, 2023, Agarwal et al, 2019). Table 12.3 represents the utilization of numerous algal species for synthesizing nanoparticles and techniques used for characterization. Metal and metal oxide both types of nanoparticles are synthesized by researchers utilizing algae. Intracellular and extracellular are the two methods used to synthesize algal nanoparticles described in Figure 12.2 (Khanna et al, 2019). The mechanism to synthesize algae-mediated nanoparticles is described in Figure 12.3. Figure 12.4. presents the procedure to synthesize algal nanoparticles in a flowchart form.

12.5 APPLICATION OF ALGAE-MEDIATED NANOPARTICLES IN THE REMEDIATION OF CONTAMINANTS FROM WASTEWATER

Algae-mediated nanoparticles have high efficiency for the remediation of contaminants such as dye and heavy metals (Bhukal et al, 2022). Heavy metals like Cd, Zn, Pb, and

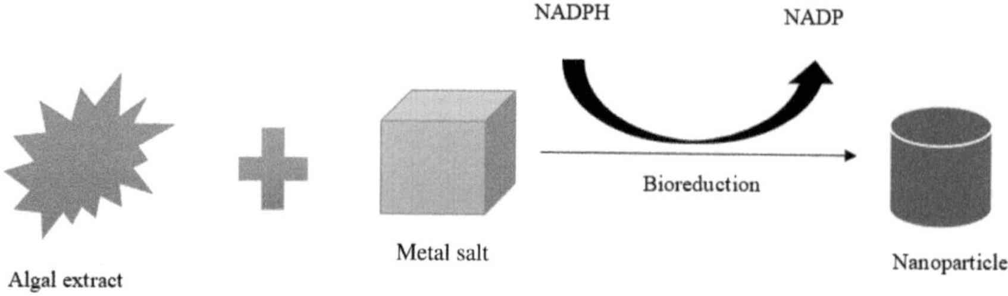

FIGURE 12.3 Simplified process for synthesizing algal nanoparticles (Guleri et al, 2020).

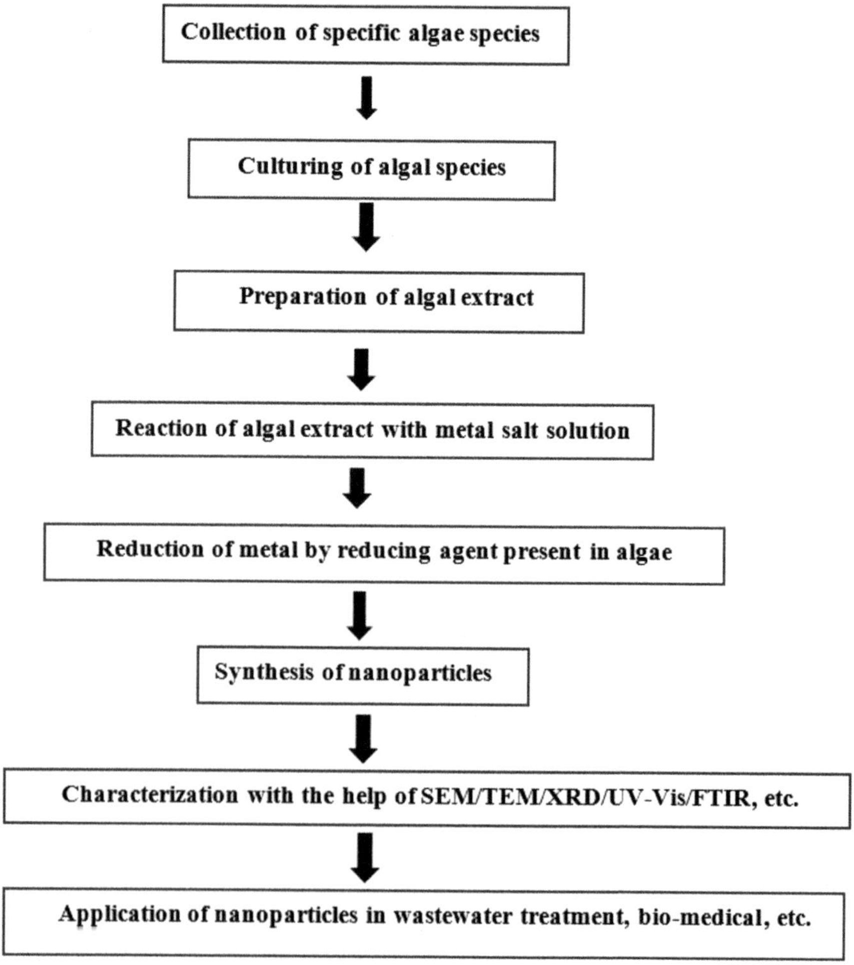

FIGURE 12.4 Procedure to synthesize algal nanoparticle.

TABLE 12.3
Synthesis of algal nanoparticles using different algal species and techniques used for their characterization

Algae	Nanoparticles	Shape and size	Characterization	References
Spirulina platensis	TiO2	Spherical	UV–VIS, XRD, SEM	(Vasanth et al, 2022)
Spirulina	FeO	Non-uniform structure	XRD, FTIR, FESEM, UV–VIS	(Bhukal et al, 2022)
Spirulina platensis	FeO	Non-uniform structure	SEM with EDX, TEM	(Shalaby et al, 2021)
Spirulina platensis	FeO	NA	XPS, BET, VSM	(Mohan et al, 2021)
Amphiroa fragilissima	Ag	Crystalline	FTIR, XRD, UV–VIS	(Khanna et al, 2019)
Microchaete sp. NCCU-342	Ag	Spherical, 60–80 nm	UV–VIS, TEM	(Husain et al, 2019)
Spirogyra submaxima	Au	Spherical, 2–50 nm	FT-IR, TEM, UV–VIS	(Khanna et al, 2019)
Scenedesmus vacuolatus	rGO	Sheet-like ultrathin, flexible structure	SEM, TEM, UV–VIS	(Ahmad et al, 2019)
Turbinaria ornate	Ag	Spherical, 20–32 nm	XRD, FTIR, SEM	(Deepak et al, 2017)
Galaxaura elongate	Au	3.85–77.13 nm	FT-IR, TEM, GC-MS	(Abdel-Raouf et al, 2017)
Cystoseira baccata	Au	Spherical, 3.4 nm	UV–VIS, TEM	(González-Ballesteros et al, 2017)
Acanthophora specifera	Ag	Cubic, 33–81 nm	XRD, FTIR	(IBM Ibraheem et al, 2016)
Anabaena sp. (SAG 12.82)	Au	Spherical, 10 nm	XRD, TEM	(Rösken et al, 2016)
Turbinaria ornate	Ag	Spherical, 22 nm	FTIR, XRD	(Krishnan et al, 2015)

TABLE 12.4
Algae-mediated nano-adsorbents for heavy metal remediation

Algae	Nanoparticles	Heavy metal removed	Adsorption efficiency (%)	References
Oedogonium sp.	Zn-O	Cr	54	(Khilji et al, 2022)
		Cd	57	
		Pb	59	
Spirulina platensis	FeO	Cr	96	(Mohan et al, 2021)
Spirulina platensis	Pd	Pb	90	(Sayadi et al, 2018)
Chlorella vulgaris	Ag	Zn	75	(Adenigba et al, 2020)
Nannochloropsis sp.	Ag	Zn	70.35	(Adenigba et al, 2020)
Chlorella	Au	Pb	58	(Sharma et al, 2023)
Nannochloropsis sp.	Au	Pb	66.53	(Adenigba et al, 2020)
Scenedesmus vacuolatus	rGO	Cu	91	(Ahmad et al, 2019)
		Pb	95	

Ni can be removed using algae-mediated nanoparticles (Mohan et al, 2021). Along with application in the remediation of pollutants, nanoparticles synthesized using algae are proven to be applicable in the biomedical sector (Mohseniazar et al, 2011). Table 12.4 presents the remediation of heavy metals using algal nanoparticles. Cr^{6+} is reduced to Cr^{4+} by iron nanomaterials that are synthesized using *Chlorococcum* sp. MM11 (Subramaniyam et al, 2015).

12.6 CONCLUSION AND FUTURE OUTLOOK

Worldwide, clean and contaminate-free water is in huge demand with the rapid increase in population, industrialization, and urbanization. There are many areas on this planet where clean water assets are inadequate to meet the domestic and economic demands. Scientists are continuously working to find the best technique to overcome these limitations. Many efficient techniques are utilized for water purification. However, most methods require engineering expertise, intensive energy, and chemical involvement. So, there is a requirement for developing novel techniques for the treatment of wastewater. Nanotechnology could be one of the best options in this aspect of wastewater treatment. Nanomaterial has unique properties such as high sensitivity and reactivity, and easy functionalization. As nano-based techniques are very beneficial for the removal of pollutants, enough attention should be paid to the selection of appropriate types of nanoparticles. Nanotechnology has already proven the advancement of nanomaterials by taking water purification challenges. Nanomaterial is very efficient, time-friendly, and less energy is required with nanostructured catalytic membranes.

However, all these methods are not economically viable for large-scale purification of water. Furthermore, there is a requirement for synthesized nanoparticles which is more efficient, user-friendly, and environment friendly. Additionally important are the cost and commercialization of these strategies for the treatment of wastewater. The tremendous application of nanoparticles can help in providing a healthy and cost-effective supply of freshwater to every corner of the world.

REFERENCES

Abboud, Y, Saffaj, T, Chagraoui, A, El Bouari, A, Brouzi, K, & Tanane, O (2014) Biosynthesis, characterization and antimicrobial activity of copper oxide nanoparticles (CONPs) produced using brown alga extract (*Bifurcaria bifurcata*). *Applied Nanosciences* 4: 571–576. doi: 10.1007/s13204-013-0233-x

Abdel-Raouf N, Al-Enazi NM, & Ibraheem IBM (2017) Green biosynthesis of gold nanoparticles using *Galaxaura elongata* and characterization of their antibacterial activity. *Arabian Journal of Chemistry* 10: S3029–S3039. https://doi.org/10.1016/j.ara bjc.2013.11.044.

Adenigba VO, Omomowo IO, Oloke JK, Fatukasi BA, Odeniyi MA, & Adedayo AA (2020, March) Evaluation of microalgal-based nanoparticles in the adsorption of heavy metals from wastewater. In *IOP Conference Series: Materials Science and Engineering* (Vol. 805, No. 1, p. 012030). IOP Publishing.

Adriano DC (2001) *Bioavailability of trace metals. In: Adriano DC, ed. Trace Elements in Terrestrial Environments.* New York: Springer, pp. 61–89. doi: 10.1007/978-0-387-21510-5

Agarwal P, Gupta R, & Agarwal N (2019) Advances in synthesis and applications of microalgal nanoparticles for wastewater treatment. *Journal of Nanotechnology*. 2019(1), 7392713. https://doi.org/10.1155/2019/7392713.

Ahmad S, Ahmad A, Khan S, Ahmad S, Khan I, Zada S, & Fu P (2019) Algal extracts based biogenic synthesis of reduced graphene oxides (rGO) with enhanced heavy metals adsorption capability. *Journal of Industrial and Engineering Chemistry* 72: 117–124.

Al Aji B, Yavuz Y, & Koparal AS (2012) Electrocoagulation of heavy metals containing model wastewater using monopolar iron electrodes. *Separation and Purification Technology* 86: 248–254.

Alcamo J, Henrichs T, & Rösch T (2017) *World Water in 2025: Global Modeling and Scenario Analysis for the World Commission on Water for the 21st Century. Report A0002.* Kassel, Germany: Centre for Environmental Systems Research, University of Kassel.

Alsaggaf MS, Diab AM, ElSaied BE, Tayel AA, & Moussa SH (2021) Application of ZnO nanoparticles phycosynthesized with *Ulva fasciata* extract for preserving peeled shrimp quality. *Nanomaterials* 11: 385. doi: 10.3390/nano11020385

Amin MT, Alazba AA, & Manzoor U (2014) A review of removal of pollutants from water/ wastewater using different types of nanomaterials. *Advances in Materials Science and Engineering* 1–24.

Arora V & Khosla B (2021) Conventional and contemporary techniques for removal of heavy metals from soil. In Biodegradation Technology of Organic and Inorganic Pollutants. IntechOpen.

Bhukal S, Sharma A, Kumar S, Deepak B, Pal K, & Mona S 2022. *Spirulina* based iron oxide nanoparticles for adsorptive removal of crystal violet dye. *Topics in Catalysis* 1–11. https://doi.org/10.1007/s11244-022-01640-3

Brayner R, Barberousse H, Hemadi M, Djedjat C, Yéprémian C, Coradin T, & Couté A (2007) Cyanobacteria as bioreactors for the synthesis of Au, Ag, Pd, and Pt nanoparticles via an enzyme-mediated route. *Journal of Nanoscience and Nanotechnology* 7(8): 2696–2708.

Chahal C, Van Den Akker B, Young F, Franco C, Blackbeard J, & Monis P (2016) Pathogen and particle associations in wastewater: Significance and implications for treatment and disinfection processes. *Advances in Applied Microbiology* 97: 63–119.

Chakraborty N, Banerjee A, Lahiri S, Panda A, Ghosh AN, & Pal R (2009). Biorecovery of gold using cyanobacteria and an eukaryotic alga with special reference to nanogold formation – a novel phenomenon. *Journal of Applied Phycology* 21: 145–152.

da Mota IDO, de Castro JA, de Góes Casqueira R, & de Oliveira Junior AG (2015) Study of electroflotation method for treatment of wastewater from washing soil contaminated by heavy metals. *Journal of Materials Research and Technology* 4(2): 109–113.

Deepak P, Sowmiya R, Ramkumar R, Balasubramani G, Aiswarya D, & Perumal P (2017) Structural characterization and evaluation of mosquito-larvicidal property of silver nanoparticles synthesized from the seaweed, *Turbinaria ornata* (Turner). *Journal of Agardh 1848. Artificial Cells, Nanomedicine, and Biotechnology* 45(5): 990–998.

Emamjomeh MM & Sivakumar M (2009) Review of pollutants removed by electrocoagulation and electrocoagulation/flotation processes. *Journal of Environmental Management* 90(5): 1663–1679.

EPA (2002) Oxyfluorfen RED facts. https://archive.epa.gov/pesticides/reregistration/web/html/oxyfluorfen_red_fs.html

Esmaeilzadeh M, Jaafari J, Mohammadi AA, Panahandeh M, Javid A, & Javan S. (2019) Investigation of the extent of contamination of heavy metals in agricultural soil using statistical analyses and contamination indices. *Human and Ecological Risk Assessment: An International Journal* 25(5): 1125–1136. doi: 10.1080/10807039.2018.1460798

González-Ballesteros N & Rodríguez-Argüelles MC (2020) Seaweeds: A promising bionanofactory for ecofriendly synthesis of gold and silver nanoparticles. *Sustainable Seaweed Technologies* 507–541. doi: 10.1016/b978-0-12-817943-7.00018-4

González-Ballesteros N, Prado-López S, Rodríguez-González JB, Lastra M, & Rodríguez-Argüelles, M (2017) Green synthesis of gold nanoparticles using brown algae *Cystoseira baccata*: Its activity in colon cancer cells. *Colloids and Surfaces B: Biointerfaces* 153: 190–198.

Gu J, Zhou W, Jiang B, Wang L, Ma Y, Guo H., Schulin R, Ji R, & Evangelou MW (2016) Effects of biochar on the transformation and earthworm bioaccumulation of organic pollutants in soil. *Chemosphere* 145: 431–437. https://doi.org/10.1016/j.chemosphere.2015.11.106.

Guleri S, Saxena A, Singh KJ, et al (2020) Phycoremediation: A novel and synergistic approach in wastewater remediation. Journal of Microbiology, Biotechnology and Food Sciences 10: 98–106. https://doi.org/10.15414/jmbfs.2020.10.1.98-106.

Hlongwane GN, Sekoai PT, Meyyappan M, & Moothi K (2019) Simultaneous removal of pollutants from water using nanoparticles: A shift from single pollutant control to multiple pollutant control. *Science of the Total Environment* 656: 808–833.

Huang X, Sun S, Gao B, Yue Q, Wang Y, & Li Q. (2015). Coagulation behavior and floc properties of compound bioflocculant–polyaluminum chloride dual-coagulants and polymeric aluminum in low temperature surface water treatment. *Journal of Environmental Sciences* 30: 215–222.

Hunger K (Ed.) (2007) *Industrial Dyes: Chemistry, Properties, Applications.* John Wiley & Sons.

Husain S, Afreen S, Hemlata, Yasin D, Afzal B, & Fatma T (2019) Cyanobacteria as a bioreactor for synthesis of silver nanoparticles-an effect of different reaction conditions on the size of nanoparticles and their dye decolorization ability. *Journal of Microbiological Methods* 162: 77–82. https://doi.org/10.1016/j.mimet.2019.05.011.

Ibraheem IBM, Abd Eliaziz BEE, Saas WF, & Fathy WA (2016) Green biosynthesis of silver nanoparticles using marine red algae *Acanthophora specifera* and its antibacterial activity. *Journal of Nanomedicine & Nanotechnology* 07. https://doi.org/10.4172/2157-7439.1000409.

Jacob JM, Ravindran R, Narayanan M, Samuel SM, Pugazhendhi A, & Kumar G (2021) Microalgae: A prospective low cost green alternative for nanoparticle synthesis. *Current Opinion in Environmental Science & Health* 20: 100–163. doi: 10.1016/j.coesh.2019.12.005

Khan F, Shahid A, Zhu H, Wang N, Javed M R, Ahmad N, & Mehmood MA (2022) Prospects of algae-based green synthesis of nanoparticles for environmental applications. *Chemosphere* 293: 133571.

Khanna P, Kaur A, & Goyal D (2019) Algae-based metallic nanoparticles: Synthesis, characterization, and applications. *Journal of Microbiological Methods* 163: 105656.

Khilji SA, Munir N, Aziz I, Anwar B, Hasnain M, Jakhar AM, & Yang H H (2022) Application of algal nanotechnology for leather wastewater treatment and heavy metal removal efficiency. *Sustainability* 14(21): 13940.

Krishnan M, Sivanandham V, Hans-Uwe D, Murugaiah SG, Seeni P, Gopalan S, & Rathinam AJ (2015) Antifouling assessments on biogenic nanoparticles: A field study from a polluted offshore platform. *Marine Pollution Bulletin* 101: 816–825. https://doi.org/10.1016/j.marpolbul.2015.08.033.

LewisOscar F, Vismaya S, Arunkumar M, Thajuddin N, Dhanasekaran D, & Nithya C (2016) Algal nanoparticles: Synthesis and biotechnological potentials. Algae – Organisms for Imminent Biotechnology 7: 157–182.

LewisOscar F, Vismaya S, Arunkumar M, Thajuddin N, Dhanasekaran D, & Nithya C (2016) Algal nanoparticles: synthesis and biotechnological potentials. *Algae – Organisms for Imminent Biotechnology* 7: 157–182.

Li K, Li P, Cai J, Xiao S, Yang H, & Li A (2016) Efficient adsorption of both methyl orange and chromium from their aqueous mixtures using a quaternary ammonium salt modified chitosan magnetic composite adsorbent. *Chemosphere* 154: 310–318. https://doi.org/10.1016/j.chemosphere.2016.03.100.

Liu C, Wang Y, Chen G, Yu D, Zhang X, Wang X, & Xu, A (2022) A novel stable nitritation process: Treating sludge by alternating free nitrous acid/heat shock. *Bioresource Technology* 347: 126753.

Lyu S, Chen W, Zhang W, Fan Y, & Jiao W (2016) Wastewater reclamation and reuse in China: Opportunities and challenges. *Journal of Environmental Science* 39: 86–96. https://doi.org/10.1016/j.jes.2015.11.012.

Ma X, Yang L, & Wu H (2021) Removal of volatile organic compounds from the coal-fired flue gas by adsorption on activated carbon. *Journal of Cleaner Production* 302: 126925.

Mohan S, Govindankutty G, Sathish A, & Kamaraj N (2021) *Spirulina platensis*-capped mesoporous magnetic nanoparticles for the adsorptive removal of chromium. *Canadian Journal of Chemical Engineering* 99(1): 294–305.

Mohseniazar M, Barin M, Zarredar H, et al (2011) Potential of microalgae and lactobacilli in biosynthesis of silver nanoparticles. *Bioimpacts* 1: 149–152. https://doi.org/10.5681/bi.2011.020.

Molazadeh M, Pourianfar HR, Lyon S, & Rampelotto PH (2019) The use of microalgae for coupling wastewater treatment with CO_2 biofixation. *Frontiers in Bioengineering and Biotechnology* 7: 42. https://doi.org/10.3389/fbioe.2019.00042/full.

Mukherjee A, Sarkar D, & Sasmal S (2021) A review of green synthesis of metal nanoparticles using algae. *Frontiers in Microbiology* 12: 693899.

Pandey BW, Ranjan OJ, Srivastava A, & Prasad AS (2017) Water pollution and its impact on human health: A case study of Allahabad City, Uttar Pradesh. *International Journal of Interdisciplinary Research in Science Society and Culture* (IJIRSSC) 3(1).

Parga JR, Cocke DL, Mencer DE, & Morkovsky P (2002) *Electrocoagulation of heavy metals and characterization of the process.* Proceedings of the TMS Fall 2002 Extraction and Processing Division Meeting 1: 671–680.

Ponnuchamy K & Jacob JA (2016) Metal nanoparticles from marine seaweeds – a review. *Nanotechnology Reviews* 5(6): 589–600.

Rajesh S, Raja DP, Rathi JM, & Sahayaraj K 2012. Biosynthesis of silver nanoparticles using *Ulva fasciata* (Delile) ethyl acetate extract and its activity against *Xanthomonas campestris* pv. m*alvacearum. Journal of Biopesticides* 5: 119–128.

Razzak SA, Faruque MO, Alsheikh Z, Alsheikhmohamad L, Alkuroud, D, Alfayez A, & Hossain MM (2022) A comprehensive review on conventional and biological-driven heavy metals removal from industrial wastewater. *Environmental Advances* 7: 100168.

González-Ballesteros N & Rodríguez-Argüelles MC (2020) Seaweeds: A promising bionanofactory for ecofriendly synthesis of gold and silver nanoparticles. *Sustainable Seaweed Technologies* 507–541. doi: 10.1016/b978-0-12-817943-7.00018-4

Rösken LM, Cappel F, Körsten S, et al (2016) Time-dependent growth of crystalline Au0-nanoparticles in cyanobacteria as self-reproducing bioreactors: 2. *Anabaena cylindrica. Beilstein Journal of Nanotechnology* 7: 312–327. https://doi.org/10.3762/bjnano.7.30

Saleh IA, Zouari N, & Al-Ghouti MA (2020) Removal of pesticides from water and wastewater: Chemical, physical and biological treatment approaches. *Environmental Technology & Innovation* 19: 101026.

Sathishkumar RS, Sundaramanickam A, Srinath R, et al (2019) Green synthesis of silver nanoparticles by bloom forming marine microalgae *Trichodesmium erythraeum* and its applications in antioxidant, drug-resistant bacteria, and cytotoxicity activity. *Journal of Saudi Chemical Society* 23: 1180–1191. https://doi.org/10.1016/j.jscs.2019.07.008

Sayadi MH, Salmani N, Heidari A, & Rezaei MR (2018) Bio-synthesis of palladium nanoparticle using *Spirulina platensis* alga extract and its application as adsorbent. *Surfaces and Interfaces* 10: 136–143.

Schrofel A, Kratošová G, Krautová M, Dobročka E, & Vávra I (2011) Biosynthesis of gold nanoparticles using diatoms – silica-gold and EPS-gold bionanocomposite formation. *Journal of Nanoparticle Research* 13: 3207–3216.

Shalaby SM, Madkour FF, El-Kassas HY, et al (2021) Green synthesis of recyclable iron oxide nanoparticles using *Spirulina platensis* microalgae for adsorptive removal of cationic and anionic dyes. *Environmental Science and Pollution Research* 28: 65549–65572. https://doi.org/10.1007/s11356-021-15544-4

Sharma A, Pal K, Saini N, Kumar S, Bansal D, & Mona S (2023) Remediation of contaminants from wastewater using algal nanoparticles via green chemistry approach: An organized review. *Nanotechnology* 34(35). https://doi.org/10.1088/1361-6528/acd45a

Singh JP, Kumar M, Sharma A, Pandey G, Chae KH, & Lee S (2020) Bottom-up and top-down approaches for MgO. In: Karakus S (Ed.), Sonochemical Reactions. IntechOpen

Subramaniyam V, Subashchandrabose SR, Thavamani P, et al (2015) *Chlorococcum* sp. MM11 – a novel phyco-nanofactory for the synthesis of iron nanoparticles. *Journal of Applied Phycology* 27: 1861–1869. https://doi.org/10.1007/s10811-014-0492-2

Symonds EM & Breitbart M (2014) Affordable enteric virus detection techniques are needed to support changing paradigms in water quality management. *CLEAN – Soil, Air, Water* 43: 8–12.

Tang WW, Zeng GM, Gong JL, Liu Y, Wang XY, Liu YY, Liu ZF, Chen L, Zhang XR, & Tu DZ (2012) Simultaneous adsorption of atrazine and Cu(II) from wastewater by magnetic multi-walled carbon nanotube. *Chemical Engineering Journal* 211: 470–478. https://doi.org/10.1016/j.cej.2012.09.102

Thunugunta T, Reddy AC, & Reddy L (2015) Green synthesis of nanoparticles: Current prospectus. *Nanotechnology Reviews* 4: 303–323. https://doi.org/10.1515/ntrev-2015-0023/html

Ting H, Haifeng L, Shanshan M, et al (2017) Progress in microalgae cultivation photobioreactors and applications in wastewater treatment: A review. *International Journal of Agricultural and Biological Engineering* 10: 1–29.

Tran TK, Chiu KF, Lin, CY, & Leu HJ (2017) Electrochemical treatment of wastewater: Selectivity of the heavy metals removal process. *International Journal of Hydrogen Energy* 42(45): 27741–27748.

Vasanth V, Murugesh KA, & Susikaran S (2022) Synthesis of titanium dioxide nanoparticles using Spirulina platensis algae extract. *The Pharma Innovation Journal* 11(7S), 266–269.

Wang, L, Gao, Y, Xiong, J, Shao, W, Cui, C, Sun, N, & He, J (2022) Biodegradable and high-performance multiscale structured nanofiber membrane as mask filter media via poly (lactic acid) electrospinning. *Journal of Colloid and Interface Science* 606: 961–970.

Wu W, Wu P, Yang F, Sun D, Zhang D-X, & Zhou Y-K. (2018) Assessment of heavy metal pollution and human health risks in urban soils around an electronics manufacturing facility. *Science of the Total Environment* 630: 53–61. doi: 10.1016/j.scitotenv.2018.02.183

Zhu Y, Sun Z, Deng Y, Liu F, Ruan W, & Xie L (2022) Mn_2O_3/Mn_3O_4-$Cu_{1.5}Mn_{1.5}O_4$ spinel as an efficient Fenton-like catalyst activating persulfate for the degradation of bisphenol A: Superoxide radicals dominate the reaction. *Science of the Total Environment* 839: 156075.

13 From Crisis to Opportunity

Advancing Plastic Waste Management in the Age of Nanowaste and COVID-Related Waste

*Muhammad Izzul Fahmi Mohd Rosli,
Mohd Saiful Samsudin, and
Siti Norabiatulaiffa Mohd Yamen*

13.1 INTRODUCTION

Plastics have transformed our daily lives over the last few decades. Globally, we use more than 260 million tonnes of plastic each year, accounting for nearly 8% of global oil production (Thompson et al., 2009). Meanwhile, plastic commercial production reached 330 million metric tonnes as the global yearly production in 2016 (Lebreton & Andrady, 2019). On a global scale, plastic usage decreased during the COVID-19 pandemic due to decreased economic activity. However, complete lockdowns and the shutdown of restaurants caused a rise in demand for food take-away packaging and plastic medical equipment leading to an increase in unorganized plastic waste (Sharma et al., 2020). Figure 13.1 shows the improper disposal of plastic waste becoming increasingly evident, exacerbating environmental concerns during the COVID-19 pandemic. Plastic pollution in the soil and water has been caused by inappropriate disposal and handling of plastics (Yukeswaran & Moni Philip Jacob, 2022). Plastic waste has long been a debated issue, both internationally and locally, due to its detrimental effects on the environment and human health (Van Deursen et al., 2023).

Plastic does not biodegrade and breaks down into microplastics, persisting in the environment because of their corrosion-resistant characteristics; most plastics are classified as hard-to-degrade materials that can endure for up to a century in the environment (Li et al., 2016). Figure 13.2 illustrates that unorganized plastic waste poses significant risks to health, as microplastics can enter the human body through various mediums, leading to potential health risks. These microplastics contaminate soil, water, and the food chain, endangering wildlife and ecosystems. Animals mistake

DOI: 10.1201/9781003543176-13

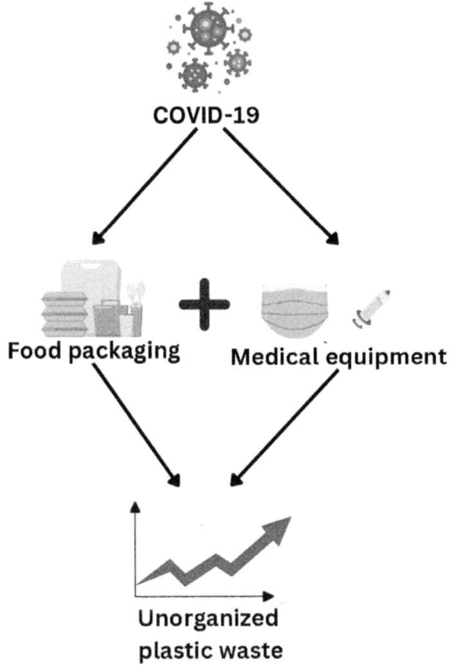

FIGURE 13.1 Unorganized plastic waste during COVID-19 pandemic.

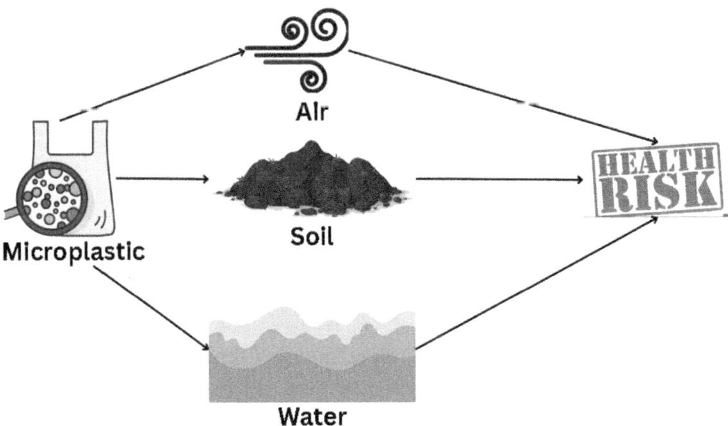

FIGURE 13.2 Medium of microplastic toward health risk.

plastic for food and suffer from suffocation, internal injuries, and death (Susanti et al., 2020). Moreover, harmful chemicals present in certain plastics, such as phthalates and bisphenol A (BPA), leach out and pose health risks to humans, including reproductive problems, developmental disorders, and cancer (Alberghini et al., 2022; Silva et al., 2022).

Efficient waste management techniques reduce plastic pollution in oceans, rivers, and landfills, thereby protecting wildlife and habitats. Recycling and reusing plastic help conserve resources and reduce the need for new plastic production. Plastic production and incineration contribute to greenhouse gas emissions; waste management methods can reduce emissions. Inadequate plastic waste management poses health hazards and contaminates water sources and soil. Effective waste management helps prevent the spread of diseases and minimizes exposure to toxic chemicals. Proper management of plastic waste creates economic opportunities through recycling, employment generation, and the emergence of new industries focused on waste reduction (Na et al., 2021; Thachnatharen et al., 2021).

Implementing effective plastic waste management techniques is crucial for various reasons. It helps mitigate environmental pollution and conserves valuable resources. By reducing the influx of plastic waste into oceans, rivers, and landfills, it safeguards wildlife and habitats. Recycling and reusing plastic materials not only conserve resources but also diminish the need for new plastic production (Huang et al., 2022; Samuel & Ahlawat, 2021). Additionally, waste management methods can reduce greenhouse gas emissions generated by plastic production and incineration. Inadequate plastic waste management poses health hazards, contaminating water sources and soil. Implementing effective waste management practices can prevent the spread of diseases and minimize exposure to toxic chemicals present in plastic waste. Furthermore, proper management of plastic waste creates economic opportunities by promoting recycling, generating employment, and stimulating the growth of industries focused on waste reduction (Al-Salem et al., 2021; da Silva et al., 2022).

This chapter provides a comprehensive overview of plastic waste as a significant environmental concern. It emphasizes the imperative need for implementing effective plastic waste management techniques. Addressing the adverse consequences of plastic waste on the environment and human health, it highlights the interconnected benefits of protecting the environment, public health, and the economy. Collaborative efforts among governments, communities, and individuals are crucial for successfully establishing and supporting sustainable practices to tackle the plastic waste crisis and pave the way for a better, more sustainable future.

13.2 PLASTIC WASTE MANAGEMENT TECHNIQUES

Plastic waste management encompasses a range of strategies aimed at mitigating the environmental impact of plastic waste. Recycling, pyrolysis, mechanical and biological treatment methods, and waste-to-energy (WtE) conversion are among the key techniques employed to tackle this pervasive issue (Na et al., 2021). Figure 13.3 shows that effective plastic waste management strategies are essential to mitigate these risks and ensure environmental sustainability. Through these approaches, plastic waste can be transformed into valuable resources, energy, or biodegradable materials, reducing its harmful effects on ecosystems and promoting a more sustainable future.

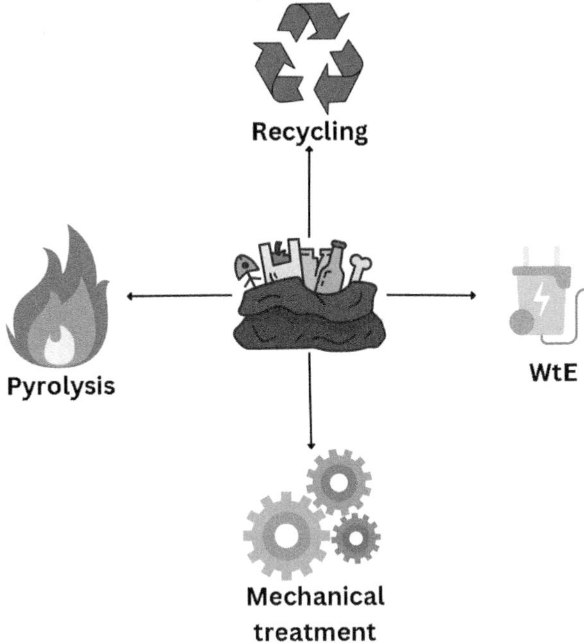

FIGURE 13.3 Plastic waste management.

13.3 RECYCLING

Plastic recycling, as a traditional method, involves a series of steps: collection, sorting, cleaning, shredding, melting, and processing of the waste. The process commences with the collection of plastic waste from various sources, including households, businesses, and other establishments (Girma, 2020). Figure 13.4 shows the key component in the recycling process to efficiently convert waste into reusable materials. This collection can be facilitated through curbside recycling programs, designated drop-off centers, or specialized collection systems. Once the plastic waste is gathered, it undergoes sorting based on its type or resin identification code. This step is vital as diverse types of plastics have distinct properties and necessitate specific recycling processes. Following the sorting stage, the plastic waste is thoroughly cleaned to eliminate any contaminants, such as dirt, labels, or residual contents. The cleaning process ensures that the recycled plastic is free from impurities. After cleaning, the plastic waste is shredded or granulated into smaller pieces. This fragmentation makes the plastic waste more manageable and facilitates further processing. Lastly, the shredded or granulated plastic undergoes a melting process, transforming it into pellets or flakes. These pellets or flakes can then be utilized as raw materials for manufacturing new plastic products using techniques like injection molding or extrusion.

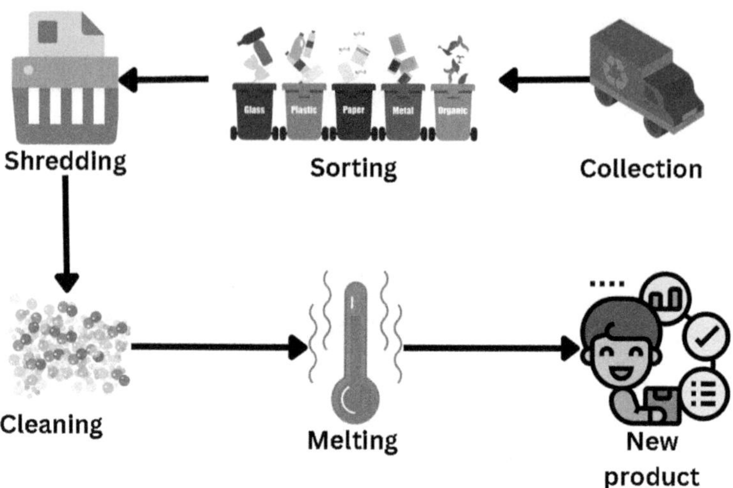

FIGURE 13.4   Recycling process flow.

Plastic recycling technologies have advanced significantly in recent years, aiming to improve the efficiency and effectiveness of recycling. One major advancement is chemical recycling, which breaks down plastic waste into chemical components for producing new materials. This expands the range of recyclable plastics and overcomes some limitations of mechanical recycling. Biodegradable plastics have also emerged, naturally degrading under specific conditions, reducing environmental impact, and providing an alternative to persistent traditional plastics. Advanced sorting technologies like optical sorting and automated systems use sensors and artificial intelligence (AI) to detect and separate different plastics accurately and efficiently, improving recycling processes. Upcycling transforms plastic waste into higher-value products, reducing the need for virgin materials and promoting sustainability. Closed-loop systems promote a circular economy, designing products and packaging to be easily recyclable and reintegrated into production. These advancements offer promise in addressing plastic waste challenges, boosting recycling rates, reducing environmental impact, and fostering a sustainable approach to plastic waste management (Beghetto et al., 2021; Chen et al., 2021).

13.4 PYROLYSIS

Pyrolysis is a process that effectively manages plastic waste by converting it into valuable products through high-temperature treatment without oxygen. Figure 13.5 shows the pyrolysis in plastic waste management by breaking down plastics into simpler compounds through thermal decomposition. The process involves preparing the plastic waste, heating it within a reactor to trigger decomposition, condensing the released gases into liquid products, treating the remaining gases to remove impurities, and utilizing or releasing them. The solid residue, called char or carbon black, can be further processed or used for energy generation. Pyrolysis offers benefits such

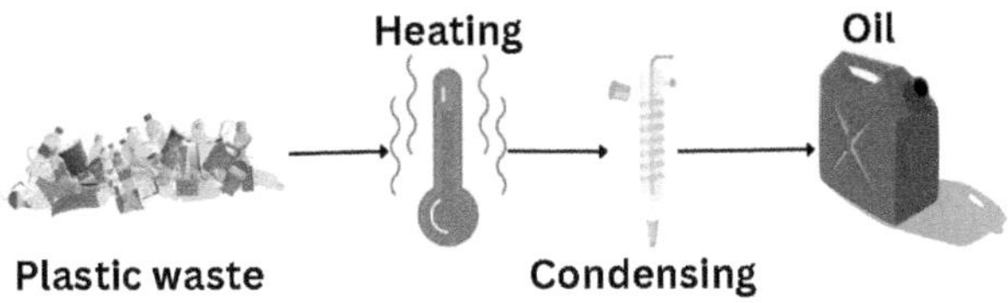

FIGURE 13.5 The pyrolysis concept for plastic waste management.

as resource recovery, waste reduction, environmental preservation, and renewable energy generation. It extracts valuable resources from plastic waste, reduces reliance on fossil fuels, and contributes to a circular economy (Alqarni et al., 2022; Farooq et al., 2022). It also decreases greenhouse gas emissions and provides an alternative to incineration, while generating renewable energy for various applications (Ridwan et al., 2022). Overall, pyrolysis is a comprehensive solution for plastic waste management.

Several companies have successfully implemented pyrolysis technology in plastic waste management, demonstrating its potential for value creation and environmental impact reduction. Agilyx Corporation, based in the USA, has developed a pyrolysis technology that converts mixed waste plastics into high-quality crude oil. Their commercial-scale facility in Oregon processes various plastic waste types, including polystyrene foam, to produce crude oil that can be further refined into fuels. Pyrolysis is used to convert trash polymers into chemical compounds for the creation of new plastics and fuels. They are focusing on complicated polymer mixes that have previously been difficult to recycle. In 2014, this company handled 3.6 million kg of plastic garbage (Sakthipriya, 2022). Another company, Plastic Energy, based in the UK, operates multiple pyrolysis plants worldwide. They specialize in converting mixed plastic waste into high-quality oils, which can be used as feedstock to produce new plastics or as a substitute for fossil fuels. By practicing Stirred-Tank Reactor (STR) technology, they can produce raw diesel, light oil, and synthetic gas components from plastic waste (Qureshi et al., 2020). Biofabrik, a company based in Germany, has developed a decentralized pyrolysis system called "WASTX Plastic." This innovative system converts plastic waste into synthetic crude oil, providing a community-level solution for local waste processing and oil production. The synthetic crude oil produced can be used for various applications, such as electricity generation or chemical processes (Braithwaite et al., 2017). These successful case studies demonstrate the practical applications of pyrolysis technology in plastic waste management. They showcase how this technology can contribute to both value creation and environmental impact reduction by transforming plastic waste into valuable resources.

13.5 MECHANICAL TREATMENT

Waste management systems use mechanical methods to process and treat solid waste. Mechanical treatment involves physical processes such as shredding, screening,

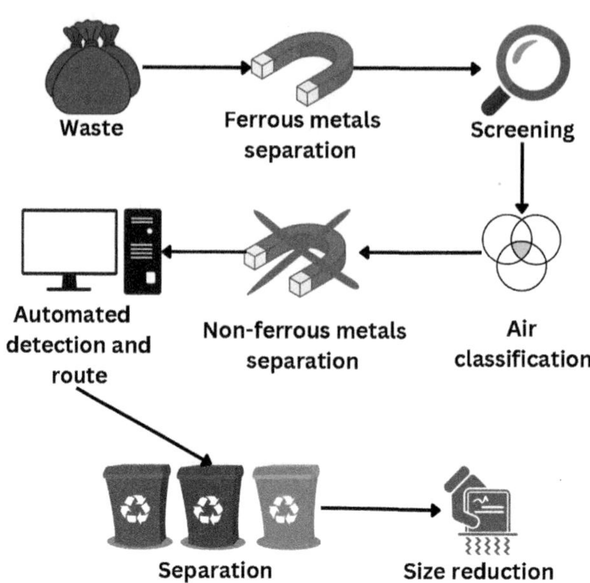

FIGURE 13.6   Process flow of waste.

magnetic separation, classification, and optical sorting. Figure 13.6 shows the process flow of waste from generation to disposal. These methods aim to recover valuable materials, reduce waste volume, and prepare them for further treatment or disposal. Shredding breaks waste into smaller pieces, screening removes large objects, magnetic separation separates ferrous materials, air classification separates lighter materials, and optical sorting sorts of materials based on their optical properties (Cesaro et al., 2016; Trulli et al., 2018). These treatment methods play a crucial role in waste reduction, resource recovery, and sustainable energy production. They are used in recycling facilities, WtE plants, composting facilities, and biogas plants. By combining physical and biological processes, these methods help reduce the environmental impact of solid waste and promote a more sustainable waste management system.

Material Recovery Facilities (MRFs) are a great example of successful mechanical treatment methods in waste management. MRFs use techniques such as shredding, screening, and optical sorting to separate valuable materials like paper, plastic, and metals from mixed waste, which helps in recycling and reducing the amount of waste sent to landfills. Another application is seen in WtE plants, where mechanical treatment is used to shred and sort waste, removing non-combustible materials. The remaining combustible fraction is then used as fuel to generate sustainable electricity or heat, reducing waste volume (Pal et al., 2001).

Large scale composting facilities demonstrate successful biological treatment methods. These facilities process organic waste by controlled decomposition with the help of microorganisms. Food waste, yard trimmings, and other organic materials are converted into nutrient-rich compost. This compost can be used in agriculture and landscaping, improving soil health and reducing the need for chemical fertilizers.

Biogas plants also utilize biological treatment through anaerobic digestion. Organic waste is treated in oxygen-free environments, broken down by anaerobic bacteria, and produces biogas as a valuable by-product. This biogas, mainly methane and carbon dioxide, can be used for electricity generation or as a renewable natural gas substitute, effectively managing organic waste (Bojan Plavac et al., 2017; Nayak & Bhushan, 2021).

These examples show how mechanical and biological treatment methods have been successfully applied in waste management. They contribute to waste reduction, resource recovery, and sustainable energy production. By highlighting the importance of these methods, we can work toward creating a more environmentally friendly and efficient waste management system.

13.6 WASTE-TO-ENERGY CONVERSION

WtE conversion is a process that turns waste materials into electricity or heat, aiming to minimize environmental harm while extracting value from waste. Figure 13.7 shows that the approach within this system is the waste-to-energy process, which not only reduces waste but also generates energy, providing a dual benefit of waste reduction and energy production. This approach offers an alternative to traditional waste management practices like landfilling and incineration without energy recovery. WtE technologies contribute to sustainable waste management and help meet energy needs. There are various methods of WtE conversion, including incineration, anaerobic digestion, and gasification (Abdelateef Mostafa et al., 2023). Incineration involves burning waste at high temperatures to generate heat for electricity or heating. Anaerobic digestion uses microorganisms to break down organic waste in an oxygen-free environment, producing biogas for fuel. Gasification converts solid waste into synthetic gas (syngas), which can be used for power generation or in chemical processes (Ong & Schideman, 2013). Implementing WtE technologies requires careful consideration of environmental impacts, controlling emissions, and proper

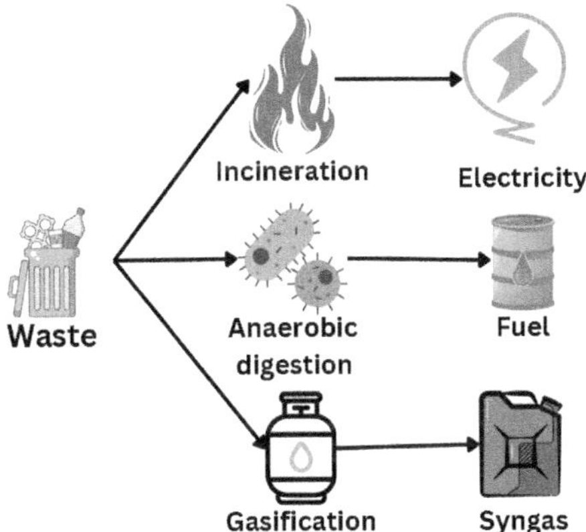

FIGURE 13.7 Waste-to-energy process.

waste management practices. Regulations and guidelines may differ across regions, influencing the suitability and viability of different conversion methods.

WtE conversion is an important part of managing plastic waste. It diverts plastic waste from landfills and uses it to produce energy. This reduces the amount of plastic waste and makes landfills last longer. It also creates clean and sustainable energy, reducing our reliance on non-renewable sources and helping the environment. Another benefit is that WtE conversion helps reduce greenhouse gas emissions. When plastic waste decomposes in landfills, it releases methane, which is a powerful greenhouse gas. By using WtE conversion, we can minimize methane emissions and lessen the impact of climate change.

WtE conversion also aligns with the circular economy concept. It integrates recycling and recovery systems by using non-recyclable plastic waste as a source of energy. This ensures that we get the most value from the waste before disposing of it. However, it is important to remember that WtE conversion should not be seen as the only solution. It should be combined with waste reduction strategies, improved recycling systems, and sustainable production practices for a comprehensive and sustainable approach to plastic waste management.

13.7 CHALLENGES AND DRAWBACKS OF EXISTING PLASTIC WASTE MANAGEMENT TECHNIQUES

Existing plastic waste management techniques face several challenges and drawbacks. One major issue is the low recycling rates, as conventional techniques, such as mechanical recycling, struggle to effectively manage the complexity of plastic composition and contaminants (Roy et al., 2021). Consequently, a significant amount of plastic waste ends up in landfills or is incinerated. In addition, the lack of effective sorting and separation methods leads to contamination and diminished recycling quality. Insufficient infrastructure, including collection systems, recycling facilities, and technologies, further hampers the proper management of plastic waste. Furthermore, certain techniques such as incineration contribute to air pollution and greenhouse gas emissions (Benny, 2022). Meanwhile, landfilling plastic waste poses environmental risks due to the leaching of harmful chemicals into the soil and water. It is important to address these challenges and drawbacks in plastic waste management to develop more effective and sustainable solutions.

13.8 INTRODUCTION TO CHEMICAL RECYCLING AS AN INNOVATIVE APPROACH

Chemical recycling is an innovative approach to plastic waste management that involves converting plastic waste into raw materials or chemicals through various chemical processes. It addresses several challenges faced by existing techniques. One key aspect is the conversion of complex plastic polymers into basic building blocks, such as monomers or small molecules (Roy et al., 2021). These can be used as feedstock for producing new plastics or other chemicals, offering a more comprehensive solution. Chemical recycling has the potential to handle a wider range of plastic waste compared to mechanical recycling. It can effectively process plastics, contaminated plastics, and difficult-to-recycle plastics.

Additionally, chemical recycling enables the transformation of plastic waste into higher-value products such as fuels, waxes, or specialty chemicals (Ragaert et al., 2017). This allows for the recovery of embedded energy and resources, promoting a circular economy. By diverting plastic waste from landfills and incineration, chemical recycling also contributes to reducing the environmental impact of plastic waste management. It helps in reducing carbon emissions and conserving natural resources. Chemical recycling represents a promising avenue for the future of plastic waste management, providing an innovative solution to the challenges posed by traditional recycling methods.

13.9 BIODEGRADABLE PLASTICS AND THEIR POTENTIAL IMPACT ON WASTE MANAGEMENT

Biodegradable plastics offer an innovative approach to plastic waste management, breaking down more rapidly under specific conditions compared to conventional plastics. The degradation process involves microorganisms such as bacteria or fungi, occurring in composting facilities or natural environments. They have the potential to reduce waste accumulation in landfills and oceans, contributing to the circular economy through composting or conversion into organic matter. However, specific conditions and proper disposal are crucial for successful management, as incorrect disposal can still lead to pollution. The challenge lies in distinguishing biodegradable from conventional plastics, which can cause contamination and inefficiencies in recycling. Proper labeling and education are necessary to ensure effective waste management. In conclusion, while biodegradable plastics offer benefits, their integration into waste systems requires careful consideration of disposal methods, infrastructure, and consumer education.

13.10 CASE STUDIES HIGHLIGHTING SUCCESSFUL IMPLEMENTATION OF INNOVATIVE APPROACHES

Plastic waste management has emerged as a critical challenge in Malaysia, necessitating the adoption of innovative solutions. This chapter examines successful case studies that demonstrate the implementation of creative approaches to address plastic waste management issues. These examples underscore the significance of innovation, community engagement, and cross-sector collaborations in fostering a sustainable and circular economy. Moreover, to encourage plastic waste recycling among communities, SWCorp Malaysia, the country's solid waste management authority, introduced the "Trash for Cash" program. This initiative incentivizes individuals and households to separate and collect recyclable plastic waste, which can be exchanged for cash or vouchers at designated recycling centers. The program has yielded promising results, including increased recycling rates and heightened awareness regarding proper waste management practices. At the same time, it will reduce the cost of waste management and waste treatment. Ultimately, operating costs are significantly reduced, and the benefits will be returned to the community in the form of improved quality of life, local infrastructure changes, and reduced taxes at the PB level. By instilling the notion of waste as a valuable resource, the program not only addresses plastic waste but also provides participants with an additional income source (Din, 2019).

In response to the plastic pollution affecting rivers, the Department of Irrigation and Drainage (DID) launched the "River of Life Public Outreach Program." This comprehensive initiative employs community engagement and educational campaigns to raise awareness about the adverse impacts of plastic waste on water bodies and marine ecosystems. The program encourages individuals and communities to actively reduce their use of single-use plastics and participate in river clean-up activities. By fostering collaboration with local businesses and industries, the program promotes sustainable practices and reduces plastic waste generation at its source. The integrated approach has resulted in significant improvements in river cleanliness and increased community involvement in preserving vital water resources (Ravindran, 2020). Zero Waste Malaysia, a non-profit organization, spearheads the "Plastic-Free Community Initiative" to empower communities in their efforts to reduce plastic consumption and adopt sustainable alternatives. This initiative collaborates with local communities, businesses, and government agencies to implement plastic-free campaigns, organize awareness events, and provide resources and training on waste reduction. Through active community engagement, the initiative fosters behavioral change and promotes the adoption of eco-friendly practices such as the use of reusable packaging and shopping bags (Baba-Nalikant et al., 2023).

13.11 EXAMPLES OF PLASTIC WASTE MANAGEMENT PRACTICES FROM DIFFERENT REGIONS

Several case studies highlight the successful implementation of innovative approaches in plastic waste management. There are numerous companies from various countries going after the closed-loop recycling of plastic. One notable example is Loop Industries, which has developed a patented technology to depolymerize waste PET plastics and convert them back into base monomers to produce new PET plastics. This process reduces the reliance on virgin materials and promotes a more sustainable approach to plastic production (Ügdüler et al., 2020). Adidas has also made significant strides in plastic waste management through its partnership with Parley for the Oceans. They collect plastic waste from beaches and coastal areas and transform it into yarn for sustainable shoe manufacturing (Joshi & Temgire, 2021). This initiative not only raises awareness about ocean pollution but also highlights the potential of recycling to address environmental challenges.

Furthermore, Dell, a computer technology company, has implemented a closed-loop recycling program. They collect old electronic equipment, disassemble it, and utilize the recovered materials to produce new products. In 2014, they succeeded in recycling 4.2 million pounds of closed-loop plastic as part of their new products (He, 2017). This approach minimizes waste generation, reduces the environmental impact of electronic waste, and decreases the demand for virgin materials. Not only that, but recognizing the need for innovative solutions, H&M, a global fashion retailer, introduced recycling reverse vending machines in their stores. These machines enable customers to return used garments and plastic bottles, receiving store discounts or vouchers in return. The collected items are subsequently recycled into new products or materials. This initiative not only encourages customers to actively participate in recycling efforts but also highlights the importance of closing the loop on textile and plastic waste, promoting the concept of circular fashion (Shen, 2014).

These case studies serve as inspiring examples of successful implementation of innovative approaches in plastic waste management. They demonstrate the potential of these approaches to promote sustainable practices, advance the circular economy, and address the pressing issue of plastic waste. Key lessons learned and best practices from these case studies highlight the importance of collaboration among governments, businesses, communities, and individuals in achieving successful plastic waste management. Investing in essential infrastructure, such as recycling facilities and waste management systems, plays a crucial role in enabling efficient collection, sorting, and processing of plastic waste. The implementation of extended producer responsibility (EPR) policies, which hold producers accountable for the entire lifecycle of their products, serves as an effective catalyst for promoting recyclability and driving recycling efforts.

Furthermore, raising public awareness, fostering behavior change, and educating individuals about proper waste management practices emerge as vital elements in reducing plastic waste and increasing recycling rates. By disseminating knowledge and encouraging responsible waste management habits, significant progress can be made in mitigating the plastic waste crisis. These valuable lessons and practices derived from the case studies serve as indispensable insights for the development of effective strategies aimed at the sustainable management of plastic waste.

13.12 CASE STUDIES FROM VARIOUS INDUSTRIES (E.G., MANUFACTURING, PACKAGING, AND RETAIL)

Various industries have highlighted successful case studies in effectively managing plastic waste. In the manufacturing sector, the automotive industry has made significant progress by implementing practices such as recycling plastic components, incorporating recycled plastics into new vehicles, exploring alternative materials, and adopting measures to reduce plastic packaging throughout the manufacturing process (Schultmann et al., 2006). Within the packaging industry, leading companies like Nestlé have made substantial commitments to use more sustainable packaging materials and improve product recyclability. Nestlé is also targeting to remove all non-recyclable or difficult-to-recycle plastic products worldwide between 2020 and 2025 (Cheng et al., 2022). They have embraced strategies such as lightweight packaging, promoting the use of reusable packaging, and making significant investments in recycling infrastructure. Additionally, retailers have taken proactive steps to minimize plastic waste by offering reusable bags to customers, encouraging the use of customer-provided containers for bulk purchases, and implementing strategies to reduce packaging materials. Notably, some retailers have introduced refill stations for cleaning and personal care products, effectively reducing the need for single-use plastic packaging.

13.13 SUCCESSFUL STRATEGIES AND APPROACHES FOR MANAGING PLASTIC WASTE

Successful strategies and approaches for managing plastic waste across industries encompass several key aspects. Firstly, designing products and packaging with recyclability in mind by using fewer plastic materials, ensuring compatibility with existing recycling systems, and incorporating recycled content can significantly contribute to effective waste management. Embracing a circular economy model is another pivotal approach, involving the reduction of plastic waste generation, the promotion of recycling and reuse, and the exploration of innovative methods to recover and reintegrate plastic materials into the production cycle. Collaboration with stakeholders, such as suppliers and waste management companies, fosters a collective effort to develop comprehensive plastic waste management strategies. Furthermore, investing in research and development of alternative materials, sustainable packaging solutions, and advanced recycling technologies can lead to more efficient practices in managing plastic waste. By examining industry-specific case studies and best practices, valuable insights can be gained to inform the development of policies, strategies, and initiatives aimed at reducing plastic waste and fostering a more sustainable approach to plastic utilization.

13.14 OVERVIEW OF NANOWASTE AND COVID-RELATED WASTE

Nanowaste and COVID waste are two distinct categories of waste with different origins and compositions. Nanowaste specifically refers to the waste generated during the production, use, and disposal of nanomaterials and nanoproducts. These materials are engineered at the nanoscale, typically ranging from 1 to 100 nanometers (Suman & Pei, 2022), and possess unique properties and applications. However, the disposal of nanowaste presents challenges that require proper management to mitigate potential environmental and health risks (Dermatas et al., 2018). This waste category encompasses various forms, including nanoparticles, nanocomposites, nanotubes, and other materials at the nanoscale (Zahra et al., 2022).

However, COVID waste pertains to the waste produced as a consequence of the COVID-19 pandemic (Haque et al., 2021). This global health crisis has led to a significant increase in the production and use of personal protective equipment (PPE) such as masks and gloves, which are often discarded after single use (Mahmoudnia et al., 2022). Furthermore, medical waste generated from COVID-19 testing, treatment, and vaccination efforts, such as syringes and contaminated materials, contributes to the growing volume of COVID waste (Das et al., 2021). It is imperative to adopt proper handling and disposal practices for COVID waste to prevent the spread of the virus and ensure the safety of waste management workers and the general public.

While both nanowaste and COVID waste pose unique challenges, they stem from different contexts and involve distinct waste streams. Nanowaste arises from the production and use of nanomaterials, which require specialized disposal methods to minimize potential environmental and health impacts. COVID waste, however, results from the global response to the COVID-19 pandemic, necessitating proper management and disposal of PPE, medical waste, and other related materials. Understanding the differences between these waste categories is crucial in developing effective waste management strategies and safeguarding the environment and public health. By implementing proper disposal practices and raising awareness about these waste streams, we can mitigate their potential negative effects and promote a sustainable future.

13.15   CHALLENGES AND RISKS ASSOCIATED WITH MANAGING THESE TYPES OF WASTE

Managing nanowaste and COVID waste presents various challenges and risks. One significant concern is the potential environmental risks associated with nanowaste. Certain nanomaterials can be toxic and have the potential to harm ecosystems, plants, animals, and microorganisms if not effectively managed. Additionally, both nanowaste and COVID waste pose health risks to waste handlers and the public. Exposure to nanomaterials, with their unique properties, can lead to respiratory, dermal, or other health issues. COVID waste may contain infectious materials, highlighting the need for proper handling and disposal to prevent the spread of diseases. Furthermore, the lack of standardized guidelines and regulations for the safe management of nanowaste and COVID waste presents an additional challenge. This makes it difficult for waste management systems to effectively address these specific waste types.

13.16   TECHNIQUES FOR SAFE HANDLING AND DISPOSAL OF NANOWASTE AND COVID WASTE

Safe handling and disposal techniques for nanowaste and COVID waste involve several key practices. Proper segregation and labeling of waste at the source are essential. Dedicated waste bins or containers should be clearly labeled to ensure accurate identification and separation of these waste streams. Waste handlers should use proper personal protective equipment (PPE), including gloves, masks, and protective clothing, to minimize exposure risks while handling nanowaste and COVID waste. Containment and packaging are crucial, with nanowaste requiring sealed containers to prevent dispersion and potential environmental contamination. COVID waste, especially infectious waste, should be properly packaged, double-bagged, and sealed to prevent transmission of pathogens.

Safe disposal methods should be followed, with nanowaste disposed of in accordance with established waste management practices such as landfilling or recycling. This should take into consideration the specific characteristics and risks associated with the nanomaterials. COVID waste, particularly infectious waste, should undergo proper treatment methods such as incineration, autoclaving, or chemical disinfection, depending on waste regulations and local guidelines.

13.17　CASE STUDIES DEMONSTRATING EFFECTIVE MANAGEMENT OF NANOWASTE AND COVID WASTE

Several case studies have shown effective management strategies for nanowaste and COVID waste. One notable example is in Jordan where their healthcare waste management during the pandemic practices the reduction of unneeded healthcare waste. They also isolate regular waste from hazardous waste. The contaminated waste was disposed of rapidly daily (Das et al., 2021). In South Korea, a comprehensive waste management system was successfully implemented during the COVID-19 pandemic. This system involved the establishment of dedicated collection points, strict segregation and packaging guidelines, and the use of advanced treatment technologies (Roy et al., 2021). The European Union's Horizon 2020 program has also played a significant role by funding multiple projects focused on nanowaste and COVID waste management. For instance, the ENNaBle project concentrates on developing safe and sustainable methods across the entire value chain (Schwab et al., 2023). These case studies highlight the crucial role of risk assessment in developing effective waste management approaches for nanowaste and COVID waste.

13.18　CONCLUSION AND RECOMMENDATIONS

This chapter delves into the challenges and opportunities surrounding the management of plastic waste, with a specific focus on nanowaste and COVID-related waste. We explore the risks associated with these waste types and examine techniques and best practices for their safe handling and disposal. Case studies are presented to highlight successful waste management approaches. Throughout the chapter, the critical importance of adopting sustainable waste management practices and addressing the global issue of plastic waste is emphasized. To effectively address the environmental and health risks associated with nanowaste and COVID-related waste, it is crucial to implement sustainable waste management practices. By doing so, we can minimize the release of hazardous substances, protect ecosystems, and ensure human well-being. Additionally, sustainable waste management plays a vital role in conserving resources, reducing greenhouse gas emissions, and promoting a circular economy. For policymakers, prioritizing the development and implementation of tailored regulations and guidelines for the management of nanowaste and COVID waste is essential. These policies should encompass proper segregation, labeling, and safe disposal practices, while also fostering research and innovation in sustainable waste management technologies. Waste management practitioners have a pivotal role in ensuring the safe handling and disposal of nanowaste and COVID waste. Investing in infrastructure and facilities that support effective management, along with training programs and awareness campaigns, is crucial to educating waste handlers about the associated risks and proper procedures. Researchers, however, should focus on advancing our understanding of the environmental and health impacts of nanowaste and COVID-related waste. Their efforts should concentrate on developing innovative waste management technologies, including improved detection and monitoring methods, eco-friendly disposal techniques, and sustainable recycling processes. Collaboration among researchers, policymakers, and waste management practitioners is paramount to address knowledge gaps and drive effective solutions. Addressing the global challenge of plastic waste requires immediate action from all stakeholders, including governments, industries, communities, and individuals. Comprehensive strategies for plastic waste management must be implemented, encompassing recycling, waste reduction, and the development of alternative materials. Promoting awareness and educational campaigns is crucial to fostering responsible plastic consumption, proper waste segregation, and recycling practices among individuals and communities. Moreover, investing in research and development for sustainable alternatives to single-use plastics is imperative.

REFERENCES

Abdelateef Mostafa, M., El-Hay, E. A., & ELkholy, M. M. (2023). Recent trends in wind energy conversion system with grid integration based on soft computing methods: Comprehensive review, comparisons and insights. *Archives of Computational Methods in Engineering*, *30*(3), 1439–1478. https://doi.org/10.1007/s11831-022-09842-4

Alberghini, L., Truant, A., Santonicola, S., Colavita, G., & Giaccone, V. (2022). Microplastics in fish and fishery products and risks for human health: A review. *International Journal of Environmental Research and Public Health*, *20*(1), 789. https://doi.org/10.3390/ije rph20010789

Alqarni, A. O., Nabi, R. A. U., Althobiani, F., Naz, M. Y., Shukrullah, S., Khawaja, H. A., Bou-Rabee, M. A., Gommosani, M. E., Abdushkour, H., Irfan, M., & Mahnashi, M. H. (2022). Statistical optimization of pyrolysis process for thermal destruction of plastic waste based on temperature-dependent activation energies and pre-exponential factors. *Processes*, *10*(8), 1559. https://doi.org/10.3390/pr10081559

Al-Salem, S. M., El-Eskandarani, M. S., & Constantinou, A. (2021). Can plastic waste management be a novel solution in combating the novel Coronavirus (COVID-19)? A short research note. *Waste Management & Research: The Journal for a Sustainable Circular Economy*, *39*(7), 910–913. https://doi.org/10.1177/0734242X20978444

Baba-Nalikant, M., Syed-Mohamad, S. M., Husin, M. H., Abdullah, N. A., Mohamad Saleh, M. S., & Abdul Rahim, A. (2023). A zero-waste campus framework: Perceptions and practices of University Campus Community in Malaysia. *Recycling*, *8*(1), 21. https://doi. org/10.3390/recycling8010021

Beghetto, V., Sole, R., Buranello, C., Al-Abkal, M., & Facchin, M. (2021). Recent advancements in plastic packaging recycling: A mini-review. *Materials*, *14*(17), 4782. https://doi.org/ 10.3390/ma14174782

Benny, V. (2022). Practices and challenges of household plastic waste disposal: An evaluation of waste management system in Kerala. *Orissa Journal of Commerce*, 27–44. https://doi. org/10.54063/ojc.2022.v43i02.03

Braithwaite, J., Mannion, R., Matsuyama, Y., Shekelle, P., Whittaker, S., Al-Adawi, S., Ludlow, K., James, W., Ting, H. P., Herkes, J., Ellis, L. A., Churruca, K., Nicklin, W., & Hughes, C. (2017). Accomplishing reform: Successful case studies drawn from the health systems of 60 countries. *International Journal for Quality in Health Care*, *29*(6), 880–886. https:// doi.org/10.1093/intqhc/mzx122

Cesaro, A., Russo, L., Farina, A., & Belgiorno, V. (2016). Organic fraction of municipal solid waste from mechanical selection: Biological stabilization and recovery options. *Environmental Science and Pollution Research*, *23*(2), 1565–1575. https://doi.org/ 10.1007/s11356-015-5345-2

Chen, A., Yang, M.-Q., Wang, S., & Qian, Q. (2021). Recent advancements in photocatalytic valorization of plastic waste to chemicals and fuels. *Frontiers in Nanotechnology, 3.* https://doi.org/10.3389/fnano.2021.723120

Cheng, P., Ji, G., Zhang, G., & Shi, Y. (2022). A closed-loop supply chain network considering consumer's low carbon preference and carbon tax under the cap-and-trade regulation. *Sustainable Production and Consumption, 29,* 614–635. https://doi.org/10.1016/j.spc.2021.11.006

da Silva, L. F., Resnitzkyd, M. H. C., Santibanez Gonzalez, E. D. R., de Melo Conti, D., & da Costa, P. R. (2022). Management of plastic waste and a circular economy at the end of the supply chain: A systematic literature review. *Energies, 15*(3), 976. https://doi.org/10.3390/en15030976

Das, A. K., Islam, Md. N., Billah, Md. M., & Sarker, A. (2021). COVID-19 pandemic and healthcare solid waste management strategy – a mini-review. *Science of the Total Environment, 778,* 146220. https://doi.org/10.1016/j.scitotenv.2021.146220

Dermatas, D., Mpouras, T., & Panagiotakis, I. (2018). Application of nanotechnology for waste management: Challenges and limitations. *Waste Management & Research, 36*(3), 197–199. https://doi.org/10.1177/0734242X18758820

Din, F. (2019, December 26). Dasar Kebersihan Negara: Kemampuan dan Persepsi "Kata di Kotakan." *Majalah Sains.*

Farooq, A., Shiung Lam, S., Hoon Rhee, G., Lee, J., Ali Khan, M., Jeon, B.-H., & Park, Y.-K. (2022). Technical benefits of using methane as a pyrolysis medium for catalytic pyrolysis of Kraft lignin. *Bioresource Technology, 353,* 127131. https://doi.org/10.1016/j.biortech.2022.127131

Girma, G. (2020). Barriers influencing implementation of plastic recycling process: A case of Dire Dawa City, Ethiopia. *Industrial Engineering Letters, 10*(3). https://doi.org/10.7176/IEL/10-3-01

Haque, Md. S., Uddin, S., Sayem, S. Md., & Mohib, K. M. (2021). Coronavirus disease 2019 (COVID-19) induced waste scenario: A short overview. *Journal of Environmental Chemical Engineering, 9*(1), 104660. https://doi.org/10.1016/j.jece.2020.104660

He, Y. (2017). Supply risk sharing in a closed-loop supply chain. *International Journal of Production Economics, 183,* 39–52. https://doi.org/10.1016/j.ijpe.2016.10.012

Huang, S., Wang, H., Ahmad, W., Ahmad, A., Ivanovich Vatin, N., Mohamed, A. M., Deifalla, A. F., & Mehmood, I. (2022). Plastic waste management strategies and their environmental aspects: A scientometric analysis and comprehensive review. *International Journal of Environmental Research and Public Health, 19*(8), 4556. https://doi.org/10.3390/ijerph19084556

Joshi, S., & Temgire, S. (2021). Life below water and the fate of humanity: With special reference to the efforts by adidas towards reducing the effects of plastic waste on the marine life. *Sustainability, Agri, Food and Environmental Research, 10*(1). https://doi.org/10.7770/safer-V10N1-art2559

Lebreton, L., & Andrady, A. (2019). Future scenarios of global plastic waste generation and disposal. *Palgrave Communications, 5*(1), 6. https://doi.org/10.1057/s41599-018-0212-7

Li, W. C., Tse, H. F., & Fok, L. (2016). Plastic waste in the marine environment: A review of sources, occurrence and effects. *Science of the Total Environment, 566–567,* 333–349. https://doi.org/10.1016/j.scitotenv.2016.05.084

Mahmoudnia, A., Mehrdadi, N., Golbabaei Kootenaei, F., Rahmati Deiranloei, M., & Al-e-Ahmad, E. (2022). Increased personal protective equipment consumption during the COVID-19 pandemic: An emerging concern on the urban waste management and strategies to reduce the environmental impact. *Journal of Hazardous Materials Advances, 7,* 100109. https://doi.org/10.1016/j.hazadv.2022.100109

Normalina, Hatta, M., Hafizianoor, & Hamdani. (2021). Plastic waste management through the role of leadership adaptation of environmental inspected habits. *Webology, 18*(Special Issue 4), 268–277. https://doi.org/10.14704/WEB/V18SI04/WEB18127

Nayak, A., & Bhushan, B. (2021). *Sustainable Solid Waste Management via Biological Treatment* (pp. 248–271). https://doi.org/10.4018/978-1-7998-4921-6.ch012

Ong, M. D., & Schideman, L. C. (2013). Hydrothermal catalytic gasification (HCG) as an alternative to anaerobic digestion. 2013 Kansas City, Missouri, July 21–July 24, 2013. https://doi.org/10.13031/aim.20131599238

Pal, K. K., Tilak, K. V. B. R., Saxcna, A. K., Dey, R., & Singh, C. S. (2001). Suppression of maize root diseases caused by *Macrophomina phaseolina, Fusarium moniliforme* and *Fusarium graminearum* by plant growth promoting rhizobacteria. *Microbiological Research, 156*(3), 209–223. https://doi.org/10.1078/0944-5013-00103

Plavac, B., Filipan, V., & Sutlović, I., & Svetičič, J. (2017). Sustainable waste management with mechanical biological treatment and energy utilization. *Tehnicki Vjesnik – Technical Gazette, 24*(4). https://doi.org/10.17559/TV-20150709113428

Qureshi, M. S., Oasmaa, A., Pihkola, H., Deviatkin, I., Tenhunen, A., Mannila, J., Minkkinen, H., Pohjakallio, M., & Laine-Ylijoki, J. (2020). Pyrolysis of plastic waste: Opportunities and challenges. *Journal of Analytical and Applied Pyrolysis, 152*, 104804. https://doi.org/10.1016/j.jaap.2020.104804

Ragaert, K., Delva, L., & Van Geem, K. (2017). Mechanical and chemical recycling of solid plastic waste. *Waste Management, 69*, 24–58. https://doi.org/10.1016/j.wasman.2017.07.044

Ravindran, S. (2020, August 7). Communities stake claim to waterways. *The Star.*

Ridwan, A., Ambarwaty, T. R., Wahyuni, N., Trenggonowati, D. L., Bahauddin, A., Umyati, A., & Kurniawan, B. (2022). Evaluating economic and environmental impact of a plastic waste processing industry nased on circular economy using benefit–cost analysis. *Jurnal Ilmiah Teknik Industri, 21*(2), 232–239. https://doi.org/10.23917/jiti.v21i2.19751

Roy, P., Mohanty, A. K., Wagner, A., Sharif, S., Khalil, H., & Misra, M. (2021). Impacts of COVID-19 outbreak on the municipal solid waste management: Now and beyond the pandemic. *ACS Environmental Au, 1*(1), 32–45. https://doi.org/10.1021/acsenviro nau.1c00005

Roy, P. S., Garnier, G., Allais, F., & Saito, K. (2021). Strategic approach towards plastic waste valorization: Challenges and promising chemical upcycling possibilities. *ChemSusChem, 14*(19), 4007–4027. https://doi.org/10.1002/cssc.202100904

Sakthipriya, N. (2022). Plastic waste management: A road map to achieve circular economy and recent innovations in pyrolysis. *Science of the Total Environment, 809*, 151160. https://doi.org/10.1016/j.scitotenv.2021.151160

Samuel, M., & Ahlawat, S. (2021). EPR in plastic waste management: Policy tool yet to be realised. *Asian Journal of Management, 12*(4), 394–400. https://doi.org/10.52711/2321-5763.2021.00059

Schultmann, F., Zumkeller, M., & Rentz, O. (2006). Modeling reverse logistic tasks within closed-loop supply chains: An example from the automotive industry. *European Journal of Operational Research, 171*(3), 1033–1050. https://doi.org/10.1016/j.ejor.2005.01.016

Schwab, F., Rothen-Rutishauser, B., Scherz, A., Meyer, T., Karakoçak, B. B., & Petri-Fink, A. (2023). The need for awareness and action in managing nanowaste. *Nature Nanotechnology, 18*(4), 317–321. https://doi.org/10.1038/s41565-023-01331-4

Sharma, H. B., Vanapalli, K. R., Cheela, V. S., Ranjan, V. P., Jaglan, A. K., Dubey, B., Goel, S., & Bhattacharya, J. (2020). Challenges, opportunities, and innovations for effective solid waste management during and post COVID-19 pandemic. *Resources, Conservation and Recycling, 162*, 105052. https://doi.org/10.1016/j.resconrec.2020.105052

Shen, B. (2014). Sustainable fashion supply chain: Lessons from H&M. *Sustainability*, *6*(9), 6236–6249. https://doi.org/10.3390/su6096236

Silva, J. dos S., Rodrigues, J. R. P., Sá, R. A. de Q. C. de, Oliveira, M. B. M. de, Silva, S. M. da, Souza, K. S., Silva, M. R. F. da, Araújo, L. C. A. de, Pereira, E. S., Bezerra, A. Â., & Barros, A. V. de. (2022). Environmental pollution by microplastics and its consequences on human health. *Research, Society and Development*, *11*(13), e520111335863. https://doi.org/10.33448/rsd-v11i13.35863

Suman, T. Y., & Pei, D.-S. (2022). Nanomaterial waste management. In *Nanomaterials Recycling* (pp. 21–36). Elsevier. https://doi.org/10.1016/B978-0-323-90982-2.00002-0

Susanti, N. K. Y., Mardiastuti, A., & Wardiatno, Y. (2020). Microplastics and the impact of plastic on wildlife: A literature review. *IOP Conference Series: Earth and Environmental Science*, *528*(1), 012013. https://doi.org/10.1088/1755-1315/528/1/012013

Thachnatharen, N., Shahabuddin, S., & Sridewi, N. (2021). The waste management of polyethylene terephthalate (PET) plastic waste: A review. *IOP Conference Series: Materials Science and Engineering*, *1127*(1), 012002. https://doi.org/10.1088/1757-899X/1127/1/012002

Thompson, R. C., Swan, S. H., Moore, C. J., & vom Saal, F. S. (2009). Our plastic age. *Philosophical Transactions of the Royal Society B: Biological Sciences*, *364*(1526), 1973–1976. https://doi.org/10.1098/rstb.2009.0054

Trulli, E., Ferronato, N., Torretta, V., Piscitelli, M., Masi, S., & Mancini, I. (2018). Sustainable mechanical biological treatment of solid waste in urbanized areas with low recycling rates. *Waste Management*, *71*, 556–564. https://doi.org/10.1016/j.wasman.2017.10.018

Ügdüler, S., Van Geem, K. M., Denolf, R., Roosen, M., Mys, N., Ragaert, K., & De Meester, S. (2020). Towards closed-loop recycling of multilayer and coloured PET plastic waste by alkaline hydrolysis. *Green Chemistry*, *22*(16), 5376–5394. https://doi.org/10.1039/D0GC00894J

Van Deursen, C., Suwan, T., Laosuwan, S., Wongmatar, P., Kaewmoracharoen, M., & Suwan, P. (2023). Development of polymeric binder from expanded polystyrene (EPS) foam waste as construction materials. *IOP Conference Series: Earth and Environmental Science*, *1146*(1), 012007. https://doi.org/10.1088/1755-1315/1146/1/012007

Yukeswaran, L., & Moni Philip Jacob, K. (2022). A review on microplastics – an indelible ubiquitous pollutant. *Biointerface Research in Applied Chemistry*, *13*(2), 126. https://doi.org/10.33263/BRIAC132.126

Zahra, Z., Habib, Z., Hyun, S., & Sajid, M. (2022). Nanowaste: Another future waste, its sources, release mechanism, and removal strategies in the environment. *Sustainability*, *14*(4), 2041. https://doi.org/10.3390/su14042041

14 *Pinus roxburghii* Sarg: Needle Management to Combat Forest Fire
Status, Prospects, and Constraints

*Prashant Sharma, Kamlesh Verma,
Daulat Ram Bhardwaj, Manoj Kumar Singh,
Dhirender Kumar, and Pankaj Thakur*

14.1 INTRODUCTION

Forests are essential for supporting the livelihoods of local communities by providing a diverse range of goods and services. These include tangible products such as fuel, timber, food, and bioproducts and intangible benefits such as regulation of greenhouse gases, air and water supply, carbon storage, and nutrient cycling. Additionally, forests contribute to greater species and genetic diversity, which is crucial for sustaining life (de Groot et al., 2002, Joshi and Negi, 2011; Joshi and Joshi, 2019). According to Langat et al. (2016), a significant proportion of the global population, ranging from 1.09 to 1.74 billion individuals, rely on forest resources to fulfill their daily needs, with approximately 200 million Indigenous communities completely depending on forests for their sustenance. However, in recent times, forest fires have emerged as a significant natural calamity and a potential hazard to forest ecosystems worldwide (Bowman et al., 2009; Bargali et al., 2022), resulting in substantial biodiversity loss and alterations in multiple ecosystems (Pew and Larsen, 2001). Each year, forest fires destroy millions of hectares of forests, costing the country around Rs. 1101 crores annually (Ministry of Environment, Forest and Climate Change, 2018). This has ultimately led to the loss of both human property and wildlife and changes in forest structure and function as well. Additionally, these fires release significant amounts of carbon into the atmosphere and require substantial costs for fire suppression and prevention (Davidenko and Eritsov, 2003) (Figure 14.1).

Forest fires have been present since the inception of forests. It is widely believed that natural forest fires have been occurring since prehistoric times, primarily due to lightning strikes. In India, the likelihood of forest fires resulting from lightning is considerably lower than those caused by human activities (Babu et al., 2016). The

DOI: 10.1201/9781003543176-14

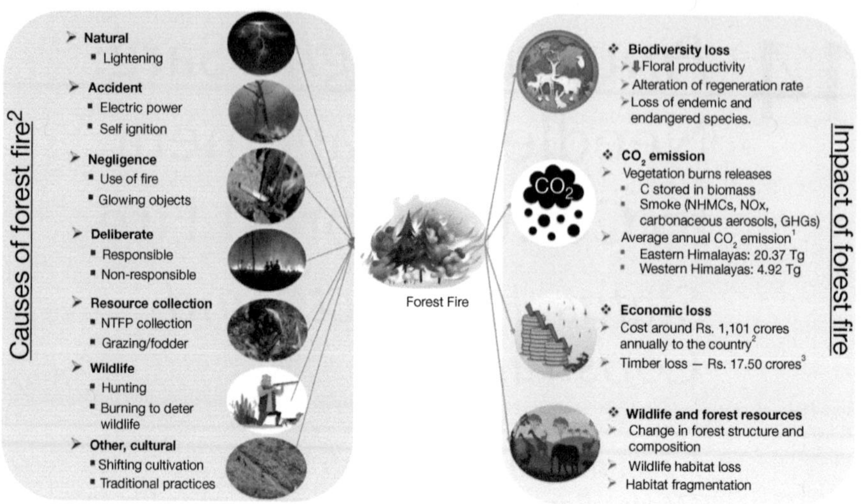

FIGURE 14.1   Potential causes and impact of forest fire.

incidence and spread of forest fires are influenced by a multifaceted interplay of various factors, including meteorological conditions (air temperature, relative humidity, wind speed, previous day rainfall, dew point temperature, air pressure, potential evapotranspiration, land surface temperature, precipitation rate, and albedo), topographical features (slope, aspect, and elevation), forest composition, human-related factors (road and rail networks, and human activity), and the characteristics of the fuel in a given landscape (Tonini et al., 2020; Bargali et al., 2022). The movement of fire is also determined by the combined effects of conduction, convection, and radiation (Bhatia et al., 2020).

14.2   INDIAN HIMALAYAN REGION AND FOREST FIRE

The forests of the Indian Himalayan region are rich and diverse, making it a significant natural repository on a global scale (Kumar et al., 2022a, 2022b). Simultaneously, it is imperative to conserve and manage the Himalayan Forest belt for posterity, as it is one of the most fragile ecosystems (Negi et al., 2012; Panda et al., 2021, 2022; Kumar et al., 2022c). Fire has become a crucial component of the mountainous terrain, particularly in the vicinity of human settlements. Specifically, the Western Himalaya region of India has been identified as a highly susceptible area to forest fires (Dobriyal and Bijalwan, 2017). The occurrence of uncontrolled forest fires in this region is a recurrent phenomenon mostly during the late spring and summer seasons and occasionally during extended periods of drought in the winter (Babu et al., 2016; Dobriyal and Bijalwan, 2017). Furthermore, these incidents are expected to become more frequent in the future due to the effects of global warming and rising temperatures (Bargali et al., 2022). In Uttarakhand, a significant proportion exceeding 50% of mountainous forests are susceptible to frequent fire occurrences with more than 4500 forest fires annually, resulting in the destruction of over 0.6 million hectares of forest (Babu, 2019). Over the course of the past century, approximately 23 years have experienced significant major forest fire incidences in Uttarakhand (Dobriyal and Bijalwan, 2017). According to Bar et al. (2021), there was a notable rise in the burn area of Uttarakhand and Himachal Pradesh at a rate of 72.94 km^2 per year from 2001 to 2019. The reoccurrence of fires has had a devastating impact on the local population and caused irreparable harm to valuable natural resources (Dobriyal and Bijalwan, 2017) in terms of forest degradation, reduced ecosystem productivity, and loss of biodiversity (Kumar and Kumar 2022). The primary factors contributing to forest fires in the Himalayan region are anthropogenic activities, including the harvesting of non-timber forest products and agricultural practices, as well as the extraction of honey from bee hives (Negi, 2019). Additionally, the region's prolonged dry weather and insufficient moisture in the forests, as well as the traditional practice of burning forests by local villagers to eliminate unwanted shrubs and better regeneration of the grass in the next year for the improved forage for livestock (Bahuguna and Upadhay, 2002), are significant contributors to the occurrence of forest fires.

14.3 ROLE OF CHIR PINE IN FOREST FIRE

Anthropogenic activities have been identified as the primary cause of forest fires in the Himalayas. However, the spread of these fires within the forest can be attributed to the significant accumulation of chir pine needles, which are a highly inflammable fuel source. This dominant forest-forming tree, scientifically known as *Pinus roxburghii* Sarg., has been identified as a key contributor to the severity of wildfires in the region (Ahmad et al., 2018; Bargali et al., 2022). The chir pine is a species that possesses a significant amount of resin, making it economically valuable. Nonetheless, the afore-mentioned characteristic renders the needle litter of the tree species susceptible to forest fires (Bargali et al., 2020), because it affects the accumulation and flammability of fuel load (Bargali et al., 2022). Chir pine, also known as three-needled Indian pine, is indigenous to the inter-ranges and valleys of the Himalayas, ranging from Afghanistan in the west to Bhutan in the east. Specifically, chir pine is the dominant species of the Subtropical Pine Forest-Group 9, (Champion and Seth, 1968) spread around 1.78 million ha area (0.54 % of the country's total geographical area) in India, comprising 0.18, 0.91, and 0.69 million ha area under very dense, medium dense, and open forest, respectively, with a carbon stock of 139.47 ton ha^{-1} (ISFR, 2021) (Figure 14.2). However, the area occupied by this specific type of forest has been observed to exhibit a declining trend over time. For instance, Reddy et al. (2015) indicated that the area under the sub-tropical pine forest is 2.29 million ha (0.70 % of the country's total geographical area), which is 22.20% higher than the recent obser-vation reported by ISFR (2021). In India, particularly chir pine is distributed in the states of Himachal Pradesh, Jammu and Kashmir, Uttarakhand, parts of Sikkim, West Bengal, and Arunachal Pradesh with elevation varying from 450 m to 2300 m above mean sea level and an estimated area of 8,90,000 ha (Sanwal et al., 2016; Bhardwaj et al., 2021). This large evergreen tree with an elongated crown attains a height of up to 50 m, with about 3.5 m in girth, and forms a straight cylindrical bole.

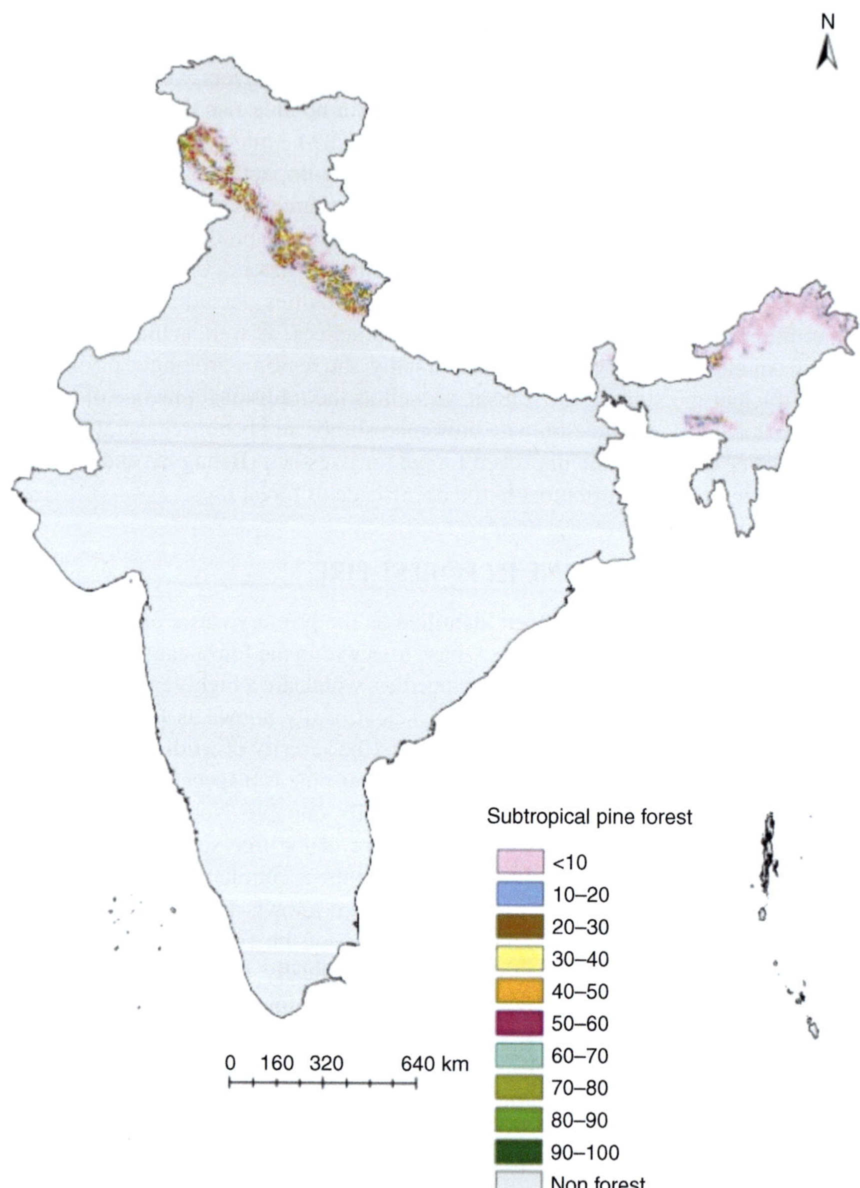

FIGURE 14.2 Extent of sub-tropical pine forest-Group 9 in India.

Source: Reddy et al. (2015); taken from Bhuvan Portal.

The chir pine species are utilized for various purposes, such as timber, fuel wood, torchwood, funeral wood, and furniture. The foliage of this plant is utilized to provide bedding for livestock and for mulching fields. The chir pine's bark serves as a significant source of charcoal, resin, and coal tar. According to Kala and Subbarao (2018), on an annual basis, a hectare of pine forest undergoes the shedding of approximately 1.2 tons of pine needles during the period ranging from April to June. In this way, approximately 4 lakh tons of pine needles are deposited in the forests of Uttarakhand each year (Negi, 2019), while in Himachal Pradesh, the annual deposition of pine needles is approximately 2 lakh tons. The high combustibility of pine needles leads to devastating forest fires during the summer season, resulting in significant economic losses in terms of timber, resin, plantations, wildlife, and other valuable biodiversity. Simultaneously, the pine needles in the forests become slippery, making it difficult for both humans and cattle to navigate through them at the same time (Chandran et al., 2011). In addition, the repetitive occurrences of forest fires facilitate the proliferation of pine forests that are conducive to fire, while simultaneously displacing ecologically significant and culturally esteemed oak (*Q. leucotrichophora*) forests (Singh et al., 1984). These oak forests are known to provide valuable resources such as fuel wood, animal feed, bedding leaves for livestock, as well as medicinal plants, among others. The phenomenon of fire-assisted invasion by pine in oak forests results in a significant alteration of microclimatic conditions, thereby rendering them more conducive to the invasion of chir pine, which is characterized by its early successional nature and limited utility for the local populace. Consequently, its territorial expansion occurs progressively with each passing year (Negi et al., 2012).

14.4   CHIR PINE NEEDLE MANAGEMENT

The Forest Department began fire prevention for the first time in the chir pine forests in 1912. One of the primary methods for managing pine needles involves either in-situ controlled burning of needles or the collection of pine needles for processing into various products for further consumption. In controlled burning, a significant quantity of carbon and greenhouse gases is released into the atmosphere, but the impact is less than a wildfire. Simultaneously, controlled burning is a labor-intensive endeavor that necessitates substantial resources. Concurrently, the collection and subsequent use of pine needles for a variety of products is an alternative solution to this issue (Figure 14.3). The chir pine needles typically comprise 43% lignin, 52% holocellulose, and 5.8% extractive content (Lal et al., 2013). The cellulose content of lignocellulose derived from pine needles is comparable to that of softwood (42 %) (Chauhan et al., 2018). Simultaneously, the lignin proportion is greater than that of softwood (Lal et al., 2013) and due to their high lignin content, fallen needles exhibit a prolonged resistance to decomposition (Kumar et al., 2021). Moreover, converting the pine needle into various products not only fulfills the diversified requirements of rural areas but also creates job opportunities for the local population, at the same time safeguarding the forests against the threat of wildfires. The process of converting pine needles into fuel gases, electricity, and paper pulp confers an economic worth upon them, thereby incentivizing individuals to gather the needles. The removal of

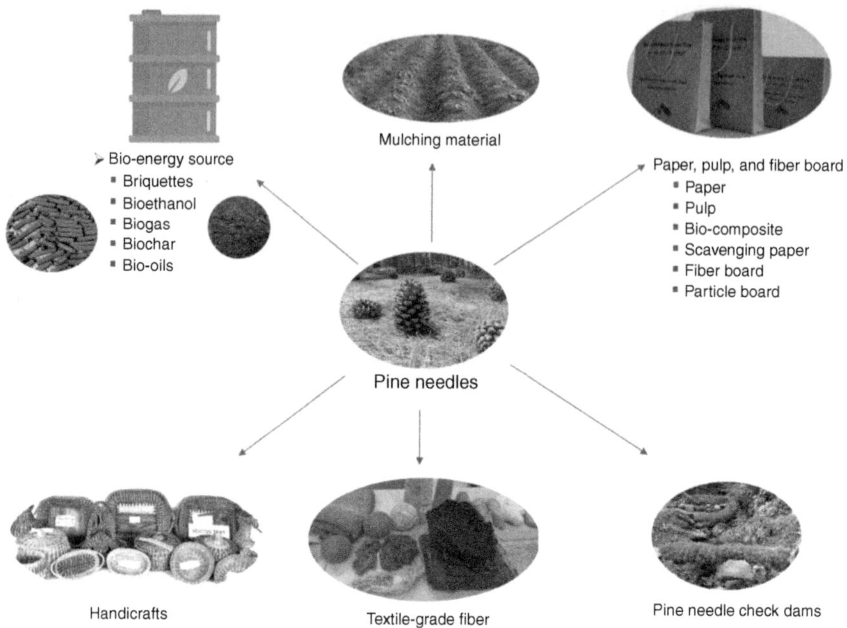

FIGURE 14.3   Utilization of chir pine needles in diversified products.

pine needles results in a reduction of fire hazards, preservation of biodiversity, and improvement of livestock feed availability (Dhaundiyal and Gupta, 2014).

14.4.1 Bio-Energy Source

The utilization of forest wood in gasification has been a longstanding practice; however, there exists the possibility of alternative resources emerging as viable fuel sources. The utilization of pine needles can benefit the renewable energy sector due to their cost-effectiveness as an energy source (Dhaundiyal and Tewari, 2016). Typically, the needle of the chir pine is utilized for direct combustion for domestic purposes; however, due to their low density and heating capacity, they are not a viable option for heating and cooking applications (Bisht and Thakur, 2016). The pine needle can be utilized for the production of various derivative products (Table 14.1), such as gasification for producer gas, anaerobic digestion for biogas production, high-pressure briquetting, fermentation for ethanol, biochar production, and thermochemical conversion for bio-oil (Roy and Kundu, 2023) that possess superior calorific value, energy efficiency, ease of use, and can be stored in a confined space for an extended duration. Pine needles possess a high content of cellulose and hemicelluloses, rendering them a viable substrate for biodegradation (Lal et al., 2013). Moreover, the pine needle exhibits a calorific value of 18.57 MJ kg^{-1}, surpassing other forms of loose biomass, including rice husk, bagasse, coconut fibers, and rice straw (Dhaundiyal and Gupta, 2014). So, pine needles can be utilized for the production of valuable by-products

TABLE 14.1
Utilization of the chir pine needle for the production of bio-energy resources

Bio-energy resource	Usage	Source
Producer gas in gasifier	In gas turbines to produce electricity	Kumar and Randa (2014); Minhas and Banshtu (2022); Varma and Mondal (2018)
Pine needle as a substrate for gasification	Alternative fuel for a downdraft gasifier; electricity production	Dhaundiyal and Gupta (2014); Dhaundiyal and Tewari (2016); Bisht and Thakur (2016, 2020); Agrawal and Sood (2021)
Biomethane from pine needle	Electricity production	Dwivedi et al. (2016)
Biochar briquettes	Cooking and household heating	Roy and Kundu (2023)
Pine needles briquettes	Energy generation	Joshi et al. (2015); Singh and Kaur (2018); Sengar et al. (2020, 2022)
Bioethanol	Alternate fuel	Vaid et al. (2018); Kumar et al. (2018); Dwivedi et al. (2022)
Biochar	Lead remediation	Choudhary et al. (2020)
Bioenergy	Energy generation	Azad et al. (2022)

which have the potential to not only mitigate the risk of forest fires, thus addressing the pressing concern of global warming caused by burning (Dwivedi et al., 2016), as well as alleviating the demand for fuel wood within forested areas (Dhaundiyal and Gupta, 2014).

From the chir pine needles, the producer gas can be generated through a biomass gasifier that can be utilized in the kiln with enhanced operational parameter control (Kumar and Randa, 2014). Similarly, the briquettes can be produced from desiccated pine needles; however, they exhibit reduced density and calorific value. Nevertheless, the biochar that is generated through torrefaction can be utilized for the production of biochar briquettes which have increased density and calorific value, as well as improved storage and transportation convenience (Roy and Kundu, 2023). For instance, each hectare of chir pine forest have the capacity to generate 8 MWh of electricity which can provide cooking fuel for a single family and create employment opportunities for one individual within a year (Dhaundiyal and Tewari, 2016). Furthermore, if 40% of the annual pine needle produced (0.82 million tons) in the Uttarakhand state is allocated for bio-gasification, it has the potential to generate 2.98×10^{10} cubic meters of biogas annually, which is equivalent to an annual electricity production of 2.98×10^{10} kWh (Dwivedi et al., 2016).

14.4.2 PULP, PAPER, AND FIBER BOARDS

The rise in environmental apprehensions regarding the utilization of synthetic petroleum-based plastics has led to a significant surge in the requirement for

TABLE 14.2
Utilization of the chir pine needle in the paper and pulp industry

Paper, pulp, and fiber boards	Usage	Source
Pine needle fiber board	Fruit packing cases	Chawla (1977)
Pine needles composite boards bonded with isocyanate prepolymer	Strong composite board appropriate for outdoor usage, particularly in damp environments.	Gupta et al. (2010); Chauhan et al. (2012);
Pine needle paper and pulp	Semi-bleached grade pulp, unbleached kraft paper, filler or paper/board.	Lal et al. (2013)
Polylactic (PLA)-based pine needle composites	Can be used as polymer composites reinforced with natural fibers	Sinha et al. (2016)
Pine needles lignocellulosic ethylene scavenging paper impregnated with nanozeolite	Food packaging application	Kumar et al. (2021)
Pine needle waste-based ethylene scavenging paper	Banana fruit packaging	Kumar et al. (2022); Rana et al. (2023)
Lignocellulosic pine needle fiber-reinforced styrene ethylene butylene styrene (SEBS) composites	Automobiles and engineering applications	Dinesh et al. (2023)
Particle board	Diversified purposes	Singh and Vinay (2023)
Bio-composites	Diversified purposes	Gairola et al. (2019)

eco-friendly packaging materials (Kumar et al., 2021). Pine needles possess a holocellulose content exceeding 50%, rendering them a viable resource for manufacturing pulp and paper (Table 14.2), despite their high lignin content, which leads to an increased requirement for bleaching chemicals or required harsh pulping conditions (Jan et al., 2006). Rather than employing the bleached variant, it can be utilized for the production of semi-bleached grade pulp, unbleached kraft paper, filler, or paper/board (Lal et al., 2013). During the late 1970s, Chawla (1977) prepared the pine needle-based fiberboard for use as a fruit packaging case. Furthermore, Kumar et al. (2021) made a type of paper from pine needles containing 30% zeolite that exhibited noteworthy efficacy in scavenging ethylene (62% efficiency), thereby providing a promising avenue for the efficient utilization of pine needles in active food packaging applications which can be further utilized in actual food systems to prolong the shelf life of climacteric fresh produce. Similarly, Kumar et al. (2022) prepared nanocomposite paper by incorporating chir pine needles with nanomaterial halloysite nanotube and micro-fibrillated cellulose to store and extend the shelf life of the banana. In addition to their use in paper and pulp production, pine needles can be effectively utilized as a reinforcement material through the creation of composite boards that meet the exterior grade standards outlined in IS: 3087-2005/EN 312-2003, with regard to both physico-mechanical properties and dimensional stability (Gupta et al., 2010; Chauhan et al., 2012; Sinha et al., 2016; Dinesh et al., 2023).

14.4.3 SMALL-SCALE SUPPLEMENTARY UTILIZATION OF CHIR PINE NEEDLES

In addition to its widespread use in energy production and the paper pulp industry, the pine needle has a multitude of applications at the local or medium scale. Traditionally, owing to their limited biodegradability, pine needles have been utilized as a bedding material for livestock (Roy and Kundu, 2023), diminutive hand brooms, and as a packaging material for fruits (Azad et al., 2022). However, in recent times, a notable shift in the utilization of pine needles was observed, attributed to the increased accessibility of cost-effective and superior packaging materials. Consequently, an abundance of chir pine needles has been left in the forest, leading to recurrent fires in the region. Concurrently, pine needles are also utilized to construct check dams as an in-situ soil and water conservation technique (Shah et al., 2022). The pine needles have a C : N ratio of more than 75 making them one of the best alternatives as a bio-mulch (Chhetri et al., 2012), and thus previously used as mulching material in cultivation land primarily for root crops such as ginger, turmeric, and Colocasia crops that not only serves to maintain/improve fertility but also promotes the growth and yield of agricultural crops and trees (Bhardwaj et al., 2017; Sharma et al., 2022). Furthermore, the initiatives from the state governments have motivated the local communities, particularly women through self-help groups (SHGs), to gather and employ pine needles to produce various decorative items, including table mats, multipurpose baskets, chapati boxes, and trays.

The Northern India Textile Research Association (NITRA) has recently created a textile-grade fiber from pine needles. This fiber has a tenacity of 10 g den^{-1}, an elongation of 5.94%, and a bundle strength of 5.64 g tex^{-1}. This development has the potential to serve as a substitute for flax fabric imported from European countries, which could lead to a decrease in imports and generate income for individuals residing in the Himalayan region (Basu et al., 2019). In a similar way, Forest Research Institute (FRI) has also developed a technology that is capable of extracting fiber from pine needles that are cream-yellow in color, possess a length of up to 20 cm, and exhibit reasonable bundle strength and good water absorption ability. The resulting fiber can be spun into handloom cloth and used to create various products, such as jackets, coats, purses, wall curtains, lampshades, mats, and ropes (Shah et al., 2022).

14.5 CONSTRAINTS IN THE UTILIZATION OF CHIR PINE NEEDLES

The exploration of the potential use of chir pine needles for various product development presents a promising opportunity for local communities. This initiative emphasizes empowering youth and women by creating employment opportunities that improve their livelihood security in their respective regions. There are several significant factors that require effective management and governance at the community level. Sengar et al. (2020) identified five primary obstacles that hinder the

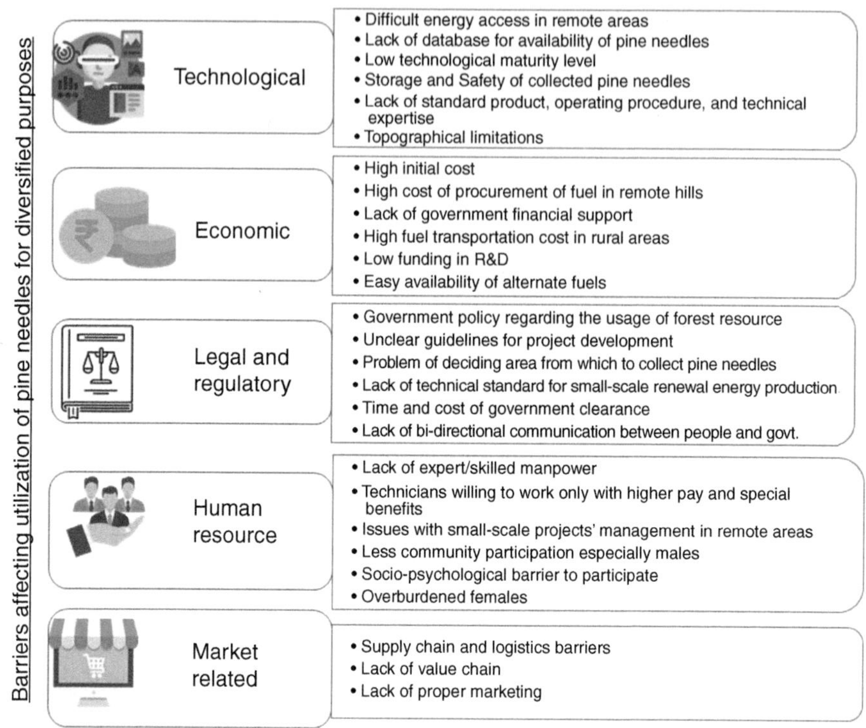

FIGURE 14.4   Barrier affecting the utilization of pine needles for diversified purposes.

Source: Adapted from Sengar et al. (2020).

widespread implementation of energy generation using chir pine needles, namely technological, economic, legal and regulatory, human resources, and market-related impediments (Figure 14.4). At the individual level, the utilization of pine needles faces two main obstacles, specifically, the economic viability and the feasibility of collecting, transporting, and storing them (Bisht and Thakur 2016). The acquisition of economic benefits is a crucial objective; however, it is vital that these benefits are enduring and that the approaches suggested for their realization are socially suitable (Joshi et al., 2015). The manufacturing of diverse goods utilizing pine needles entails significant initial investment and operating expenses, particularly in energy generation initiatives. These costs are further exacerbated by the accessibility of various fuel sources, including mustard bio-briquettes, liquefied petroleum gas, hydropower-generated electricity, imported coal, and bagasse (Sengar et al., 2020).

Furthermore, despite the considerable efforts undertaken by governmental authorities to incentivize the proficient utilization of pine needles among Indigenous populations, these individuals are not receiving sufficient compensation to support their livelihoods solely through pine needle utilization initiatives. Furthermore, the process of gathering pine needles, particularly in the rugged terrain of the Himalayan area, presents a significant challenge. Advanced tools and machines cannot be employed for the collection of the needle in this type of topography. Concurrently, the individuals who are engaged in the act of gathering needles are not receiving adequate compensation. In Himachal Pradesh, people currently receive a remuneration of Rs. 10 per quintal, which is inadequate for their sustenance. Furthermore, owing to the substantial size of pine needles, transportation costs tend to be elevated. Additionally, given the limited availability of pine needles, a significant storage capacity is necessary to ensure a consistent supply of the raw material to the facility year-round. Simultaneously the lack of market infrastructure is another major hindrance to selling their products, especially briquets and handicrafts. In general, the determination of the community plays a significant role in preventing forest fires and effectively utilizing pine needles. Sometimes, the indigenous population chooses not to gather needles or put fire themselves ultimately resulting in a new flush of grasses for their livestock in the subsequent year. Concurrently, the government is offering incentives to the local panchayats where no fire incidents are documented in adjoining forests within a given year. Typically, it has been observed that local residents participate actively in the program for 1 or 2 years. However, after 3–4 years, a significant fire outbreak occurred in the area, attributed to the accumulation of heavy fuel load in the form of pine needles on the forest floor.

14.6   FUTURE PERSPECTIVES

Amid the contemporary context of climate change, researchers and policymakers are expending significant efforts toward optimizing the utilization of pine needles. Nonetheless, there are certain issues that require immediate attention and further investigation. At the national level, there exists a robust network of fire monitoring systems that disseminate forest fire alerts to subscribed local residents. However, the country currently lacks a methodology for accurately estimating the potential fuel load, specifically pine needles, which can facilitate the local population in obtaining information regarding the availability of needles in the area that can be utilized for their small-scale projects. Furthermore, lifting the ban on green felling and the continued use of outdated management techniques by the forest department are additional factors that require modernization. Moreover, the combination of pine needles and cow dung can be employed in producing vermicompost as an alternative cost-effective, and easy means of utilizing pine needles.

The state government and forest departments of the respective states are effectively raising awareness and providing training to local communities on the utilization of pine needles by spending crores of rupees. However, additional funding is still required to support research and development activities. The researchers must prioritize their attention on the three primary components in an expeditious manner. Initially, it is imperative to investigate the impact of various chir pine pruning and thinning regimes on the yield and regrowth of the understory species. In the absence of felling operations, mature chir pines are generating a significant quantity of needles despite having already surpassed their rotation period. From a technical standpoint, trees show minimal annual biomass accumulation and carbon sequestration. Furthermore, it is imperative to diversify chir pine forests by incorporating indigenous multi-purpose tree species to make them mixed stands. This approach not only satisfies the varied needs of local communities but also enhances the rate of pine needle decomposition. Simultaneously, the extraction of chir pine needles from forested regions may result in a prolonged insufficiency of nutrients and organic carbon within the chir pine forest, ultimately leading to soil infertility. Therefore, the in-situ decomposition of pine needles should be prioritized over exploring alternative uses. From a perspective of practicality, it is recommended that researchers and policymakers focus their efforts on the creation of a bio-decomposer for pine needles, akin to the bio-decomposer developed for rice straw to mitigate the issue of rice straw burning. This approach not only allows a reduction in annual fire control expenditures but also promotes the on-site decomposition of pine needles, thereby contributing to the enrichment of soil with nutrients and organic carbon. The crux of the matter is determining the means to utilize the pine needle in an efficient way which can be an ecologically beneficial endeavor that also generates economic gains, thereby promoting the socio-economic advancement of the populace.

14.7 CONCLUSION

Forest fires pose a significant threat to the forests, resulting in the loss of biodiversity, economic resources, and environmental degradation. Moreover, forest fire and climate change are mutually reinforcing phenomena that possess significant potential to exacerbate future conditions, particularly in the vulnerable Himalayan region. To address this issue, it is crucial to effectively and efficiently manage the pine needles. Fortunately, several ex-situ alternatives exist for the utilization of chir pine needles, ranging from traditional uses such as livestock bedding to novel alternatives like bio-ethanol. The generation of biomass energy has the potential to function as a mechanism for generating employment opportunities, which can facilitate the empowerment of youth and women residing in rural areas, leading to a favorable impact on their standard of living. However, it is crucial to thoroughly evaluate the potential ecological ramifications of these interventions for safeguarding biodiversity and maintaining the ecosystem's health. Nonetheless, the extraction of pine needles from the forest floor can be proved as an inappropriate approach in the long run as it may result in unfertile soil. Therefore, researchers, local communities, and policymakers need to collaborate toward identifying and developing sustainable and context-specific strategies and mechanisms for dealing with the problem of chir pine needles and simultaneously maintaining the ecological equilibrium.

REFERENCES

Agrawal, A., & Sood, D. (2021). Development and performance analysis of pine needle based downdraft gasifier system. In: Baredar, P.V., Tangellapalli, S., & Solanki, C.S. (eds), *Advances in Clean Energy Technologies*. Springer Proceedings in Energy. Springer, Singapore. https://doi.org/10.1007/978-981-16-0235-1_13

Ahmad, F., Goparaju, L., & Qayum, A. (2018). Himalayan forest fire characterization in relation to topography, socio-economy and meteorology parameters in Arunachal Pradesh, India. *Spatial Information Research*, 26(3), 305–315. https://doi.org/10.1007/s41324-018-0175-1

Azad, D., Pateriya, R.N., & Sharma, R.K. (2022). Feasibility study of pine needles as a potential source of bio-energy. *Pantnagar Journal of Research*, 20(3), 519–523.

Babu, K.V.S. (2019). *Developing Forest Fire Danger Index Using Geo-spatial Techniques.* Doctoral's thesis. International Institute of Information Technology, Hyderabad.

Babu, K.V.S., Roy, A., & Prasad, P.R. (2016). Forest fire risk modeling in Uttarakhand Himalaya using TERRA satellite datasets. *European Journal of Remote Sensing*, 49(1), 381–395. https://doi.org/10.5721/EuJRS20164921

Bahuguna, V.K., & Upadhay, A. (2002). Forest fires in India: Policy initiatives for community participation. *International Forestry Review*, 4(2), 122–127. https://doi.org/10.1505/IFOR.4.2.122.17446

Bar, S., Parida, B.R., Pandey, A.C., & Kumar, N. (2022). Pixel-based long-term (2001–2020) estimations of forest fire emissions over the Himalaya. *Remote Sensing*, 14(21), 5302. https://doi.org/10.3390/rs14215302

Bar, S., Parida, B.R., Roberts, G., Pandey, A.C., Acharya, P., & Dash, J. (2021). Spatio-temporal characterization of landscape fire in relation to anthropogenic activity and climatic variability over the Western Himalaya, India. *GIScience & Remote Sensing*, 58(2), 281–299. https://doi.org/10.1080/15481603.2021.1879495

Bargali, H., Calderon, L.P.P., Sundriyal, R.C., & Bhatt, D. (2022). Impact of forest fire frequency on floristic diversity in the forests of Uttarakhand, western Himalaya. *Trees, Forests and People*, 9, 100300. https://doi.org/10.1016/j.tfp.2022.100300

Bargali, H., Singh, P., & Bhatt, D. (2020). Role of chir pine (*Pinus roxburghii* sarg.) in the forest fire of Uttarakhand Himalaya. *Envis Bulletin Himalayan Ecology*, 28, 82–85.

Basu, A., Pandey, A., Parmar, M., & Chauhan, S. (2019). Value added products from natural fibers of Indian Himalayan region. *International Journal of Engineering Research & Technology*, 8(3), 118–122. http://dx.doi.org/10.17577/IJERTV8IS030092

Bhardwaj, D.R., Kikon, G., Thakur, C.L., & Kumar, N. (2017). Effect of bamboo species and mulch materials on turmeric crop production and curcumin contents. *Indian Journal of Agroforestry*, 19(2), 13–22.

Bhardwaj, D.R., Tahiry, H., Sharma, P., Pala, N.A., Kumar, D., & Kumar, A. (2021). Influence of aspect and elevational gradient on vegetation pattern, tree characteristics and ecosystem carbon density in Northwestern Himalayas. *Land, 10*(11), 1109. https://doi.org/10.3390/land10111109

Bhatia, A.K., Sharma, K., & Sharma, D. (2020). Forest fire as an evil or necessity. *Journal of Pharmacognosy and Phytochemistry*, 9(5), 138–141.

Bisht, A.S., & Thakur, N.S. (2016, December). Pine needle biomass is a potential energy source for Himalayan region. In: *2016 7th India International Conference on Power Electronics, Delhi (IICPE)*. IEEE (Paper Presentation), India. https://doi.org/10.1109/IICPE.2016.8079505

Bisht, A.S., & Thakur, N.S. (2020). Pine needles biomass gasification based electricity generation for Indian Himalayan region: Drivers and barriers. In: Druck, H., Mathur, J., Panthalookaran, V., Sreekumar, V. (eds), *Green Buildings and Sustainable Engineering. Springer Transactions in Civil and Environmental Engineering*. Springer, Singapore. https://doi.org/10.1007/978-981-15-1063-2_4

Bowman, D.M.J.S., Balch, J.K., Artaxo, P., Bond, W.J., Carlson, J.M., Cochrane, M.A., D'Antonio, C., Defries, R., Doyle, J., Harrison, S., Johnston, F., Keeley, J., Krawchuk, M., Kull, C., Marston, J., Moritz, M., Prentice, I., Roos, C., Scott, A., & Pyne, S. (2009). Fire in the earth system. *Science*, 324(5926), 481–484. https://doi.org/10.1126/science.1163886

Champion, H.G., & Seth, S.K. (1968). *A Revised Forest Types of India*. Manager of Publications, Government of India, Delhi.

Chandran, M., Sinha, A.R., & Rawat, R.B.S. (2011, May 9–13). Replacing controlled burning practice by alternate methods of reducing fuel load in the Himalayan Long leaf Pine (*Pinus roxburghii* Sarg.) forests. In: *5th International Wildland Fire Conference, South Africa*, Sun City, South Africa.

Chauhan, L., Omre, P.K., Singh, T.P., Kumar, A., & Verma, A.K. (2018). Utilization of pine needles for development of biobased packaging material. *International Journal of Current Microbiology and Applied Sciences*, *7*(10), 1274–1279. https://doi.org/10.20546/ijcmas.2018.710.143

Chauhan, M., Gupta, M., Singh, B., Singh, A.K., & Gupta, V.K. (2012). Pine needle/isocyanate composites: Dimensional stability, biological resistance, flammability, and thermoacoustic characteristics. *Polymer Composites*, *33*(3), 324–335. https://doi.org/10.1002/pc.22151

Chawla, J.S. (1977). Studies on pine needle (leaves) fibre: *Pinus roxburghii*. *Indian Pulp & Paper Technical Association*, *14*(3), 205–208.

Chhetri, R., Toomsan, B., Kaewpradit, W., & Limpinuntana, V. (2012). Mass loss, nitrogen, phosphorus and potassium release patterns and non-additive interactions in a decomposition study of chir pine (*Pinus roxburghii*) and oak (*Quercus griffithii*). *International Journal of Agricultural Research*, *7*(7), 332–344. http://dx.doi.org/10.3923/ijar.2012.332.344

Choudhary, V., Patel, M., Pittman Jr, C.U., & Mohan, D. (2020). Batch and continuous fixed-bed lead removal using Himalayan pine needle biochar: Isotherm and kinetic studies. *ACS Omega*, *5*(27), 16366–16378. https://doi.org/10.1021/acsomega.0c00216

Davidenko, E.P., & Eritsov, A. (2003). The fire season in 2002 in Russia. Report of the Aerial Forest Fire Service, Avialesookhrana. *International Forest Fire News*, *28*, 15–17.

De Groot, R.S., Wilson, M.A., & Boumans, R.M.J. (2002). A typology for the classification, description and valuation of ecosystem functions, goods and services. *Ecological Economics*, *41*(3), 393–408. https://doi.org/10.1016/S0921-8009(02)00089-7

Dhaundiyal, A., & Gupta, V.K. (2014). The analysis of pine needles as a substrate for gasification. *Hydro Nepal: Journal of Water, Energy and Environment*, *15*, 73–81. https://doi.org/10.3126/hn.v15i0.11299

Dhaundiyal, A., & Tewari, P.C. (2016). Performance evaluation of throatless gasifier using pine needles as a feedstock for power generation. *Acta Technologica Agriculturae, 19*(1), 10–18. https://doi.org/10.1515/ata-2016-0003

Dinesh, Kumar, B., & Kim, J. (2023). Mechanical and dynamic mechanical behavior of the lignocellulosic pine needle fiber-reinforced SEBS composites. *Polymers*, *15*(5), 1225. https://doi.org/10.3390/polym15051225

Dobriyal, M.J.R., & Bijalwan, A. (2017). Forest fire in western Himalayas of India: A review. *New York Science Journal*, *10*(6), 39–46. https://doi.org/10.7537/marsnys100617.06

Dwivedi, D., Rathour, R.K., Sharma, V., Rana, N., Bhatt, A.K., & Bhatia, R. K. (2022). Co-fermentation of forest pine needle waste biomass hydrolysate into bioethanol. *Biomass Conversion and Biorefinery*, *14*, 1–13. https://doi.org/10.1007/s13399-022-02896-1

Dwivedi, R.K., Singh, R.P., & Bhattacharya, T.K. (2016). Studies on bio-pretreatment of pine needles for sustainable energy thereby preventing wild forest fires. *Current Science*, *111*(2), 388–394. https://doi.org/10.18520/cs/v111/i2/388-394

Gairola, S., Gairola, S., Sharma, H., & Rakesh, P. K. (2019, February 18–22). Impact behavior of pine needle fiber/pistachio shell filler based epoxy composite. In: *Journal of Physics: Conference Series 2nd International Conference on New Frontiers in Engineering, Science & Technology* (NFEST), Kurukshetra, Haryana, India. http://dx.doi.org/0.1088/1742-6596/1240/1/012096

Gupta, M., Chauhan, M., Khatoon, N., & Singh, B. (2010). Composite boards from isocyanate bonded pine needles. *Journal of Applied Polymer Science*, *118*(6), 3477–3489. https://doi.org/10.1002/app.32703

ISFR. (2021). *Indian State of Forest Report-201*. Forest Survey of India, Dehradun, India.

Jan, A., Shah, J., Khan, F.U., & Rahman, A.U. (2006). Investigation of pine needles for pulp/paper industry. *Biological Sciences-PJSIR*, *49*(6), 407–409.

Joshi, A.K., & Joshi, P.K. (2019). Forest ecosystem services in the central Himalaya: Local benefits and global relevance. *Proceedings of the National Academy of Sciences, India Section B: Biological Sciences*, *89*, 785–792. https://doi.org/10.1007/s40011-018-0969-x

Joshi, G., & Negi, G.C. (2011). Quantification and valuation of forest ecosystem services in the western Himalayan region of India. *International Journal of Biodiversity Science, Ecosystem Services & Management*, *7*(1), 2–11. https://doi.org/10.1080/21513732.2011.598134

Joshi, K., Sharma, V., & Mittal, S. (2015). Social entrepreneurship through forest bioresidue briquetting: An approach to mitigate forest fires in Pine areas of Western Himalaya, India. *Renewable and Sustainable Energy Reviews*, *51*, 1338–1344. https://doi.org/10.1016/j.rser.2015.07.057

Kala, L.D., & Subbarao, P.M.V. (2018). Estimation of pine needle availability in the Central Himalayan state of Uttarakhand, India for use as energy feedstock. *Renewable Energy*, *128*, 9–19. https://doi.org/10.1016/j.renene.2018.05.054

Kumar, A., Deshmukh, R.K., & Gaikwad, K.K. (2022). Quality preservation in banana fruits packed in pine needle and halloysite nanotube-based ethylene gas scavenging paper during storage. *Biomass Conversion and Biorefinery*, *14*, 1–10. https://doi.org/10.1007/s13399-022-02708-6

Kumar, A., Gupta, V., Singh, S., Saini, S., & Gaikwad, K.K. (2021). Pine needles lignocellulosic ethylene scavenging paper impregnated with nanozeolite for active packaging applications. *Industrial Crops and Products*, *170*, 113752. https://doi.org/10.1016/j.indcrop.2021.113752

Kumar, A., & Randa, R. (2014). Experimental analysis of a producer gas generated by a chir pine needle (leaf) in a downdraft biomass gasifier. *International Journal of Engineering Research and Applications*, *4*(10), 122–130.

Kumar, D., Bhardwaj, D.R., Sharma, P., Bharti, Sankhyan, N., Al-Ansari, N., & Linh, N.T.T. (2022a). Population dynamics of *Juniperus macropoda* Bossier forest ecosystem in relation to soil physico-chemical characteristics in the cold desert of North-Western Himalaya. *Forests*, *13*(10), 1624. https://doi.org/10.3390/f13101624

Kumar, D., Bhardwaj, D.R., Thakur, C.L., Sharma, P., & Ayele, G.T. (2022b). Vegetation shift of *Juniperus macropoda* Boisser forest in response to climate change in North-Western Himalayas, India. *Forests*, *13*(12), 2088. https://doi.org/10.3390/f13122088

Kumar, D., Thakur, C.L., Bhardwaj, D.R., Sharma, N., Sharma, P., & Sankhyan, N. (2022c). Biodiversity conservation and carbon storage of *Acacia catechu* Willd. Dominated northern tropical dry deciduous forest ecosystems in north-western Himalaya: Implications of different forest management regimes. *Frontiers in Environmental Science*, *10*, 1638. https://doi.org/10.3389/fenvs.2022.981608.

Kumar, S., & Kumar, A. (2022). Hotspot and trend analysis of forest fires and its relation to climatic factors in the western Himalayas. *Natural Hazards*, *114*, 3529–3544. https://doi.org/10.1007/s11069-022-05530-5

Kumar, V., Nanda, M., Verma, M., & Singh, A. (2018). An integrated approach for extracting fuel, chemicals, and residual carbon using pine needles. *Biomass Conversion and Biorefinery*, *8*, 447–454. https://doi.org/10.1007/s13399-018-0304-z

Lal, P.S., Sharma, A., & Bist, V. (2013). Pine needle – an evaluation of pulp and paper making potential. *Journal of Forest Product and Industries*, *2*(3), 42–47.

Langat, D.K., Maranga, E.K., Aboud, A.A., & Cheboiwo, J.K. (2016). Role of forest resources to local livelihoods: The case of east mau forest ecosystem, Kenya. *International Journal of Forestry Research*, 10. https://doi.org/10.1155/2016/4537354

Minhas, A., & Banshtu, R.S. (2022). Pine needle gasification: An approach towards renewable energy production and life cycle assessment of chir pine. *International Journal for Research in Applied Science & Engineering Technology*, *10*(8), 1381–1387. https://doi.org/10.22214/ijraset.2022.46427

Ministry of Environment, Forest and Climate Change, Government of India; World Bank. (2018). *Strengthening Forest Fire Management in India*. World Bank, Washington DC. https://doi.org/10.1596/30013

Negi, G.C.S. (2019). Forest fire in Uttarakhand: Causes, consequences and remedial measures. *International Journal of Ecology and Environmental Sciences*, *45*(1), 31–37. https://nieindia.org/Journal/index.php/ijees/article/view/1578/480

Negi, G.C.S., Rawal, R.S., Dhyani, P.P., & Palni, L.M.S. (2012). Twenty priority issues for forestry research with particular reference to Indian Himalayan region in the RIO+ 20 era. *Glimpses of Forestry Research in the Indian Himalayan Region*. Envis Centre on Himalayan Ecology, G.B. Pant Institute of Himalayan Environment & Development Kosi-Katarmal, Almora, 1.

Panda, S., Bhardwaj, D.R., Sharma, P., Handa, A.K., & Kumar, D. (2021). Impact of climatic patterns on phenophase and growth of multi-purpose trees of north-western mid-Himalayan ecosystem. *Trees, Forests and People*, *6*, 100143. https://doi.org/10.1016/j.tfp.2021.100143

Panda, S., Bhardwaj, D.R., Thakur, C.L., Sharma, P., & Kumar, D. (2022). Growth response of seven multipurpose tree species to climatic factors: A case study from northwestern Himalayas, India. *Journal of Forest Science*, *68*(3), 83–95. https://doi.org/10.17221/159/2021-JFS

Parashar, A., & Biswas, S. (2003, September 21–28). The impact of forest fire on forest biodiversity in the Indian Himalayas (Uttaranchal). In: *XII World Forestry Congress* (Paper presentation), Quebec, Canada.

Pew, K.L., & Larsen, C.P.S. (2001). GIS analysis of spatial and temporal patterns of human-caused wildfires in the temperate rain forest of Vancouver Island, Canada. *Forest Ecology and Management*, *140*(1), 1–18. https://doi.org/10.1016/S0378-1127(00)00271-1

Rana, A.K., Guleria, S., Gupta, V.K., & Thakur, V.K. (2023). Cellulosic pine needles-based biorefinery for a circular bioeconomy. *Bioresource Technology*, *367*, 128255. https://doi.org/10.1016/j.biortech.2022.128255

Reddy, C.S., Jha, C.S., Diwakar, P.G., & Dadhwal, V.K. (2015). Nationwide classification of forest types of India using remote sensing and GIS. *Environmental Monitoring and Assessment*, *187*, 1–30. https://doi.org/10.1007/s10661-015-4990-8

Roy, M., & Kundu, K. (2023). Production of biochar briquettes from torrefaction of pine needles and its quality analysis. *Bioresource Technology Reports*, *22*, 101467.

Sanwal, C.S., Kumar, R., & Bhardwaj, S.D. (2016). Integration of *Andrographis paniculata* as potential medicinal plant in chir pine (*Pinus roxburghii* Sarg.) plantation of North-Western Himalaya. *Scientifica*, 2016(1), 2049532. https://doi.org/10.1155/2016/2049532

Sengar, A., Sharma, V., Agrawal, R., Dwivedi, A., Dwivedi, P., Joshi, K., & Barthwal, M. (2020). Prioritization of barriers to energy generation using pine needles to mitigate climate change: Evidence from India. *Journal of Cleaner Production*, *275*, 123840. https://doi.org/10.1016/j.jclepro.2020.123840

Sengar, A., Sharma, V., Joshi, K., Agrawal, R., Dwivedi, A., Dwivedi, P., & Barthwal, M. (2022). A fuzzy Analytic Hierarchy Process based analysis for prioritization of enablers to pine briquettes-based energy generation in alignment with the United Nations' sustainable development goals: Evidence from India. *Biomass and Bioenergy, 165*, 106580. https://doi.org/10.1016/j.biombioe.2022.106580

Shah, N., Dhiman, I.R., & Verma, I.R. (2022) Utilisation of Pine needles (*Pinus roxburghii*) to reduce the forest fire in Nowshera Forest Division, J&K. *Epitome: International Journal of Multidisciplinary Research, 8*(12), 12–18.

Sharma, U., Bhardwaj, D.R., Sharma, S., Sankhyan, N., Thakur, C.L., Rana, N., & Sharma, S. (2022). Assessment of the efficacy of various mulch materials on improving the growth and yield of ginger (*Zingiber officinale*) under bamboo-based agroforestry system in NW-Himalaya. *Agroforestry Systems, 96*(5–6), 925–940. http://dx.doi.org/10.1007/s10457-022-00753-8

Singh, J.S., Rawat, Y.S., & Chaturvedi, O.P. (1984). Replacement of oak forest with pine in the Himalaya affects the nitrogen cycle. *Nature, 311*(5981), 54–56. http://dx.doi.org/10.1038/311054a0

Singh, S., & Vinay, D. (2023). Preparation and properties of particle board from dry pine needles. *The Pharma Innovation Journal, 3*(6), 173–176.

Singh, T.P., & Kaur, L. (2018). Effect of operational parameters on properties of pine needle cattle dung briquettes. *Agricultural Engineering Today, 42*(2), 28–36.

Sinha, P., Mathur, S., Sharma, P., & Kumar, V. (2016). Potential of pine needles for PLA-based composites. *Polymer Composites, 39*(4), 1339–1349. https://doi.org/10.1002/pc.24074

Tonini, M., D'Andrea, M., Biondi, G., Degli Esposti, S., Trucchia, A., & Fiorucci, P. (2020). A machine learning-based approach for wildfire susceptibility mapping. The case study of the Liguria region in Italy. *Geosciences, 10*(3), 105. https://doi.org/10.3390/geosciences10030105

Vaid, S., Nargotra, P., & Bajaj, B.K. (2018). Consolidated bioprocessing for biofuel-ethanol production from pine needle biomass. *Environmental Progress & Sustainable Energy, 37*(1), 546–552. https://doi.org/10.1002/ep.12691

Varma, A. K., & Mondal, P. (2018). Pyrolysis of pine needles: effects of process parameters on products yield and analysis of products. *Journal of Thermal Analysis and Calorimetry, 131*, 2057–2072. http://dx.doi.org/10.1007/s10973-017-6727-0

15 Waste Management Practices of Riverside Communities at the Limbang River Sarawak, Malaysia

Siti Noor Aliza Apandi and Norli Ismail

15.1 INTRODUCTION

In today's rapidly growing world, waste management has become a critical issue for communities worldwide. The effective management of waste is crucial not only for environmental sustainability but also for public health and the overall well-being of communities. This chapter provides an overview of waste management practices specifically implemented in riverside communities. By examining the unique challenges and strategies employed by these communities, we can gain valuable insights into sustainable waste management practices that can be applied in similar settings.

Malaysia is facing a significant challenge in dealing with urbanization and modernization; as the population increases, more waste is deposited into the environment. Mixed paper, plastic, wood, and metals waste are the types of waste generated and disposed of, or so-called municipal solid waste (MSW) (Jayawardhana et al., 2016), generated by Malaysian households (Cheng et al., 2022). Humans use river water for consumptive and non-consumptive purposes (Abdullah, 2012). Consumptive use of river water is included irrigation in agriculture, and water supply, while non-consumptive use is as a transportation path, producing hydroelectric dams and other recreational activities by humans (Goi, 2020). Riverbank communities use both consumptive and non-consumptive uses of the river water daily, such as washing, swimming, bathing, and aquaculture (Wagiono et al., 2022). Domestic sewage, agriculture runoff, and industrial waste are among the leading causes of changes in river water quality (Gyawali et al., 2012; Suthar et al., 2010). Pathogenicity, chemical composition, contaminants, and inefficient sewage systems among riverside communities result in human waste to be disposed into the river, which causes degraded river water quality (Pandey et al., 2014; Picardal & Marababol, 2012).

The catastrophe of using the river water for a bath and swimming while using an active direct toilet in the Sarawak River is poorly understood and under-reported.

DOI: 10.1201/9781003543176-15

Installing the complete toilet with a septic system in stilt houses is laborious, costly, and time-consuming compared to landed homes due to the septic tank's location on the land, which is away from the houses themselves. The early prediction is that septic tanks are only used by households with higher monthly incomes because the price of septic tanks can be extremely expensive. However, improper sewage systems produce olfactory pollution, introducing life-threatening pathogens carried into the river. This poses a risk to humans, particularly children under five years, who might come into contact with the contaminated river water through bathing and swimming. The second initial prediction of this study included fewer healthy children in the downstream (DS) zone. It is crucial to know the status of the human waste system in a community and its relationship with human health and the environment near the river to improve awareness among the community and have proper waste management.

This study's objective is to better understand the relationship between the riverside community and waste management practices by determining the socio-demographic profile, river water utilization, and waste disposal practices at the Limbang River.

15.2 METHODS

15.2.1 STUDY AREA

The study was conducted from August to September 2022 at Kampung Seberang Kedai (KSK), a village located on the riverside of the Limbang River. The KSK, Sarawak, Malaysia, lies between latitude 4°45′9.96″N and longitude 115°0′23.81″E, and has a length of 1.15 km. KSK is one of the river bank communities in the Limbang River, which has an area of 0.10 km².

Limbang River is located in the northern part of Sarawak near the Brunei border of west and east and is approximately 88 km long and 150 m wide, making it the third longest gazette river in the state of Sarawak, and it drains into the South China Sea. The main water supply at Limbang comes from the Lembaga Air Kawasan Utara (LAKU), where raw water sources come from the Berawan and the Pandaruan rivers (LAKU, 2022). The KSK has a total of 116 houses and a population of 903. Three zones were divided as the study sites: DS, midstream (MS), and upstream (US). The KSK has a 5- to 7-m elevation from sea level and is surrounded by palm and mangrove trees. The KSK is only accessible by boat during the high tide, and there is only a 2 ft wide concrete bridge connecting the US to DS zones.

15.2.2 RESPONDENTS

A total of 45 representatives from 45 houses were selected, specifically having at least 1 child younger than 5. The respondents were 15 men and 30 women. The questionnaires contained details about the community household waste management, sanitation, hygiene practices, and health status. Most respondents were women, and most were homemakers who look after their children at home while the men go to work. Of the respondents, 33.3% finished secondary school, and no respondent continued their study at higher learning.

15.2.3　Data Collections

This study's qualitative data collection methods used modified socio-health survey (SHS) questions, as Sanchez et al. (2022) suggested. The questionnaire consisted of three parts: respondent's background, waste management, and health status. The respondent's backgrounds included age, gender, educational background, monthly household income, religion, race, and nationality.

The waste management part asked about the toilet type, their hygiene after visiting the toilet, and waste disposal practices. The last part is the children's health status, including the health impact of hygiene practices and behaviors. Before the survey, a short meeting with the village chief was held on August 2022 to gain information on the current waste management system in the KSK and the effort to have proper waste management in the KSK.

15.2.4　Data Analysis

The collected data were tabulated by frequencies and percentages in pie charts and bar charts. The results were further analyzed using a correlation test at a 95% confidence level using the statistical package for the social sciences (SPSS) software by International Business Machines (IBM) version 27.

15.2.5　Ethical Considerations

Before the survey, all respondents were asked to consent to participate in this study. No respondents were forced to participate in this study and all respondents were allowed to withdraw at any time they wanted without penalty. The name and house location were not to be published in any journal, article, presentation, or publication to protect the respondent's identity and privacy without prior permission. The outcomes of this study determined the status of waste management at the KSK and eventually will help the low-income community to have a better life quality by using the study outcomes to propose the dumpster and scheduled waste collection from the city council in the future.

15.3　RESULTS AND DISCUSSION

15.3.1　Demographic Profile of Respondents

The demographic profile of respondents along the riverside of the Limbang River is presented in Table 15.1. Accommodations such as a mosque, retail shops, tailors, and assembly area are located at the central of the KSK, clear evidence that the MS zone is the management center of the KSK.

In all these 3 study sites, the gender is similar in frequency; 5 are male, and 10 are female. The respondents in the 26–30 year group had completed their secondary

TABLE 15.1
Demographic profile of representatives from houses selected along the Limbang River

Demographic profile		Downstream (DS) $n = 15$		Midstream (MS) $n = 15$		Upstream (US) $n = 15$	
		Frequency	(%)	Frequency	(%)	Frequency	(%)
Age	26–30 years old	5	33.3	3	20.0	2	13.3
	31–35 years old	1	6.7	0	0.0	0	0.0
	36–40 years old	1	6.7	0	0.0	0	0.0
	41–45 years old	4	26.7	3	20.0	3	20.0
	46–50 years old	3	20.0	2	13.3	2	13.3
	>51 years old	1	6.7	7	46.7	8	53.3
Gender	Male	5	33.3	5	33.3	5	33.3
	Female	10	66.7	10	66.7	10	66.7
Household monthly	RM0–1,000	4	26.7	8	53.3	9	60.0
income	RM1,001–2,000	8	53.3	6	40.0	6	40.0
	RM2,001–3,000	3	20.0	1	6.7	0	0.0
Highest educational	PMR	5	33.3	3	20.0	6	40
background	SPM	5	33.3	4	26.7	6	40
	STPM/DIPLOMA	1	6.7	2	13.3	0	0.0
	No Formal education	0	0.0	1	6.7	1	6.7
	Others (UPSR)	4	26.7	5	33.3	2	13.3

TABLE 15.2
Interview outcomes with the KSK village chief

No	Questions	Answers
1	What is the main water supply at the KSK?	Our water source is entirely from LAKU.
2	How about waste management here in the KSK?	Some people dump the waste directly into the river; some are burned and gathered, then transported across the river into the dumpster.
3	How about the household sewerage system?	Most people here have free septic tanks from hospitals, and others make their own septic tank from concrete, and the rest have no septic tank, which the human waste directly into the river.
4	What is the main occupation of people here?	Primarily the KSK people work as charter boat drivers, and some are fishermen. The number of fishermen is decreasing due to the reduction of total fish catch.
5	Is there any ongoing effort that has been made to improve waste management here?	Yes. We are currently trying to appeal for free septic tanks for all houses here and a green dumpster in the backyard, but there is still no feedback from the government or the Ministry of Health, Malaysia.

school, and only 6% pursued their studies at higher learning. The variables of age groups and highest education levels are closely related, where the younger generation had a proper education compared to the older generation.

The average household member in each household is five persons with a median monthly household income in the range of RM0–RM1,000. Their source of income is mostly from charter boat drivers, selling the fish catch, making the fishing gear, making cakes, sewing cloth, and getting paid for removing anchovies' gut. Others only rely on financial aid from the government, which is RM300–RM500 per month. Overall, all the KSK people fall under the bottom 40% (B40) groups. B40 groups have a monthly household income of less than RM4,849, and 2.91 million households in Malaysia fall into this category (DOSM, 2020).

The interview outcomes with the village chief, shown in Table 15.2, clarify that the primary occupation of the KSK people is charter boat drivers and fishermen. The village chief also mentioned that most septic tanks in the KSK were donated by the Ministry of Health, while the rest either use self-made septic tanks or do not own any. There are household waste disposal practices at the KSK: disposing into the river, burning, and disposing into the dumpster across the river.

15.3.2 Anthropogenic Activity in the Limbang River

The bar chart in Figure 15.1 shows the activities of the KSK people in the Limbang River. Of the respondents, 73.3% practiced dumping household waste into the river, while the other 26.7% collected their garbage and dumped into the dumpster across

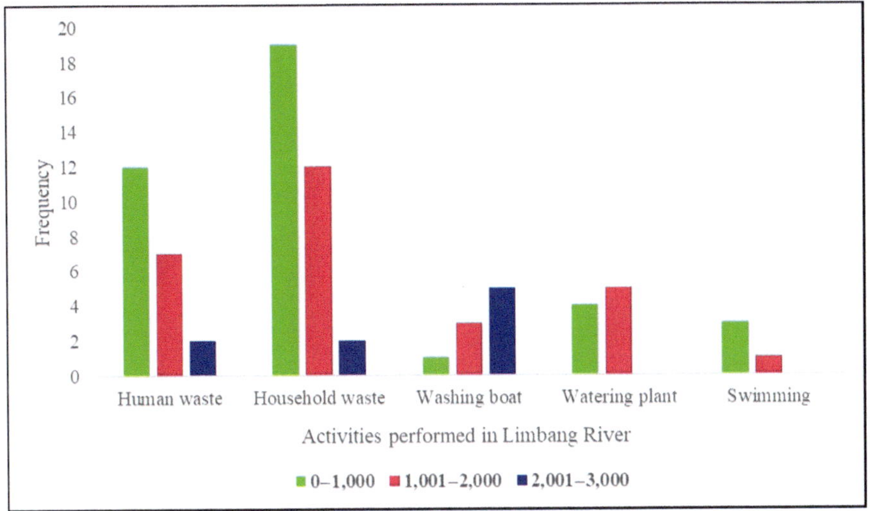

FIGURE 15.1　Activities performed by the KSK people in Limbang River.

the river or burned it down in the backyard. In the human waste disposal system, 53.3% have a septic tank installed in their toilet; however, 46.6% of respondents have an improper sewage system.

The riverside water is accessible for agricultural and domestic purposes (Picardal & Marababol, 2012); one of the domestic purposes is watering the plant, which has become the KSK people's daily routine. Only 20% of respondents come from a monthly household income group of RM0–1,000 and RM1,001–RM2,000 using the river water for their plants, while no data were recorded under an income group above RM2,000. Less than 25% of the KSK people use river water to wash the boats. Washing boats are performed by the boat owner, which can be seen in Figure 15.1, performed mainly by the income group of RM2,001–RM3,000. Swimming in the river is a less performed activity done at Limbang River. Only 8.8% of respondents swam recreationally in the Limbang River, most coming from the MS zone. Respondents from DS and US admitted they used to swim and bathe in the river but are no longer doing so because of crocodile threat and the foul smell of the river. The olfactory pollution is caused by the pile of floating plastics containing biological matter mixed with domestic wastewater and human waste stacked up between the house's stilts. The smell worsens when it is low tide, and the river does not wash the human fecal material. The average low tide in Limbang River is 0.51 m above sea level, while the high tide is 2.07 m (Tideschart, 2022). The low tides make some houses in the KSK inaccessible by boat.

Human waste disposal practice by zones in the KSK is depicted in Figure 15.2. where the people in the MS zone have more septic tanks compared to the US and DS zones, with 43% and 38%, respectively, human waste disposed directly into the river. The availability of septic tanks at a respondent's home is related to the monthly household income, where a higher number of people under the monthly household

Human waste disposal	Zones		
	DS	MS	US
Into septic tank	7	11	6
Direct to the river	8	4	9
Total	15	15	15

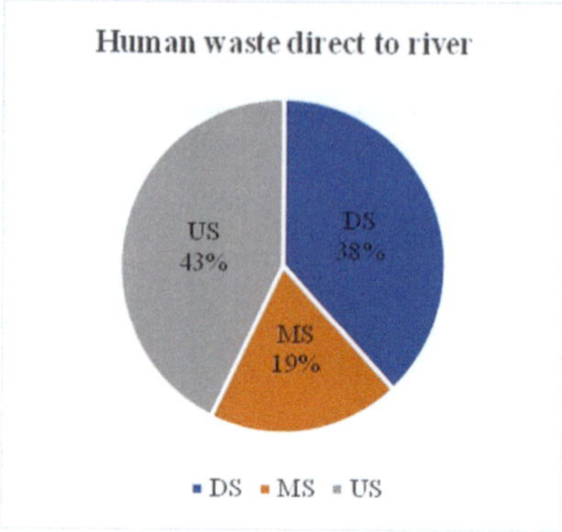

FIGURE 15.2   Human waste disposal practice by zones.

income group of RM2,001–RM3,000 has installed the proper septic tank at their residence. Septic tanks made from cement or polyethene, where the wastewater from household flow into, are a common type of septic tank that allows the biological matter to decompose by microbial agents before reaching groundwater (EPA, 2001). However, septic systems are a prevalent source of groundwater contamination, which could deteriorate the river water quality and increase waterborne disease outbreaks and other health issues (Sangodoyin, 1993). More than half the population in the KSK have septic tanks; however, some respondents claimed they had made the septic tank by themselves using concrete. The condition of self-made septic tanks in the KSK is concealed, which could be improper design, faulty, and incorrect operation and lead to groundwater contamination (EPA, 2001).

15.3.3   HOUSEHOLD DRINKING WATER SUPPLY

The main water supply in the KSK is from LAKU, and they admitted to using another water source from collected rainwater when there is a water supply interruption. The LAKU raw water sources are monitored periodically for contamination to protect the water quality (Borneo Post Online, 2020).

The drinking water supply, as illustrated in Figure 15.3, shows the boiled tap water is the most common drinking water among the KSK people, where 64% of the KSK people rely on LAKU tap water for the drinking water source, prevalently among the monthly income group of RM0–RM1,000. However, boiled tap water does not remove the chlorine and heavy metal contained in water. Chlorination is the standard method to prevent bacterial contamination; chlorine concentration is strictly maintained at 0.2–0.5 mg/L (Qaiser et al., 2014). Drinking water quality standards set a threshold of total coliform of water sample should be 0 in 100 ml (MOH, 2016). Chlorine and foreign substances can be removed by a nanowater filter (Maxwell et al., 2021). Twenty-nine percent of the respondents have installed water filters in their homes, and only 7% drink bottled mineral water. A group of monthly households with income between RM2,001 and 3,000 fully installed the nanowater filter system at their home.

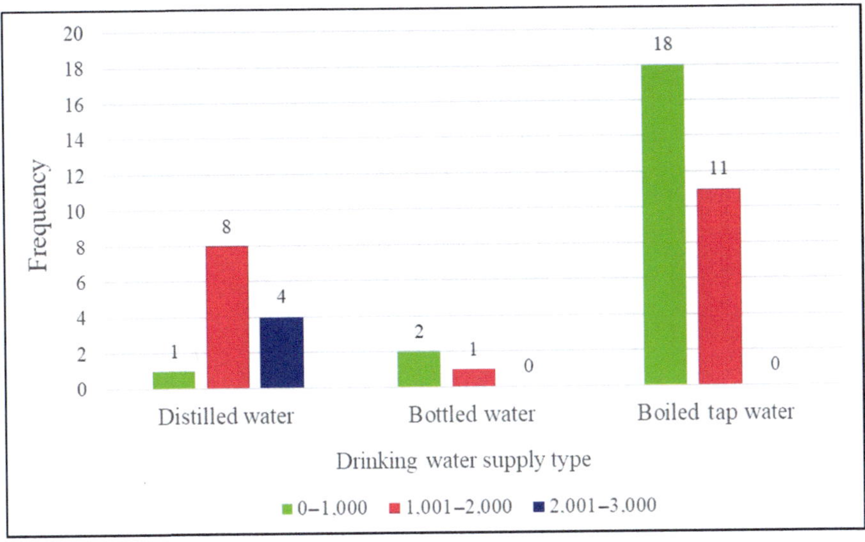

FIGURE 15.3 Drinking water supply in the KSK household.

The drinking water hygiene also depends on the storage method. The filtered water has filtered out the antibacterial agents, which makes the water viable for bacterial growth. It is recommended to store the filtered water in the refrigerator below 4°C to slow bacterial growth (Wiebe et al., 1992). The drinking water hygiene also depends on the storage method. The filtered water has filtered out the antibacterial agents, which makes the water viable for bacterial growth. It is recommended to store the filtered water in the refrigerator below 4°C to slow bacterial growth (Hess, 2018).

15.3.4 HOUSEHOLD WASTE MANAGEMENT

Household waste in the KSK includes municipal waste, and domestic waste (kitchen waste) is dumped directly into the river. According to Radhi (2020), the Malaysian waste composition is presented in Figure 15.4. Almost half of Malaysian and other countries' waste is from food waste, which is unsustainably disposed into the environment (Nguyen et al., 2022).

The practice of household waste disposal in the KSK is depicted in Figure 15.5. The absence of waste management in the KSK has led to the inevitable practice of dumping household waste into the river, burning, or transporting it across the river regardless of the type of waste. The study found that only 5% of the KSK waste was disposed into the dumpster provided by the city council. Undoubtedly it is much easier and more convenient to dump the waste into the river or burn it than throw it across the river, which has a regular garbage collection. More than 70% of the KSK household waste continues to deposit into the river, while the 22% is burned on the land. However, continuing the unsustainability of waste disposal into the river will affect the DS zone community's environment, where more waste will linger between

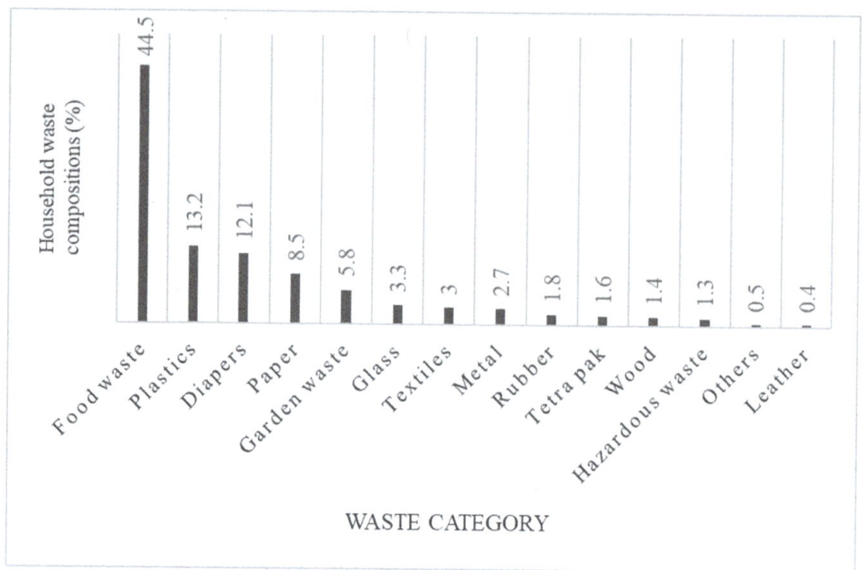

FIGURE 15.4 Malaysian waste compositions.

Source: Radhi (2020).

Household waste disposal practice	Response	Household waste
Disposed into the river	33 (73%)	
Burn	10 (22%)	
Disposed into the dumpster across the river	2 (5%)	
Total	**45 (100%)**	

FIGURE 15.5 Household waste management responses from 45 houses.

the house's stilts in the future. According to the village chief, the city council has not granted the placement of a dumpster near the KSK because the new road is still under construction by Northern Regional Development Agencies (NRDA), which has granted by Deputy Chief Minister of Sarawak (Ismail, 2020). The 3.7 km new road will connect the KSK to Bukit Lubok and is expected to be complete by the end of 2023 (RECODA, 2022).

15.3.5 HEALTH STATUS AND IMPACT OF THE UNSUSTAINABILITY WASTE DISPOSAL IN THE RIVER

The health status of children who live in DM, MS, and the US is illustrated in Figure 15.6. This study revealed that there are higher percentages of healthy children who live in the MS zone, and fever is the most common health issue among children, particularly in the DS region; the possible causes of these issues are related to activities, unsustainability, waste disposal practice, and sanitation access. The health issues worsen when there is a disruption water supply. At DS and US, more houses are not equipped with septic tanks compared to MS. The distribution of septic tanks donated by MOH seems to be focused on the MS zone (central area); as shown in Figure 15.6, 80% of human waste disposal are actively practiced at DS and US zones. In general, children under 5 are the sensitive group to environmental pollution of air, water, soil, and chemical pollution (Landrigan et al., 2019). The most common health issue among the sensitive group recorded in this study was fever. Around 46% of children in the KSK have had a fever in the past 3 months, where 20% come from DS, 15% from the US, and 11% from MS. Besides fever, the contaminated river also can lead to skin diseases, waterborne diseases, and gastrointestinal infections (Haldar et al., 2022; Khan et al., 2018).

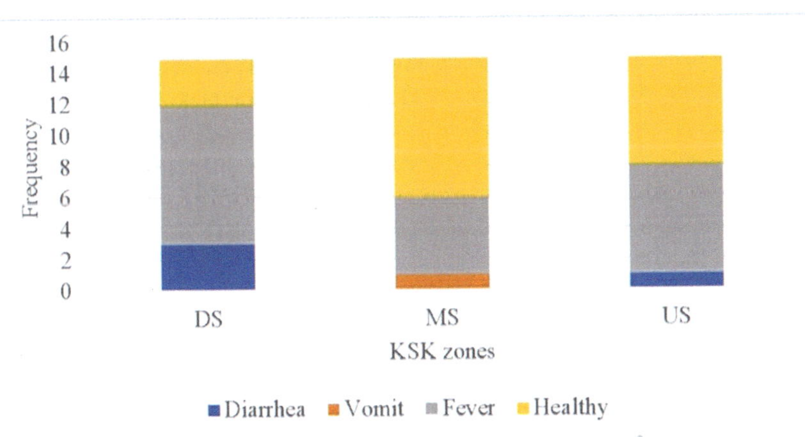

	Downstream (DM)	Midstream (MS)	Upstream (US)
Diarrhea	3	0	1
Vomit	0	1	0
Fever	9	5	7
Healthy	3	9	7
Total	15	15	15

FIGURE 15.6 Health issues occur among children under 5 years old at the DS, MS, and US zones in the KSK.

The river flow also plays a vital role in determining the accumulation of human and household waste in the DS zone. Over 13% of Malaysian household waste is plastics (Radhi, 2020). In time, plastics will degrade into smaller plastic called microplastic, which takes thousands of years to decompose (Kamsuri et al., 2022). In other parts of the Southeast Asian River, such as the Cisadane River, Indonesia, and Tebrau River, Malaysia, there is a higher abundance of microplastic DS than midstream (Sarijan et al., 2018; Sulistyowati et al., 2022). Fisheries in the Limbang River have slowed down over the past few years due to the decrement in the number of fish caught. Microplastic could be the culprit for this phenomenon, in which fish often mistake microplastic for the source of food (Wang et al., 2020), which will bring serious health problems to marine animals and eventually reduce their population.

Anthropogenic activities that have affected the river quality (Sidabutar et al., 2017) seem likely unavoidable. The Limbang River has become the KSK people's primary source of income for transportation and fishing. During the pandemic of COVID-19, the closure of the Brunei border (Laeng, 2022) left the Limbang people no choice but to commute on the Limbang River by water transportation either to Labuan or Lawas. The Limbang River quality currently falls under the class II category, clean with a WQI of 81 (DOE, 2018). However, if the practice of household waste is disposed of unsustainably into the Limbang River, the river water quality will eventually deteriorate.

15.3.6 Variables Correlation

The correlation constructs are presented in Table 15.3. There is negative correlation between monthly household incomes and owning a septic tank. One possible interpretation of this case is that the donation of septic tanks by MOH, as the village chief claimed, is inadequate for all houses in the KSK. Installing the septic tank is laborious and expensive—the price of one tank ranges from RM529.34 to RM4,562.86, depending on the material and size of the septic tank (Jomprice, 2022).

The household income does not affect the decision to install a septic tank or not, and the correlation between MI and WM shows a moderate positive correlation. The

TABLE 15.3

Correlations constructs for monthly household income (MI), human waste system (WS), household waste management (WM), and location (LC)

Correlated constructs	r-values	p-value
Monthly household income (MI) and human waste system (WS)	−0.143	0.348
Monthly household income (MI) and household waste management (WM)	0.355	0.017*
Location (LC) and human waste system (WS)	0.055	0.722
Location (LC) and household waste management (WM)	−0.369	0.013*
Location (LC) and health issue (HI)	0.251	0.096

* *p*-Values at 95% confidence level.

group with a household income between RM2,001 and RM3,000 owns the boat, enabling them to transport the waste and dispose of it into the dumpster across the river. LC and WS have a weak positive correlation, as only 53.3% of respondents have septic tanks. As for LC and WM, it has a robust negative correlation; the interpretation from this is that all respondents from any zones have similar waste disposal practices. LC and HI also have a moderate positive correlation, as DS zone children report more health issues than MS and US.

15.3.7 COMPARISON OF THE KSK COMMUNITIES' HEALTH IMPACT WITH OTHER RIVERSIDE DWELLERS IN SOUTHEAST ASIAN COUNTRIES

This study examined the waste management practice in the KSK by considering the demographic and location of the respondents at the river. Three other rivers from another journal article on waste management practice were deliberately selected to be compared with the KSK results. The summary comparison on the KSK waste management practice with other riverside dwellers from Martapura River, Indonesia; Butuanon River, Philippines; and Sapangdaku River, Philippines, by taking into account the number of respondents, gender, household income, and waste management practices is shown in Table 15.3. The result shows the outcomes are similar to the Butuanon River community, where more than 70% of the household waste is deposited into the river due to fewer disposal methods available. The Martapura and Sapangdaku river communities show that less than half its riverside dwellers choose not to dump the waste into the river.

According to Sanchez et al. (2022), the Sapangdaku riverside dwellers have midstream garbage collection, which covers more than half of the waste produced by its people. However, 98% of the Sapangdaku riverside community admitted that the river is the best site to dump waste. On the other hand, the Martapura riverside community waste management practice studied by Hanafi et al. (2018) has four methods of disposal; dump into the river, burn, landfill, and bury in the ground. About 47.5% of waste were dumped into the river, prevalently among the groups who live between 101 and 200 m (51.7%), and more than 200 m (77.9%) away from landfill choose to dump the waste into the river. The only thing that these four riverside communities had in common are the monthly household income. Most riverside communities for all four rivers fall into the lower income proportion as shown in Table 15.4.

15.3.8 CHALLENGES IN WASTE MANAGEMENT PRACTICES OF RIVERSIDE COMMUNITIES

Limited Space: Riverside communities often face challenges in terms of limited space for waste disposal. Due to their location near water bodies, these communities may have restricted land availability for waste management facilities, such as landfill sites or recycling centers. This limitation necessitates innovative and space-efficient waste management strategies. Environmental Impact: Improper waste management practices in riverside communities can have severe environmental consequences. Waste, if not effectively managed, can contaminate rivers and other water bodies,

TABLE 15.4

Summary comparison of the KSK household waste management with other journal articles on another part of Southeast Asian Countries

Authors	Title	Location	No. of households interviewed	Respondents (%)		Household income		Waste disposal practice directly into the river (%)
				Male	Female	Income	%	
Hanafi et al. (2018)	"Household Waste Management among Riverside Communities and other Determinants"	Martapura River, Banjarmasin, Indonesia	784	41.8	58.2	Lower: < RM648.88 Upper: ≥ RM648.88	64.3 35.7	47.3
Sanchez et al. (2022)	"Water Management Practices and Environmental Attitudes of Riparian Communities in Sapangdaku River, Cebu Island, Philippines"	Sapangdaku River, Toledo, Philippines	120	52.5	47.5	Lower: < RM1,792.13 Upper: ≥ RM1,79213	89.2 10.8	43.0
Picardal and Marababol (2012)	"Impacts of Waste Disposal Practices and Water Utilization of Riverside Dwellers on Physicochemical and Microbiological Properties of Butuanon River, Central Visayas"	Butuanon River, Cebu, Philippines	90	48.9	51.1	Lower: < RM641.00 Upper: ≥ RM641.00	62.2 37.8	81.7
This Study		Limbang River, Sarawak, Malaysia.	45	33.3	66.7	Lower: < RM2,000 Upper: ≥ RM2,000	91.11 0.08	73.0

causing water pollution and endangering aquatic ecosystems. It can also contribute to air pollution, soil degradation, and the release of greenhouse gases, worsening climate change.

Flooding and Erosion: Riverside communities are susceptible to flooding and erosion, which can further complicate waste management efforts. During floods, improperly disposed waste can be washed into rivers, increasing the risk of pollution. Erosion can also destabilize waste management infrastructure and disposal sites, leading to potential environmental hazards. Lack of Infrastructure: Some riverside communities, particularly those in developing regions, may lack the necessary waste management infrastructure. Inadequate waste collection systems, recycling facilities, and treatment plants make it challenging to manage waste effectively. Limited access to waste management services further increases the likelihood of improper waste disposal practices.

Public Awareness and Participation: Promoting public awareness and active participation in waste management practices can be a significant challenge. Riverside communities often consist of diverse populations, including permanent residents, tourists, and transient individuals. Engaging and educating these groups about proper waste disposal methods, recycling, and the importance of environmental stewardship requires tailored communication strategies and continuous community outreach efforts.

Hazardous Waste Management: Riverside communities may face additional challenges in managing hazardous waste generated from various activities such as industries, boating, or agricultural practices. The proper handling, storage, and disposal of hazardous materials require specialized knowledge, infrastructure, and regulatory compliance, which can pose significant challenges for these communities. Addressing these challenges and implementing effective waste management practices in riverside communities are crucial for preserving the environment, protecting public health, and ensuring sustainable development. By adopting integrated waste management approaches, including waste reduction, recycling, and safe disposal methods, riverside communities can mitigate the environmental impact of waste while promoting a cleaner and healthier living environment for their residents.

15.4 CONCLUSION

This study thoroughly understood the KSK community's environmental behaviors and attitudes and how they affect the sustainable use and management of the Limbang River. Although claimed health problems cannot be solely ascribed to poor water quality, the use of river water for basic home tasks is intimately tied to their waste disposal habits. The correlation test results do not support the hypothesis that household income does not affect septic tank availability but support the second hypothesis that health issues among children are very much in the DS community. Confounding variables could significantly affect how riverside communities deal with challenges related to water rehabilitation or how they practice sustainable use of river water. The rehabilitation of the river is thought to require comprehensive planning and management, but all community segments must engage equally in these efforts.

REFERENCES

Abdullah, S. (2012). *"Water Resource Users in Malaysia—Issues and Challenge" World Water Vision Report (2000) "A Water Secure World."* November. mywp.org.my

Borneo Post Online. (2020, November 13). Bintulu water supply tested regularly for safety—Dr Abdul Rahman. *Borneo Post Online.* www.theborneopost.com/2020/11/13/bintulu-water-supply-tested-regularly-for-safety-dr-abdul-rahman/

Cheng, K. M., Tan, J. Y., Wong, S. Y., Koo, A. C., & Amir Sharji, E. (2022). A review of future household waste management for sustainable environment in Malaysian cities. *Sustainability, 14*(11). https://doi.org/10.3390/su14116517

DOE. (2018). *Malaysia Environmental Quality Report.* Annual Report, Department of Environment, Ministry of Natural Resources And Environmental Sustainability. www.doe.gov.my/en/environmental-quality-report.

DOSM. (2020). *Household Income and Basic Amenities Survey Report.* www.dosm.gov.my

EPA. (2001). *Source Water Protection Practices Bulletin.* www.gov.epa/septic/source-water-protection-practices-bulletin

Goi, C. L. (2020). The river water quality before and during the Movement Control Order (MCO) in Malaysia. *Case Studies in Chemical and Environmental Engineering, 2,* 100027. https://doi.org/https://doi.org/10.1016/j.cscee.2020.100027

Gyawali, S., Techato, K., & Yuangyai, C. (2012). *Effects of Industrial Waste Disposal on the Surface Water Quality of U-tapao River.* Thailand.

Haldar, K., Kujawa-Roeleveld, K., Hofstra, N., Datta, D. K., & Rijnaarts, H. (2022). Microbial contamination in surface water and potential health risks for peri-urban farmers of the Bengal delta. *International Journal of Hygiene and Environmental Health, 244,* 114002. https:// doi.org/10.1016/j.ijheh.2022.114002

Hanafi, A. S., Sholihah, Q., Martina, M., & Deniati, E. N. (2018). Pengelolaan Sampah Rumah Tangga pada Masyarakat Tepi Sungai dan Faktor yang Mempengaruhinya. *Media Kesehatan Masyarakat Indonesia, 14*(4), 368. https://doi.org/10.30597/mkmi.v14i4.5091

Hess, A. (2018). How often should I change my water filter? *Home Revolution.* www.homerev.com/blogs/home-revolution/change-water-filter#:~:text=The basic rule of thumb,Brand%2Funit of filtration system

Ismail, M. (2020, August 25). DCM: Link road part of coastal road network to open up rural areas. *The Borneo Post.* www.theborneopost.com/2020/08/25/dcm-link-road-part-of-coastal-road-network-to-open-up-rural-areas

Jayawardhana, Y., Kumarathilaka, P., Herath, I., & Vithanage, M. (2016). Chapter 6—Municipal solid waste biochar for prevention of pollution from landfill leachate. *Environmental Materials and Waste: Resource Recovery and Pollution Prevention* (M. N. V. Prasad & Kaimin Shih, Eds.; pp. 117–148). Elsevier, Amsterdam. https://doi.org/10.1016/B978-0-12-803837-6.00006-8

Jomprice. (2022, November 24). The Latest Malaysia Home Septic Tank Price List 2022. *Jomprice.* https://jomprice.com/bahan-binaan/harga-tangki-septik-rumah/

Kamsuri, N., Tarmizi, N. A. A., & Mojiri, A. (2022). Occurrence, impact, toxicity, and degradation methods of microplastics in environment—a review. *Environmental Science Pollution Research International, 29*(21), 30820–30836.

Khan, K., Lu, Y., Saeed, M. A., Bilal, H., Sher, H., Khan, H., Ali, J., Wang, P., Uwizeyimana, H., Baninla, Y., Li, Q., Liu, Z., Nawab, J., Zhou, Y., Su, C., & Liang, R. (2018). Prevalent fecal contamination in drinking water resources and potential health risks in Swat, Pakistan. *Journal of Environmental Sciences, 72,* 1–12. https://doi.org/10.1016/j.jes.2017.12.008

Laeng, J. (2022, July 15). Brunei's land, sea borders fully reopen from Aug 1. *The Borneo Post*. www.theborneopost.com/2022/07/15/bruneis-land-sea-borders-fully-reopen-from-aug-1

LAKU. (2022). *Raw Water Sources*. Official Website of LAKU Management Sdn. Bhd. www.lakumanagement.com.my/page-0-91-68-Raw-Water-Sources.html

Landrigan, P. J., Fuller, R., Fisher, S., Suk, W. A., Sly, P., Chiles, T. C., & Bose-O'Reilly, S. (2019). Pollution and children's health. *Science of the Total Environment*, *650*, 2389–2394. https://doi.org/10.1016/j.scitotenv.2018.09.375

Maxwell, O., Oghenerukevwe, O. F., Adewoyin Olusegun, O., Joel, E. S., Daniel, O. Arinze., Oluwasegun, A., Jonathan, H. O., Samson, T. O., Adeleye, N., Michael, O. M., Omeje Uchechukwu, A., Akinwumi Oluwasayo, A., Akinpelu, A., L, A. M., & Oladokun, O. (2021). Sustainable nano-sodium silicate and silver nitrate impregnated locally made ceramic filters for point-of-use water treatments in sub-Sahara African households. *Heliyon*, *7*(12), e08470. https://doi.org/10.1016/j.heliyon.2021.e08470

MOH. (2016). *Drinking Water Quality Standard*. https://environment.com.my/wp-content/uploads/2016/05/Drinking-Water-MOH.pdf

Nguyen, T. T. T., Malek, L., Umberger, W. J., & O'Connor, P. J. (2022). Household food waste disposal behaviour is driven by perceived personal benefits, recycling habits and ability to compost. *Journal of Cleaner Production*, *379*, 134636. https://doi.org/10.1016/j.jclepro.2022.134636

Pandey, P. K., Kass, P. H., Soupir, M. L., Biswas, S., & Singh, V. P. (2014). Contamination of water resources by pathogenic bacteria. *AMB Express*, *4*(1), 51. https://doi.org/10.1186/s13568-014-0051-x

Picardal, J., & Marababol, M. (2012). *Impacts of Waste Disposal Practices and Water Utilization of Riverside Dwellers on Physicochemical and Microbiological Properties of Butuanon River, Central Visayas*. www.researchgate.net/publication/327631933_Impacts_of_Waste_Disposal_Practices_and_Water_Utilization_of_Riverside_Dwellers_on_Physicochemical_and_Microbiological_Properties_of_Butuanon_River_Central_Visayas

Qaiser, S., Hashmi, I., & Nasir, H. (2014). Chlorination at treatment plant and drinking water quality: A case study of different sectors of Islamabad, Pakistan. *Arabian Journal for Science and Engineering*, *39*(7), 5665–5675. https://doi.org/10.1007/s13369-014-1097-4

Radhi, N. A. M. (2020, February 23). More Households Embracing Waste Separation. *New Straits Times*. www.nst.com.my/news/nation/2020/02/568249/more-households-embracing-waste-separation

RECODA. (2022, October 3). *New 3.7km Road from Bukit Lubok to Kampung Seberang Kedai Roads Progressing*. Regional Corridor Development Authority. https://recoda.gov.my/2022/10/03/new-3-7km-road-from-bukit-lubok-to-kampung-seberang-kedai-roads-progressing/

Sanchez, J. M. P., Caturza, R. R. A., Picardal, M. T., Librinca, J. M., Armada, R. L., Pineda, H. A., Libres, M. T., Paloma, Ma. L. B., Ramayla, S. P., & Picardal, J. P. (2022). Water management practices and environmental attitudes of riparian communities in Sapangdaku River, Cebu Island, Philippines. *Biosaintifika: Journal of Biology & Biology Education*, *14*(2), 147–159. https://doi.org/10.15294/biosaintifika.v14i2.36185.

Sangodoyin, A. Y. (1993). Considerations on contamination of groundwater by waste disposal systems in Nigeria. *Environmental Technology*, *14*(10), 957–964. https://doi.org/10.1080/09593339309385370.

Sarijan, S., Azman, S., Said, M. I. M., Andu, Y., & Zon, N. F. (2018). Microplastics in sediment from Skudai and Tebrau river, Malaysia: a preliminary study. *MATEC Web of Conferences*. https://doi.org/10.1051/matecconf/201825006012.

Sidabutar, N., Namara, I., Hartono, D., & Soesilo, T. (2017). The effect of anthropogenic activities to the decrease of water quality. *IOP Conference Series: Earth and Environmental Science, 67,* 012034. https://doi.org/10.1088/1755-1315/67/1/012034.

Sulistyowati, L., Nurhasanah, Riani, E., & Cordova, M. R. (2022). The occurrence and abundance of microplastics in surface water of the midstream and downstream of the Cisadane River, Indonesia. *Chemosphere, 291,* 133071. https://doi.org/10.1016/j.chemosphere.2021.133071.

Suthar, S., Sharma, J., Chabukdhara, M., & Nema, A. K. (2010). Water quality assessment of river Hindon at Ghaziabad, India: Impact of industrial and urban wastewater. *Environmental Monitoring and Assessment, 165*(1), 103–112. https://doi.org/10.1007/s10661-009-0930-9

Tideschart. (2022). Sapo Point Brunei Bay tide times and tide charts for this week. In *Tideschart.* www.tideschart.com/Malaysia/Sarawak/Bahagian-Limbang/Sapo-Point-(Brunei-Bay)/Weekly

Wagiono, F., Shaddiq, S., Junaidy, J., Wibowo, D. E., & Yahya, M. Y. D. (2022). Community habits in floating houses (Lanting) in utilizing the river as an shower, wash, and toilet (MCK) facility in the S. Parman Down Area Neighborhood 01 Hamlet XVII Palangka Raya. *JED (Jurnal Etika Demokrasi), 7*(1), 109–121. https://doi.org/10.26618/jed.v7i1.6770.

Wang, W., Ge, J., & Yu, X. (2020). Bioavailability and toxicity of microplastics to fish species: A review. *Ecotoxicology and Environmental Safety, 189,* 109913. https://doi.org/10.1016/j.ecoenv.2019.109913

Wiebe, W. J., Sheldon, W. M., & Pomeroy, L. R. (1992). Bacterial growth in the cold: Evidence for an enhanced substrate requirement. *Applied and Environmental Microbiology, 58*(1), 359–364. https://doi.org/10.1128/aem.58.1.359-364.1992

Index

Note: Page numbers in **bold** refers to Tables.

A

Acetogenesis, 148
Acidogenesis, 148
Acid rain, 6, 64, 214
Acinetobacter, 88
Acrylic/polymethyl methacrylate, 207
Acrylonitrile butadiene styrene (ABS), 205
Activated carbon, 261, 266
Activated sludge, 253, 261, 266
Adsorption, 84, 210–212, 250, 252, 264, 266, 270
Advanced oxidation processes (AOP), 122, 266
Agricultural activities, 3, 15, 182
Airborne pollution, 64
Airborne transmission, 137
Air pollution, 6, 17, 29, 44, 49, 64, 79, 111,
    114–115, 120–121, 134, 147, 214, 247,
    284, 325
Aluminum oxide, 69, 75, 242
Ammonia, 12, 46, 134, 150
Anaerobic bacteria, 101, 217, 283
Anaerobic digestion, 4, 15–16, 20, 46, 54, 119–120,
    134, 141, 147–150, 183, 191–192, 194, 247,
    283, 300
Animal husbandry, 44, 46, 48, 53, 132, 150
Anthropogenic activities, 297, 322
Antibiotic resistance, 135, 143, 150
Antibiotic-resistant pathogens, 135
Antisense RNA technology, 220
Antisolvent technique, 251
Archaeoglobus fulgidus, 220
Arenicola marina, 209
Aspergillus spp., 136, 139, 144, 212, 219

B

Bacillus, 86–88, 90, 93, 217–218, 222
Bagasse, 5, 19, 300, 304
Bioaccumulation, 210, 244, 249
Bioactive compounds, 263
Biochar, 120, 249, 300–301
Bio-CNG, 194–195
Bio-composites, 302
Biodegradable waste, 3–4, 9, 18, 35, 37–38,
    182–183
Bioenergy, 50, 182, 301
Bioethanol, 195, 300–301
Biofilms, 219, 221, 253
Bioinformatics, 220, 222–223

Biological processes, 3, 148, 182–183, 192,
    242, 282
Biomedical waste, 1, 7, 18–19, 33
Biomethane, 301
Bioremediation, 121, 187, 219, 248, 249,
    261, 263
Biotransformation, 222, 244–245
Bisphenol A, 197–198, 277
Blastomyces dermatitidis, 136, 139
Brevibacillus, 217

C

Calorific value, 12–13, 300–301
Calothrix, 219, 264
Campylobacter, 135–136, 139, 142–143, 149
Cancer risk index, 172–175
Carbon monoxide, 10, 30, 114, 193
Carbon nanotube, 239–240, 254
Central Pollution Control Board (CPCB), 3, 35, 194
Chlorella, 210, 264, 270
Chlorinated paraffins (CPs), 116
Circular economy, 54–56, 61, 83, 100, 113, 147,
    187–188, 190–191, 195–196, 223, 280–281,
    284–285, 287–288, 290
Climate change, 15, 20, 35, 49–50, 54, 65, 120,
    147, 150, 193–194, 215, 284, 295, 305–306
Cloud point extraction, 253
Clostridium, 136, 139, 149
Coagulation, 192, 265
Coconut fibers, 300
Combustion, 5, 10, 13–14, 17–18, 34, 39,
    60–61, 63–64, 68–70, 80, 192, 213–214, 245,
    247–248, 300
Commercial waste, 1, 19, 179, 181
Composting, 2, 4, 9, 10, 15, 18, 20, 47–48, 54,
    118–119, 123, 134, 141, 145–147, 149–150,
    179–180, 182, 187–188, 190, 192, 195, 247,
    282, 285
Construction and demolition waste, 7, 62, 65, 67,
    73, 86, 115, 181–182
Construction waste, 63, 66–68, 79
Coronaviruses, 136, 139
COVID-19, 139
CRISPR, 220
Cryptococcus neoformans, 136, 139
Cryptosporidium, 137, 139, 149
Cucumaria frondose, 209
Cystoseira baccata, 269

D

Dermatophytes, 136, 139
Diatoms, 197, 219
Dichlorodiphenyldichloroethylene (DDE), 200
Dichlorodiphenyltrichloroethane (DDT), 121
Diffusion, 91, 211
Dioxins, 10, 111, 115, 200, 211
Dispersion, 211, 246, 251, 253, 289
Domestic waste, 1, 6, 8–9, 19–20, 165–166, 317, 319
Dye, 144, 202, 207, 262–263, 267

E

Earthworms, 47, 54, 211–212, 246
Ecological risk, 158–159, 162, 168, 170
Economic development, 2, 186
Ecosystem, 2, 4, 10, 64, 111, 113, 144, 150, 157,
    168, 187, 197–200, 210–214, 217, 219, 221,
    223, 240–243, 246, 276, 278, 286, 289–290,
    295–297, 306
Electricity, 6, 9, 30, 33, 79, 119–120, 147,
    191–194, 207, 214, 281–283, 299, 301, 304
Electrodeposition, 252
Electro-flotation, 264
Electronic waste, 4, 6, 34, 116, 181–183, 185–186,
    190–191, 286
Energy, 4, 6, 13, 17, 19–20, 27–31, 33, 39–40, 45,
    47–48, 52, 54–55, 65, 67, 70–71, 79–80, 82, 87,
    90–91, 93, 118–122, 133–135, 141, 147, 179,
    184–185, 187–195, 202, 209, 213–216, 237,
    240–241, 247–248, 254, 262, 270, 278,
    280–285, 300–301, 303–304, 306
Engineering, 55, 61, 62, 85, 121, 201, 207,
    219–220, 222–223, 237, 242, 270, 302
Environment Protection Act, 35, 38
Environmental contamination, 81, 116, 135, 141,
    240, 248, 289
Environmental impact, 15, 53, 62, 68–70, 80, 86,
    120, 122, 133–134, 147–188, 190–191,
    193–194, 237, 246–247, 278, 280–283,
    285–286, 323, 325
Enzymatic activity, 211
Enzymes, 16, 51, 142–143, 195, 217–220,
    222–223
Escherichia coli, 88, 132, 135–136, 139, 142–144,
    149, 220, 222

F

Fasciola hepatica, 137, 139
Flavonoids, 267
Fly ash, 17, 60–61, 63–64, 68–72, 79–80, 82–84,
    86, 101
Foodborne illnesses, 135, 142
Food security, 44, 49–50, 157
Forest Research Institute (FRI), 303

Fossil fuels, 30, 63–65, 147, 192–194, 281
Fourier-transform infrared spectroscopy (FTIR),
    84, 97, 218, 269
4R's, 41, 184

G

Galaxaura elongate, 269
Gasification, 9, 30–31, 39, 119, 121, 192–193,
    283, 300–301
Gene editing, 220, 223
Genetic engineering, 219–220, 222–223
Geopolymerization, 61–62, 71, 80–82, 84–85,
    100–101
Geopolymer concrete, 84–86
Giardia, 137, 139, 149
Global warming, 6, 15–16, 18, 38, 49, 54, 65, 112,
    214, 296, 301
Green waste, 6, 181–182, 186
Greenhouse gas, 6, 15–16, 17, 20, 29, 47, 65, 111,
    118, 120, 134, 147–148, 150, 183, 190, 193–194,
    278, 281, 284, 290, 295, 299, 325
Groundwater, 2, 4, 16–17, 37, 45, 48, 54, 63, 65,
    67, 112–113, 121–122, 132, 137, 150, 157–160,
    165–167, 172–173, 215, 246–247, 318

H

Hazardindex, 171–173
Hazardous waste, 1, 8–11, 15, 20, 33–34, 36, 67,
    81, 83, 111, 116–117, 119, 121–122, 132, 181,
    188–189, 192, 252, 290, 325
Health hazards, 63, 65, 67, 111, 113–114, 116–118,
    120–123, 165, 179, 188–189, 210, 239–242, 278
Health risks, 1, 8, 15, 37, 64, 67, 111, 113–115,
    120–123, 136, 144, 149–150, 158, 185, 191,
    240, 243–244, 247, 254, 276–277, 288–290
Heavy-metal evaluation index (HEI), 158
Heavy metal pollution index (HPI), 158, 161,
    165–166
Hemicelluloses, 300
Hepatitis E virus, 136
Histoplasma capsulatum, 136, 139
Holocellulose, 299, 302
Holothuria floridana, 209
Household waste, 9, 15, 30, 47, 116, 179–181,
    262, 313, 316, 319–320, 322–324
Hydrocarbons, 111, 121, 215–216, 219, 221, 247

I

Ideonella sakaiensis, 222
Incinerated sewage sludge ash, 84
Incineration, 7, 9, 14, 17, 20, 31, 68–69, 83, 111,
    119–120, 187, 191–192, 194, 213–215,
    247–248, 251–252, 278, 281, 284–285, 289
Index of geoaccumulation (Igeo), 158

Industrial waste, 1, 8, 60–71, 83–84, 100–101, 114–115, 122, 168, 186, 312
Influenza viruses, 136
Institutional Waste, 181
Integrated waste management, 2, 26, 39, 191–192

L

Landfill design, 120
Leachate, 4, 13, 17, 38, 77–78, 95, 113, 118, 120, 157, 259–160, 164–167, 175, 198
Leather processing waste, 144–145
Leptolyngbya, 219, 264
Leptospira spp, 136, 139
Lignin, 70, 217, 299, 302
Lignocellulose, 299
Limbang river, 313–317, 322, 324–325
Listeria monocytogenes, 136, 139
Livelihood, 295, 303, 304
Livestock industry waste, 133
Lysinibacillus, 219

M

Material recovery facilities, 282
Meat processing waste, 143
Mechanical biological treatment, 192
Medical waste, 1, 6, 33–34, 115–116, 181, 185, 288–289
Metagenomic technique, 220
Methane (CH$_4$), 10, 12, 14, 15–18, 37–38, 46, 50, 54, 65, 120, 135, 147–150, 193, 200, 214–215, 283–284
Methanogenesis, 148–149
Microalgal nanoparticles, 266
Microbes, 16, 49, 132, 140, 217–218, 220–221, 223, 266
Microbial-induced carbonate precipitation (MICP), 82
Microcystis, 219
Microemulsion, 251–252
Microplastics, 8, 199–200, 207–212, 217–219, 276
Micro-polyesters, 208
Mulching, 299–300, 303
Municipal solid waste, 1, 18–19, 33–35, 63, 69, 86, 113, 179–184, 195, 312
Mutagenesis, 220
Mycotoxins, 142

N

Nano-adsorbents, 266–267, 270
Nanoceramics, 239, 242
Nanocomposites, 243, 249, 288
Nanofibers, 239, 242
Nanofiltration, 261, 266

Nanomembranes, 239, 243, 253
Nanoparticles, 237–240, 243–255, 261–264, 266–271, 288
Nanoplastics, 208–209, 212–213, 239, 242, 245, 249–250
Nano-pollution, 246
Nanotechnology, 237–238, 254, 262, 270
Nanowaste, 237–239, 244–252, 254, 276, 288–290
Needle management, 295, 299
Nitrogen oxides (NOx), 6, 64, 114, 247, 296
Nitrous oxide, 10, 54, 65
Non-biodegradable, 4, 183
Nongovernmental organizations, 40
Norovirus, 136, 139, 149
Nuclear magnetic resonance (NMR), 84
Nutrient cycling, 113, 249, 295

O

Open dumping, 2, 15, 17, 18, 157, 188–189, 247
Optical property, 204
Organic fertilizers, 44, 50, 150
Organic matter, 3, 10, 14, 31, 34, 51, 69, 133–135, 137, 141, 144–146, 148, 149, 182, 211, 238, 245, 263, 266, 285
Organic pollutants, 112, 121, 200
Oxidative stress, 210, 244–245
Ozonation, 266

P

Palm oil fuel ash (POFA), 68, **72**
Particulate matter, 10, 17, 64–65, 111, 115, 119, 242, 247
Persistent organic pollutants (POPs), 122, 210, 212
Personal care products, 261, 288
Pesticides, 50, 112–114, 116, 181, 211, 262, 266
Petrochemicals, 215–216
Photocatalysts, 252
Phthalates, 197–198, 205, 207, 210, 212, 277
Phyco-nanotechnology, 267
Pinus roxburghii Sarg., 297
Plant growth, 48, 211, 240, 264
Plasma, 31, 39, 119, 121, 160, 192
Plasticizers, 14, 92, 202–203, 205
Plastisphere, 220–222
Pleurocapsa, 219
Policy, 18, 29–31, 39, 117–118, 122, 185, 187–188, 190–191, 254, 304
Poliovirus, 136, 139
Polybrominated diphenyl ethers (PBDE), 198, 207
Polycarbonate, 198, 201, 207
Polychlorinated biphenyls (PCBs), 7, 121
Polychlorinated dibenzofurans (PCDFs), 114–115

Polychlorinated dibenzo-p-dioxins (PCDD), 114–115, 211
Polycyclic aromatic hydrocarbons (PAHs), 111, 121, 247
Polyethylene terephthalate (PET), 200, 204, 218, 220
Polyfluorinated compounds, 197
Polymerase chain reaction, 220
Polypropylene, 242
Polytetrafluoroethylene, 201
Polyvinyl chloride (PVC), 7, 200, 205, 219
Potential ecological risk index (PERI), 158, 162, 168, 170
Prochlorothrix, 219
Pseudomonas, 88, 219
Public health, 2–3, 7, 14, 20, 33, 35–36, 40, 45, 67–68, 113, 115, 117, 121, 123, 182, 214, 278, 289, 312, 325
Public–Private Partnership, 188, 191, 195
Pyrolysis, 9, 20, 31, 39, 119–120, 192, 213, 215–216, 248, 251, 278, 280–281

Q

Quartz, 75–76, 90, 95

R

Radioactive, 1, 5, 7–10, 33–34
Red mud, 71, 83–84, 101
Renewable energy, 4, 28, 39, 70, 120, 133–134, 147, 190–194, 281, 300
Resource recovery, 44, 53, 55, 60–61, 79, 83, 86, 101–102, 118, 122, 185, 187–188, 190, 192, 281–283
Respiratory illnesses, 66, 68
Reuse, 4, 18, 28–30, 33, 38–39, 46, 53, 55–56, 117–118, 141, 145, 183–184, 187, 215, 223, 248–250, 261, 288
Reverse osmosis, 261, 266
Rice husk, 68, 70–72, 300
Rivularia, 219
Rotavirus, 136, 139, 149

S

Saccharomyces cerevisiae, 212, 220
Salinity, 16
Salmonella, 132, 135–136, 139, 142–143, 149
SARS-CoV-2, 136
Scanning electron microscopy (SEM), 84, 87 89, 219, 269
Scenedesmus vacuolatus, 269
Scytonema, 219

Segregation, 18–19, 34, 36, 38, 67, 120, 145, 185, 187–191, 289–291
Sewage sludge, 5–6, 33–34, 84, 200, 239, 252
Sewage treatment plant, 33
Smelting, 60–61
Smog, 64
Society, 1, 19, 46, 49–50, 55, 84, 111–112, 195, 213–214, 261
Socio-economic, 306
Soil fertility, 44, 49, 113, 133, 135, 211
Spirogyra submaxima, 269
Spirulina, 264, 269–270
Stabilizers, 90, 202
Stakeholders, 8, 53, 101, 180, 190, 288, 291
Sulfur dioxide (SO_2), 6, 17, 64, 247
Sustainable development goals, 49, 50
Swachh Bharat Mission, 16, 188, 194
Synechococcus, 219
Syngas, 283

T

Taenia spp., 137, 139, 143
Terrestrial ecosystem, 210, 212
Thermolysis, 266
3R's 46, 54–56, 118
Thyonella gemmate, 209
TiO_2, 240, 244–245, 248, 251–252, 254, 269
Toxoplasma gondii, 137, 139
Toxicity, 4, 10, 14, 17, 121–122, 144, 209, 238–246, 249, 254, 262
TSH, 212–213
Turbinaria ornate, 269

U

Ultrafiltration, 266
United States Environmental Protection Agency (USEPA), 10, 64, 68, 162–163, 165
Urbanization, 1–2, 5, 16–17, 19, 33–34, 41, 112, 157, 179, 186, 188–189, 199, 261, 270, 312

V

Vector-borne, 66, 68
Vermicomposting, 18, 47–48, 54, 183
Viruses, 67, 135–136, 139, 145–147
Volatile fatty acids (VFAs), 134
Volatile gases, 213
Volatile organic compounds, 6, 64, 79, 114, 134

W

Waste management and handling rules, 33
Waste minimization, 44, 101, 117

Waste-to-energy, 39–40, 118, 193, 283
Wastewater, 5–6, 8, 44, 50, 65, 83, 90, 122, 132,
    141–145, 165, 192, 208–209, 218, 246, 252,
    254, 261–267, 270–271, 317–318
Wastewater treatment, 6, 246, 261–262,
    270
Water contamination, 8, 17, 65, 137
World Bank, 1–2, 34, 179, 185–186, 188
World Health Organization (WHO), 17, 111, 136,
    161, 261

X

X-ray fluorescence (XRF), 76–77

Z

Zero waste, 48, 56, 122–123, 286
Zeta potential, 87, 92, 95–98, 101
Zika virus, 66, 68
Zinc finger proteins, 220
Zoonotic diseases, 135, 144, 150